Weighted Sobolev Spaces and Degenerate Elliptic Equations

Weighted Sobolev Spaces and Degenerate Elliptic Equations

By

Albo Carlos Cavalheiro

Cambridge Scholars
Publishing

Weighted Sobolev Spaces and Degenerate Elliptic Equations

By Albo Carlos Cavalheiro

This book first published 2023

Cambridge Scholars Publishing

Lady Stephenson Library, Newcastle upon Tyne, NE6 2PA, UK

British Library Cataloguing in Publication Data
A catalogue record for this book is available from the British Library

ISBN (10): 1-5275-5166-0
ISBN (13): 978-1-5275-5166-4

To my parents, Simone and Larissa

Contents

Introduction

The goal of this book it to introduce the reader to different topics of the theory of degenerate elliptic partial differential equations.

Let ω be a *weight* on $\mathbb{R}^N$, i.e., a locally integrable function on $\mathbb{R}^N$ such that $0 < \omega(x) < \infty$ for a.e. $x \in \mathbb{R}^N$. Let $\Omega \subset \mathbb{R}^N$ be open, $1 \leq p < \infty$ and k a nonnegative integer. The weighted Sobolev space $W^{k,p}(\Omega, \omega)$ consists of all functions u with weak derivatives $D^\alpha u$, $|\alpha| \leq k$, satisfying

$$\|u\|_{W^{k,p}(\Omega,\omega)} = \left(\sum_{|\alpha| \leq k} \int_\Omega |D^\alpha u|^p \, \omega \, dx \right)^{1/p} < \infty.$$

In the case $\omega = 1$, this space is denoted $W^{k,p}(\Omega)$ (classic Sobolev spaces). Sobolev spaces without weights occur as spaces of solutions for elliptic and parabolic partial differential equations. For degenerate partial differential equations, i.e., equations with various types of singularities in the coefficients, it is natural to look for solutions in weighted Sobolev spaces (see, e.g., [32], [35], [43], [56]).

A class of weights, which is particularly well understood, is the class of A_p weights that was introduced by B. Muckenhoupt (see [50]). These classes have found many useful applications in harmonic analysis (see [55]). Another reason for studying A_p-weights is the fact that powers of the distance to submanifolds of $\mathbb{R}^N$ often belong to A_p (see [46]). There are, in fact, many interesting examples of weights (see [43] for p-admissible weights).

In various applications, we can meet boundary value problems for elliptic equations whose ellipticity is disturbed in the sense that some degeneration or singularity appears. This bad behavior can be caused by the coefficients of the corresponding differential operator as well as by the solution itself. There are several very concrete problems form practice which lead to such differential equations, e.g form glaciology, non-Newtonian fluid mechanics, flows through porous media, differential geometry, celestial mechanics, climatology, reaction-diffusion problems, etc (see some examples of applications of degenerate elliptic equations in [8], [27] [57]).

Let us start with some well-known facts. We will consider a nonlinear differential operator of order $2k$ in the divergence form

$$(Au)(x) = \sum_{|\alpha|\leq k} (-1)^{|\alpha|} D^{\alpha} a_{\alpha}(x, u, \nabla u, ..., \nabla^{k} u) \qquad (0.1)$$

for $x \in \Omega \subset \mathbb{R}^N$, where the coefficients $a_\alpha = a_\alpha(x,\xi)$ are defined on $\Omega \times \mathbb{R}^m$, and where $\nabla^j u = \{D^\gamma u : |\gamma| = j\}$ (for $j = 0, 1, ..., k$) is the gradient of the j-th order. Here we suppose that

(i) $a_\alpha(x,\xi)$ satisfy the Carathéodory condition, i.e., $a_\alpha(.,\xi)$ is measurable in Ω for every $\xi \in \mathbb{R}^m$ and $a_\alpha(x,.)$ is continuous in $\mathbb{R}^m$ for a.e. $x \in \Omega$;

(ii) $a_\alpha(x,\xi)$ satisfy the *growth condition*

$$|a_\alpha(x,\xi)| \leq C_\alpha \left(g_\alpha(x) + \sum_{|\beta|\leq k} |\xi_\beta|^{p-1} \right)$$

for a.e. $x \in \Omega$ and every $\xi \in \mathbb{R}^m$, $C_\alpha > 0$ and $g_\alpha \in L^{p'}(\omega)$ $(\frac{1}{p} + \frac{1}{p'} = 1)$;

(iii) $a_\alpha(x,\xi)$ satisfy the *ellipticity condition*

$$\sum_{|\alpha|\leq k} a_\alpha(x,\xi)\, \xi_\alpha \geq C \sum_{|\alpha|\leq k} |\xi_\alpha|^p,$$

for every $\xi \in \mathbb{R}^m$ with constant $C > 0$ independent of ξ;

(iv) $a_\alpha(x, \xi)$ satisfy the *monotonicity condition*

$$\sum_{|\alpha| \leq k} (a_\alpha(x, \xi) - a_\alpha(x, \eta))(\xi_\alpha - \eta_\alpha) > 0$$

for every $\xi, \eta \in \mathbb{R}^m$, $\xi \neq \eta$.

A typical example of a differential operator A satisfying all the foregoing conditions is the so-called p-Laplacian Δ_p defined for $p > 1$ by

$$\begin{aligned} \Delta_p u &= \operatorname{div}(|\nabla u|^{p-2} \nabla u) \\ &= \sum_{i=1}^{N} \frac{\partial}{\partial x_i}\left(\left[\left(\frac{\partial u}{\partial x_1}\right)^2 + \dots + \left(\frac{\partial u}{\partial x_N}\right)^2\right]^{(p-2)/2} \frac{\partial u}{\partial x_i}\right), \end{aligned}$$

or its modification, the operator $\tilde{\Delta}_p$ defined by

$$\tilde{\Delta}_p u = \sum_{i=1}^{N} \frac{\partial}{\partial x_i}\left(\left|\frac{\partial u}{\partial x_i}\right|^{p-2} \frac{\partial u}{\partial x_i}\right),$$

or the little more complicated operator

$$Au = -\tilde{\Delta}_p u + |u|^{p-2} u.$$

Let us slightly change the foregoing operator into

$$(Au)(x) = -\sum_{i=1}^{N} \frac{\partial}{\partial x_i}\left(a_i(x)\left|\frac{\partial u}{\partial x_i}\right|^{p-2} \frac{\partial u}{\partial x_i}\right) + a_0(x)|u|^{p-2} u, \quad (0.2)$$

with given functions (coefficients) $a_i(x)$ $(i = 1, \dots, N)$ satisfying

$$a_i \in L^\infty \ (i = 0, 1, \dots, N), \quad (0.3)$$

$$\text{and } a_i(x) \geq C_1 > 0 \ (i = 0, 1, \dots, N) \text{ for a.e. } x \in \Omega. \quad (0.4)$$

Then all conditions (i), (ii), (iii) and (iv) remain satisfied and we look for a weak solution in the Sobolev space $W^{1,p}(\Omega)$. However, the situation changes dramatically if some of the coefficients $a_i(x)$ violate conditions (0.3) and/or condition (0.4) (i.e., with coefficients which are singular ($a_i(x)$ are unbounded) and/or degenerated ($a_i(x)$ are only positive a.e.)). The situation can be saved using the weighted Sobolev space $W^{1,p}(\Omega,\omega)$ instead of the classical Sobolev space $W^{1,p}(\Omega)$.

A typical example is the degenerate p-Laplacian

$$-\operatorname{div}(a(x)|\nabla u|^{p-2}\nabla u),$$

with $p\neq 2$.

In the linear case, we consider the second order, linear, elliptic equation with divergence structure

$$\operatorname{div}\,(A(x)\,\nabla u(x)) = 0 \tag{0.5}$$

where $A(x) = [a_{ij}(x)]_{i,j=1,\ldots,N}$ is a symmetric matrix with measurable coefficients, defined in a domain $\Omega\subset\mathbb{R}^N$ $(N\geq 2)$. We assume the following ellipticity condition

$$0 < \omega(x)|\xi|^2 \leq \sum_{i,j=1}^{N} a_{ij}(x)\xi_i\xi_j \leq |\xi|^2 v(x), \tag{0.6}$$

for all $\xi\in\mathbb{R}^N$ and a.e. $x\in\Omega$, where ω and v are measurable functions, finite and positive a.e. $x\in\Omega$. The equation (0.5) is *degenerate* if ω^{-1} is unbounded, and the equation (0.5) is *singular* if v is unbounded.

Below, there follows a short outline of the book

Chapter 1 is of an introductory character. In the first section, we present the notation and conventions that will be used in the book and we summarize some facts from the theory of weighted Sobolev spaces.

In the Chapter 2 we summarize some facts from the theory of A_p-weights and an approximation result for A_p-weights.

The theme of Chapter 3 is degenerate linear elliptic equations. In this chapter we study the existence and uniqueness of solution, the approximation of solutions and the concept of entropy solution for degenerate linear elliptic equations.

Chapter 4 is devoted to the study of some problems with degenerate semilinear elliptic equations. In Chapter 5 we study some degenerate quasilinear elliptic equations (existence, uniqueness and entropy solutions). Finally, in Chapter 6 we investigate the existence and uniqueness of solutions for some degenerate nonlinear elliptic problems using the theory of monotone operators.

A. C. Cavalheiro
Department of Mathematics
State University of Londrina
Brazil

List of symbols

Here we introduce the basic notation which will be observed throughout this book.

$\mathbb{N}$ - the set of natural numbers.
$\mathbb{R}$ - the set of real numbers.
$\mathbb{R}^N$ - the N-dimensional Euclidean space of points $x = (x_1, x_2, ..., x_N)$.
$\alpha = (\alpha_1, ..., \alpha_N)$ - an (N-dimensional) multiindex (i.e., with components α_j which are nonnegative integers).
$|\alpha| = \alpha_1 + ... + \alpha_N$ - the length of the multiindex α.
$D^\alpha = \dfrac{\partial^{|\alpha|}}{\partial x_1^{\alpha_1} .... \partial x_N^{\alpha_N}}$ - partial derivative of order $|\alpha|$.
sup E - lowest upper bound (supremum) of the set E.
inf E - greatest lower bound (infimum) of the set E.
$s^+ = \max\{s, 0\}$ - the positive part of $s \in \mathbb{R}$.
$s^- = \max\{-s, 0\}$ - the negative part of $s \in \mathbb{R}$.
X^* - the adjoint (dual) space to the space X.
$\langle ., . \rangle$ - inner product in $\mathbb{R}^N$.
$(.|.)_X$ - duality between the Banach spaces X and X^*, $(f|x) = f(x)$, $x \in X$ and $f \in X^*$.
$\|x\|_X$ - norm of x in a Banach space X.
$x_n \to x$ - convergence in norm.
$x_n \rightharpoonup x$ - weak convergence.
X^N - product space of X ($N \in \mathbb{N}$), i.e., X^N consists of all $u = (u_1, u_2, ..., u_N)$, where $u_j \in X$ for all j. If X is a Banach space over $\mathbb{R}$, then X^N is also a Banach space over $\mathbb{R}$ with the norm

$$\|u\| = \sum_{j=1}^{N} \|u_j\|.$$

$\mathcal{D}(f)$ - domain of f.
$\mathcal{R}(f)$ - range of f.
supp$f$ - support of the function f.

$\underline{\lim}$, $\overline{\lim}$ - lower, upper limit.
$\partial\Omega$ - boundary of a set Ω.
$\partial\Omega \in C^{0,1}$ - piecewise smooth boundary.
$\overline{\Omega} = \Omega \cup \partial\Omega$ - the closure of Ω.
$C^\infty(\Omega)$ - space of infinitely continuously differentiable functions $f : \Omega \to \mathbb{R}$.
$C_0^\infty(\Omega)$ - the space of infinitely differentiable functions with compact support.
$X \hookrightarrow Y$ - the space X is continuously and compactly embedded into the space Y.
∇u - gradient of the map u, $\nabla u = \left(\frac{\partial u}{\partial x_1}, ..., \frac{\partial u}{\partial x_N}\right)$.
$\text{div} f$ - divergence of the map f, $\text{div} f = \sum_{j=1}^{N} \frac{\partial f}{\partial x_j}$.
Δf - Laplace operator of the map f, $\Delta f = \sum_{j=1}^{N} \frac{\partial^2 f}{\partial x_j^2}$.
$\text{osc}(u, E)$ - oscillation of the function $u : E \to \mathbb{R}$, $\text{osc}(u, E) = \sup_E u - \inf_E u$.
$|E|$ - measure of the measurable set E with respect to the N-dimensional Lebesgue measure.
$\mu(E)$ - measure of the set E with respect to the measure μ.

$\int_\Omega f \, d\mu$ - integral with respect to measure μ.

$\int_{\partial\Omega} f \, d\sigma(x)$ - surface integral.

$L^p(\Omega)$ - Lebesgue space.
$L^p(\Omega, \omega)$ - weighted Lebesgue space.
$W^{k,p}(\Omega, \omega)$ - weighted Sobolev space.
$\mathcal{M}^p(\Omega, \omega)$ - weighted Marcinkiewicz space.

Chapter 1

Weighted Sobolev Spaces

In this first chapter we introduce the *weighted Sobolev spaces* $W^{k,p}(\Omega, \omega)$ and investigate their basic properties which are needed in chapters to come. Throughout this book Ω will denote an open subset of $\mathbb{R}^N$, $N \geq 2$.

1.1 Preliminaries

Let $\mathbb{R}^N$ denote Euclidean N-space. The norm of a point $x = (x_1, ..., x_N)$ in $\mathbb{R}^N$ is given by $|x| = (x_1^2 + ... + x_N^2)^{1/2}$. If $E \subset \mathbb{R}^N$, the boundary, the closure and the complement of E with respect to $\mathbb{R}^N$ are denoted by ∂E, $\overline{E}$ and $E^c = \mathbb{R}^N \setminus E$. Let χ_E the characteristic function of a set E, i.e.,

$$\chi_E(x) = \begin{cases} 1, & \text{if } x \in E, \\ 0, & \text{if } x \notin E. \end{cases}$$

The set $B(a, r) = \{x \in \mathbb{R}^N : |x - a| < r\}$ is an open ball in $\mathbb{R}^N$ with radius r and center a. If $\lambda > 0$, then $\lambda B(a, r) = B(a, \lambda r)$.

Let $\alpha = (\alpha_1, ..., \alpha_N) \in \mathbb{Z}_+^N$ be a multi-index. Then $|\alpha| = \alpha_1 + ... + \alpha_N$, $\alpha! = \alpha_1!...\alpha_N!$, $x^\alpha = x_1^{\alpha_1}...x_N^{\alpha_N}$ for $x \in \mathbb{R}^N$, and $D^\alpha = \dfrac{\partial_1^\alpha}{\partial x_1^{\alpha_1}} \cdots \dfrac{\partial_N^\alpha}{\partial x_N^{\alpha_N}}$.

We denote by $C(\Omega)$ the set of all continuous functions $\varphi : \Omega \to \mathbb{R}$. Moreover, supp$\varphi$ (support of φ) is the closure (in the Euclidean norm of $\mathbb{R}^N$) of the set $\{x \in \Omega : \varphi(x) \neq 0\}$, i.e., $\text{supp}\varphi = \overline{\{x \in \Omega : \varphi \neq 0\}}$. For $k \in \mathbb{N}$, then $C^k(\Omega)$ is the set of k times continuously differentiable functions $\varphi : \Omega \to \mathbb{R}$ and $C^\infty(\Omega) = \bigcap_{k=1}^{\infty} C^k(\Omega)$.

The notation $E \Subset \Omega$ means that $\overline{E}$ is a compact subset of Ω. Then, if $k \in \mathbb{N}$, $C_0^k(\Omega) = \{\varphi \in C^k(\Omega) : \text{supp}\varphi \Subset \Omega\}$ and $C_0^\infty = \{\varphi \in C^\infty(\Omega) : \text{supp}\varphi \Subset \Omega\}$. The gradient of $\varphi \in C^1(\Omega)$ if $\nabla\varphi = (\partial_1\varphi, ..., \partial_N\varphi)$.

If $1 \leq p \leq \infty$, then p' is the conjugate exponent to p given by $\dfrac{1}{p} + \dfrac{1}{p'} = 1$ with the usual conventions when $p = 1$ or $p = \infty$.

Let $\Omega \subset \mathbb{R}^N$ be open. We denote the Lebesgue space by

$$L^p(\Omega) = \left\{ u : \Omega \to \mathbb{R} : \|u\|_{L^p(\Omega)} = \left(\int_\Omega |u|^p \, dx \right)^{1/p} < \infty \right\},$$

and

$$L^\infty(\Omega) = \left\{ u : \Omega \to \mathbb{R} : \|u\|_{L^\infty(\Omega)} = \text{ess sup}\, |u(x)| < \infty \right\}.$$

The space $L^p(\Omega)$ equipped with the norm $\|u\|_{L^p(\Omega)}$ $(1 \leq p \leq \infty)$ is a Banach space. By $L^p_{\text{loc}}(\Omega)$ we will denote the set of all functions $\varphi \in L^p(K)$ for every compact set $K \subset \Omega$.

For $k \in \mathbb{N}$ and $1 \leq p < \infty$ we denote by

$$W^{k,p}(\Omega) = \{u \in L^p(\Omega) : D^\alpha u \in L^p(\Omega), |\alpha| \leq k\}.$$

The Sobolev space $W^{k,p}(\Omega)$ is a Banach space if equipped with the norm $\|u\|_{W^{k,p}(\Omega)} = \left(\sum_{|\alpha| \leq k} \|D^\alpha u\|_{L^p(\Omega)} \right)^{1/p}$. We denote by

$W^{k,p}_{\text{loc}}(\Omega)$ the set of all functions $u \in W^{k,p}(K)$ for every compact set $K \subset \Omega$. Further, the space $W^{k,p}_0(\Omega)$ is defined as the closure of $C^\infty_0(\Omega)$ with respect to the norm $\|.\|_{W^{k,p}(\Omega)}$.

We will use the abbreviation "a.e." for "*almost everywhere* " or "*almost every*" with respect to Lebesgue measure. Similarly, "*measurable*" and "*locally integrable*" always mean Lebesgue measure and locally integrable with respect to Lebesgue measure, respectively. If Ω is an open subset of $\mathbb{R}^N$, then the set of locally integrable functions on Ω will denoted by $L^1_{\text{loc}}(\Omega)$.

The Lebesgue N-measure of a measurable subset $E \subset \mathbb{R}^N$ is denoted $|E|$.

Throughout, C will denote a positive constant whose value is not necessarily the same at each occurrence; it may vary even within a line. Two quantities A and B are said *equivalent* or *comparable* if there exists two constant C_1 and C_2 such that $C_1\, A \leq B \leq C_2\, A$ (and we use the notation $A \approx B$).

1.2 Weighted Sobolev Spaces

In this section, we intend to recall some concepts and results which will play an important role in our further considerations. The present section explores some properties of weighted Sobolev spaces and investigate their basic properties which are need in chapters to come. These spaces are only a slight modification of the classical Sobolev spaces $W^{k,p}(\Omega)$ but their property are much more complicated and less transparent. Complete expositions can be found in [38], [43], [46] and [56].

Definition 1.2.1. *Let Ω be an open set in $\mathbb{R}^N$. We denote by $\mathcal{W}(\Omega)$ the set of all measurable a.e. in Ω positive, finite and locally integrable functions $\omega = \omega(x)$, $x \in \Omega$ ($0 < \omega(x) < \infty$ a.e. in Ω). Elements of $\mathcal{W}(\Omega)$ will be called weight functions.*

Every weight $\omega \in \mathcal{W}(\Omega)$ gives rise to a measure on the measurable subsets of Ω through integration. This measure will also be denoted by ω. Thus, $\omega(E) = \int_E \omega\, dx$ for measurable set $E \subset \Omega$ (or $\mu(E) = \int_E \omega(x)\, dx$).

Definition 1.2.2. *Let $\Omega \subset \mathbb{R}^N$ an open set and $\omega \in \mathcal{W}(\Omega)$. For $1 \leq p < \infty$, we define $L^p(\Omega, \omega)$ as the set of measurable functions f on Ω such that*

$$\|f\|_{L^p(\Omega,\omega)} = \left(\int_\Omega |f|^p\, \omega\, dx \right)^{1/p} < \infty. \tag{1.1}$$

For $\omega(x) \equiv 1$ we obtain the usual Lebesgue space $L^p(\Omega)$.

Theorem 1.2.1. *The space $L^p(\Omega, \omega)$ equipped with the norm (1.1) is a Banach space.*

Proof. See [33], Theorem III.6.6 □

We now determine conditions on the weight ω that guarantee that functions in $L^p(\Omega, \omega)$ are locally integrable on Ω.

Definition 1.2.3. *Let $1 \leq p < \infty$ and let ω be a weight. We say that $\omega \in \mathcal{B}_p(\Omega)$ if*

(a) $\omega^{-1/(p-1)}$ is locally integrable, when $p > 1$;

(b) $\operatorname{ess\,sup}_{x \in B} \dfrac{1}{\omega(x)} < \infty$ for all ball B, when $p = 1$.

Now, suppose that $f \in L^p(\Omega, \omega)$, $\omega \in \mathcal{B}_p(\Omega)$ and let $B \subset \Omega$ be a ball.

(i) If $1 < p < \infty$, by Hölder's inequality (with $1/p + 1/p' = 1$), we obtain

$$\begin{aligned} \int_B |f|\, dx &= \int_B |f|\, \omega^{1/p}\, \omega^{1/p'}\, \omega^{-1}\, dx \\ &\leq \left(\int_B |f|^p\, \omega\, dx \right)^{1/p} \left(\int_B \omega^{(1-p')}\, dx \right)^{1/p'} \end{aligned}$$

$$\leq \|f\|_{L^p(\Omega,\omega)}\left(\int_B \omega^{-1/(p-1)}\,dx\right)^{1/p'}.$$

(ii) If $p = 1$, we have

$$\int_B |f|\,dx = \int_B |f|\,\omega\,\omega^{-1}\,dx \leq \left(\operatorname*{ess\,sup}_{x\in B} \frac{1}{\omega(x)}\right)\left(\int_\Omega |f|\,\omega\,dx\right).$$

Its follows that $L^p(\Omega,\omega) \subset L^1_{\text{loc}}(\Omega)$ and that the convergence in $L^p(\Omega,\omega)$ implies local convergence in $L^1(\Omega)$.

Hence, we have the following result.

Theorem 1.2.2. *Let $\Omega \subset \mathbb{R}^N$ be an bounded open set and a weight $\omega \in \mathcal{B}_p(\Omega)$ $(1 \leq p < \infty)$. Then $L^p(\Omega,\omega)$ is continuously embedded in $L^1(\Omega)$*
(i.e., $L^p(\Omega,\omega) \hookrightarrow L^1_{loc}(\Omega)$).

If $\omega \notin \mathcal{B}_p(\Omega)$, the conclusion of Theorem 1.2.2 need not hold.

Example 1.2.1. *Let $\Omega = (-1/2, 1/2) \subset \mathbb{R}$, $p = 2$ and $\omega(x) = |x|$. We have that $\omega \notin \mathcal{B}_2(\Omega)$, since $\omega^{-1/(2-1)}(x) = |x|^{-1}$. Let us take*
$f(x) = |x|^{-1}|ln(x)|^{-2/3}$. We have that $f \in L^2(\Omega,\omega)$ because

$$\begin{aligned}
\|f\|^2_{L^2(\Omega,\omega)} &= \int_{-1/2}^{1/2}\left(|x|^{-1}|ln(x)|^{-2/3}\right)^2 |x|\,dx \\
&= 2\int_0^{1/2} |x|^{-1}|ln(x)|^{-4/3}\,dx \\
&= 2\int_{ln(2)}^{\infty} t^{-4/3}\,dt < \infty,
\end{aligned}$$

but $f \notin L^1_{loc}(\Omega)$, since $(-1/4, 1/4) \subset (-1/2, 1/2)$ and

$$\int_{-1/4}^{1/4} |f|\,dx = \int_{-1/4}^{1/4} |x|^{-1}|ln(x)|^{-2/3}\,dx$$

$$= \ 2\int_{ln(4)}^{\infty} t^{-2/3}\, dx = \infty.$$

If $\omega \in \mathcal{B}_p(\Omega)$ then $L^p(\Omega,\omega) \subset L^1_{\text{loc}}(\Omega)$ for every open set Ω. It thus make sense to talk about weak derivatives of functions in $L^p(\Omega,\omega)$.

Definition 1.2.4. *Let $\Omega \subset \mathbb{R}^N$ be open set, $1 \le p < \infty$ and k a nonnegative integer. Suppose that the weight $\omega \in \mathcal{B}_p(\Omega)$. We define the weighted Sobolev space $W^{k,p}(\Omega,\omega)$ as the set of functions $u \in L^p(\Omega,\omega)$ with weak derivatives $D^\alpha u \in L^p(\Omega,\omega)$ for $|\alpha| \le k$. The norm of $u \in W^{k,p}(\Omega,\omega)$ is given by*

$$\|u\|_{W^{k,p}(\Omega,\omega)} = \left(\sum_{|\alpha| \le k} \int_\Omega |D^\alpha u|\, \omega\, dx \right)^{1/p}. \qquad (1.2)$$

Theorem 1.2.3. *Let $\omega \in \mathcal{B}_p(\Omega)$, $1 \le p < \infty$. Then $W^{k,p}(\Omega,\omega)$ is a Banach space if equipped with the norm (1.2).*

Proof. For simplicity, we will deal with the case $k = 1$, i.e., with space $W^{1.p}(\Omega,\omega)$.

Step 1. If $\omega \in \mathcal{B}_p(\Omega)$ and $\varphi \in C_0^\infty(\Omega)$. Then, if $u \in L^p(\Omega,\omega)$, we have

$$L_0(u) = \int_\Omega u\, \varphi\, dx, \ \text{and}$$
$$L_j(u) = \int_\Omega u\, D_j\varphi\, dx,$$

defines a continuous linear functional L_j ($j = 0, 1, ..., n$) on $L^p(\Omega,\omega)$. In fact, if we denote $\Omega_\varphi = supp(\varphi)$ (support of φ), then $\Omega_\varphi = \overline{\Omega}_\varphi \subset \Omega$ and

$$|L_j(u)| \le \int_\Omega |u|\, |D_j\varphi|\, dx = \int_\Omega |u|\, \omega^{1/p}\, |D_j\varphi|\, \omega^{-1/p}\, dx$$

$$\le \left(\int_\Omega |u|^p \,\omega\, dx \right)^{1/p} \int_{\Omega_\varphi} |D_j\varphi|^{p'} \omega^{-p'/p}\, dx \Big)^{1/p'}$$

$$= \|u\|_{L^p(\Omega,\omega)} \left(\int_{\Omega_\varphi} |D_j\varphi|^{p/(p-1)}\, \omega^{-1/(p-1)}\, dx \right)^{(p-1)/p}$$

$$\le \|u\|_{L^p(\Omega,\omega)} \left(\sup_{\Omega_\varphi} |D_j\varphi| \right) \left(\int_\Omega \omega^{-1/(p-1)}\, dx \right)^{(p-1)/p}$$

$$= C\, \|u\|_{L^p(\Omega,\omega)},$$

where

$$C = \left(\sup_{\Omega_\varphi} |D_j\varphi| \right) \left(\int_\Omega \omega^{-1/(p-1)}\, dx \right)^{(p-1)/p} < \infty,$$

since $\varphi \in C_0^\infty(\Omega)$ and $\omega \in \mathcal{B}_p(\Omega)$.

Step 2. Let $\{u_n\}$ be a Cauchy sequence in $W^{1,p}(\Omega,\omega)$, i.e.,

$$\|u_m - u_n\|^p_{W^{1,p}(\Omega,\omega)}$$

$$= \int_\Omega |u_m - u_n|^p \,\omega\, dx + \sum_{j=1}^N \int_\Omega |D_j u_m - D_j u_n|^p \,\omega\, dx$$

$$\to 0,$$

when $m, n \to \infty$. Then $\{u_n\}$ and $\{D_j u_n\}$ $(j = 1, ..., n)$ are Cauchy sequences in $L^p(\Omega,\omega)$. By Theorem 1.2.1 there exist functions v, Ψ_j such that $v = \lim u_n$ and $\Psi_j = \lim D_j u_n$ in $L^p(\Omega,\omega)$. We have that $\Psi_j = D_j v$ $(j = 1, ..., N)$. In fact, by Step 1 the functional L_j is a continuous linear functional on $L^p(\Omega,\omega)$, consequently

$$L_j(u_n) \to L_j(v) \text{ for } n \to \infty.$$

We also have that $L_0(u) = \displaystyle\int_\Omega u\,\varphi\, dx$ $(\varphi \in C_0^\infty(\Omega))$ define a continuous linear functional on $L^p(\Omega,\omega)$. We have

$$\begin{aligned} L_j(u_n) &= \int_\Omega u_n\, D_j\varphi\, dx \\ &= -\int_\Omega \varphi\, D_j u_n\, dx \\ &= -L_0(D_j u_n). \end{aligned}$$

By a limiting process we obtain $L_j(v) = -L_0(\Psi_j)$, i.e.,

$$\int_\Omega v\, D_j\varphi\, dx = -\int_\Omega \Psi_j\, \varphi\, dx,$$

for every $\varphi \in C_0^\infty(\Omega)$. Therefore Ψ_j is the distributional derivative of v, i.e., $\Psi_j = D_j v$ $(j = 1, 2, .., N)$. Moreover, we have

$$\begin{aligned} \|u_n - v\|^p_{W^{1,p}(\Omega,\omega)} &= \int_\Omega |u_n - v|^p\, \omega\, dx \\ &\quad + \sum_{j=1}^N \int_\Omega |D_j u_n - D_j v|^p\, \omega\, dx \to 0 \end{aligned}$$

for $n \to \infty$. Therefore, the Cauchy sequence $\{u_n\}$ converges to v in $W^{1,p}(\Omega,\omega)$, i.e., $W^{1,p}(\Omega,\omega)$ is a Banach space. □

Theorem 1.2.4. *Let $\omega \in \mathcal{W}(\Omega)$. Then $C_0^\infty(\Omega) \subset W^{1,p}(\Omega,\omega)$ if and only if $\omega \in L^1_{loc}(\Omega)$.*

Proof. $(\Leftarrow)$ If $\omega \in L^1_{\text{loc}}(\Omega)$ and $\varphi \in C_0^\infty(\Omega)$ $(\overline{\Omega}_\varphi = \Omega_\varphi$ the support of $\varphi)$, we obtain

$$\int_\Omega |\varphi|^p\, \omega\, dx \leq \max_{\overline{\Omega}_\varphi}(|\varphi|^p)\left(\int_{\Omega_\varphi} \omega\, dx\right) < \infty,$$

and

$$\int_\Omega |D_j\varphi|^p\, \omega\, dx \leq \max_{\overline{\Omega}_\varphi}(|D_j\varphi|^p)\left(\int_{\Omega_\varphi} \omega\, dx\right) < \infty.$$

Hence, $C_0^\infty(\Omega) \subset W^{1,p}(\Omega, \omega)$.
($\Rightarrow$) If $C_0^\infty(\Omega) \subset W^{1,p}(\Omega, \omega)$. Let $K \subset \Omega$ be a compact set. Then there exists a function $\varphi \in C_0^\infty(\Omega)$ such that $D_j\varphi(x) = 1$ for $x \in K$ and we have

$$\begin{aligned} 0 &\leq \int_K \omega\, dx = \int_K |D_j\varphi|^p\, \omega\, dx \\ &\leq \int_\Omega |D_j\varphi|^p\, \omega\, dx \\ &\leq \|\varphi\|_{W^{1,p}(\Omega,\omega)} < \infty. \end{aligned}$$

Hence, $\omega \in L^1_{\text{loc}}(\Omega)$. $\square$

Definition 1.2.5. *Let $\Omega \subset \mathbb{R}^N$ be open and $\omega \in L^1_{loc}(\Omega)$. We define the space $W_0^{1,p}(\Omega, \omega)$ as the closure of $C_0^\infty(\Omega, \omega)$ with respect to the norm $\|.\|_{W^{1,p}(\Omega,\omega)}$ (i.e., $W_0^{1,p}(\Omega, \omega) = \overline{C_0^\infty(\Omega)}$).*

1.3 Examples of weights

Next we give some examples of weights.

1.3.1 Power-type weights

(i) Let Ω be a bounded domain in $\mathbb{R}^N$. Let M be a nonempty subset of $\overline{\Omega} = \Omega \cup \partial\Omega$, and denote $d_M(x) = \text{dist}(x, M)$ for $x \in \Omega$ (the distance of the point x from the set M). Let $\varepsilon \in \mathbb{R}$ and let us denote the power-type weight

$$\omega(x) = [d_M(x)]^\varepsilon.$$

So the singularity ($\varepsilon < 0$) or degenerations ($\varepsilon > 0$) can appear on the boundary $\partial\Omega$ of Ω as well as in the interior of the domain. The set M is very often a closed part of the boundary $\partial\Omega$ (i.e., $M \subset \partial\Omega$).

(ii) Let $s = s(t)$ be a continuous positive function defined for $t > 0$ and such that

$$\text{either } \lim_{t\to 0} s(t) = 0 \text{ or } \lim_{t\to 0} s(t) = +\infty.$$

Let us denote $\omega(x) = s\,(d_M(x))$. We have that:

(a) if $\lim_{t\to 0} s(t) = 0$ then $W^{k,p}(\Omega) \hookrightarrow W^{k,p}(\Omega, s(d_M))$;

(b) if $\lim_{t\to 0} s(t) = \infty$ then $W^{k,p}(\Omega, s(d_M)) \hookrightarrow W^{k,p}(\Omega)$.

For more information on properties of spaces with theses weights see [46].

1.3.2 A_p - weights

The class of A_p-weight was introduced by B. Muckenhoupt (see [50], where he showed that the A_p weights are precisely those weights ω for which the Hardy-Littlewood maximal operator is bounded from $L^p(\mathbb{R}^N, \omega)$ to $L^p(\mathbb{R}^N, \omega)$ $(1 < p < \infty)$, that is

$$M:\; L^p(\mathbb{R}^N, \omega) \to L^p(\mathbb{R}^N, \omega)$$
$$(Mf)(x) = \sup_{r>0} \frac{1}{|B_r(x)|} \int_{B_r(x)} |f(y)|\, dy,$$

is bounded if and only if $\omega \in A_p$ $(1 < p < \infty)$, i.e., there exists a positive constant $C = C_{p,\omega}$ such that

$$\left(\frac{1}{|B|}\int_B \omega\, dx\right)\left(\frac{1}{|B|}\int_B \omega^{-1/(p-1)}\, dx\right)^{p-1} \le C,$$

for every ball $B \subset \mathbb{R}^N$.

We say that a weight ω belongs to A_1 if there is a constant $C = C_{1,\omega}$ such that

$$\frac{1}{|B|}\int_B \omega\,dx \leq C\,\operatorname*{ess\,inf}_B \omega$$

for all balls B in $\mathbb{R}^N$.

The union of all Muckenhoupt classes A_p is denoted by A_∞, i.e., $A_\infty = \bigcup_{p>1} A_p$.

Example of A_p-weights

(i) If ω is a weight and there exist two positive constants C_1 and C_2 such that $C_1 \leq \omega(x) \leq C_2$ for a.e. $x \in \mathbb{R}^N$, then $\omega \in A_p$ for $1 \leq p < \infty$.
(ii) If $x \in \mathbb{R}^N$, $\omega(x) = |\,x|^{\alpha}$ is in A_p if and only if $-N < \alpha < N(p-1)$ (see Corollary 4.4 in [55]).
(iii) $\omega(x) = e^{\lambda\varphi(x)} \in A_2$, with $\varphi \in W^{1,N}(\Omega)$ (where Ω is a open set in $\mathbb{R}^N$) and λ is sufficiently small (see Corollary 2.18 in [38]).
(iv) $\omega(x) = |x|^{\alpha}(\max\{1, -\ln(|x|)\})^{\beta}$, $x \in \mathbb{R}^n$, is an A_1-weight if and only if $\alpha < 0$ or $\alpha = 0 \leq \beta$ (see Proposition 7.2 in [4]).
For more information about A_p-weights see [35], [38], [55] and [56].

1.3.3 p - admissible weights

Let ω be a locally integrable, nonnegative function in $\mathbb{R}^N$ and $1 < p < \infty$. We say that ω is *p-admissible* if the following four conditions are satisfied:

(I) $0 < \omega(x) < \infty$ a.e. $x \in \mathbb{R}^N$ and ω is *doubling* , i.e., there is a constant $C_1 > 0$ such that $\omega(2B) \leq C_1\,\omega(B)$, whenever $B = B(x,r)$ is a ball in $\mathbb{R}^N$ (where $2B(x,r) = B(x,2r)$).

(II) If Ω is an open set and $\varphi_k \in C^\infty(\Omega)$ is a sequence of functions such that

$$\int_\Omega |\varphi_k|^p \omega\, dx \to 0 \quad \text{and} \quad \int_\Omega |\nabla\varphi_k - \vartheta|^p \omega\, dx \to 0$$

as $k \to \infty$, where $\vartheta = (\vartheta_1, ..., \vartheta_N)$ is a vector-valued measure function with $\vartheta_j \in L^p(\Omega, \omega)$, then $\vartheta = 0$.
(III) There are constants $\theta > 1$ and $C_2 > 0$ such that

$$\left(\frac{1}{\omega(B)} \int_B |\varphi|^{\theta p} \omega\, dx\right)^{1/\theta p} \leq C_2 R \left(\frac{1}{\omega(B)} \int_B |\nabla\varphi|^p \omega\, dx\right)^{1/p}.$$

whenever $B = B(x_0, R)$ is a ball in $\mathbb{R}^N$ and $\varphi \in C_0^\infty(B)$.
(IV) There is a constant $C_3 > 0$ such that

$$\int_B |\varphi - \varphi_B|^p \omega\, dx \leq C_3 R^p \int_B |\nabla\varphi|^p \omega\, dx$$

whenever $B = B(x_0, R)$ is a ball in $\mathbb{R}^N$ and $\varphi \in C^\infty(B)$ is bounded and $\varphi_B = \dfrac{1}{\omega(B)} \displaystyle\int_B \varphi\, \omega\, dx$.

It follows immediately from condition (I) that the measure ω and the Lebesgue measure are mutually absolutely continuous. Condition (II) guarantees that the gradient of a Sobolev function is well defined. Condition (III) is the *weighted Sobolev inequality* and condition (IV) is the *weighted Poincaré inequality*.

Examples of p-admissible weights

(1) If $\omega \in A_p$ $(1 < p < \infty)$ then ω is a p-admissible weight.

(2) $\omega(x) = |x|^\alpha$, $x \in \mathbb{R}^N$, $\alpha > -N$, is a p-admissible weight for all $p > 1$.

(3) If $f:\mathbb{R}^N \to \mathbb{R}^N$ is a K - quasiconformal mapping and $J_f(x)$ is the determinant of its jacobian matrix, then $\omega(x) = J_f(x)^{1-p/N}$ is p-admissible for $1 < p < N$.

A continuous mapping $f:\mathbb{R}^N \to \mathbb{R}^N$, $f = (f_1, ..., f_N)$, is K-quasiconformal if:

(a) the coordinate functions $f_j \in W^{1,N}_{\text{loc}}(\Omega)$;

(b) f is a homeomorphism onto $f(\mathbb{R}^N)$;

(c) there is $K \geq 1$ such that the inequality $|f'(x)|^N \leq K\, J_f(x)$ is satisfied for a.e. x in Ω (here $f'(x)$ denotes the formal derivative of f at x, i.e., the $N \times N$ matrix $\left(\dfrac{\partial f_i}{\partial x_j}(x)\right)$ of the partial derivatives of the coordinate functions f_i of f, $|f'(x)| = \max\limits_{|h|=1} |f'(x)\, h|$ and $J_f(x) = \det(f'(x))$ is the Jacobian determinant of f at x.

(4) See [5] for non-A_p examples of p-admissible weights.

For more information about p-admissible weights see [43].

Remark 1.3.1. *Recently P.Hajłasz and P.Koskela (see [41]) showed that conditions (I) - (IV) can be reduced to only two: ω is a p-admissible weights ($1 < p < \infty$) if and only if ω is doubling and there are constants $C > 0$ and $\lambda \geq 1$ such that*

$$\frac{1}{\omega(B)} \int_B |\varphi - \varphi_B|\, \omega\, dx \leq C\, R\left(\frac{1}{\omega(\lambda B)} \int_{\lambda B} |\nabla\varphi|^p\, \omega\, dx\right)^{1/p},$$

(the weak $(1,p)$ -Poincaré inequality).

1.3.4 Regular weights

Let $\omega(x) \geq 0$ be a weight, with $\omega \in L^1_{\text{loc}}(\Omega)$. We consider the set of functions

$$X = \left\{ u \in W^{1,1}_{\text{loc}}(\Omega) \text{ such that } \|u\|^2_\omega = \int_\Omega (u^2 + |\nabla u|^2)\, \omega\, dx < \infty \right\}.$$

A weighted Sobolev space can be defined, in general, in two ways:
(a) $W = W(\Omega, \omega)$ is the completion of the set X with respect to the norm $\|.\|_\omega$;
(b) $H = H(\Omega, \omega)$ is the completion of the $\{u \in C^\infty(\Omega) : \|u\|_\omega < \infty\}$ with respect to the norm $\|.\|_\omega$ (the energy norm).

We have that $H \subseteq W$. By definition, functions smooth in the interior of Ω are dense in H, while the space W is known to contain all functions of finite well-defined " energy".

The classical Sobolev space corresponds to the weight $\omega(x) \equiv 1$ and is uniquely defined since the spaces $W(\Omega)$ and $H(\Omega)$ are the same for each domain Ω. Of course, if ω is bounded above and away from zero $(0 < c_1 \le \omega(x) \le c_2 < \infty)$, the spaces W and H are also the same. However, condition $\omega \in L^1_{\text{loc}}(\Omega)$ do not in general ensure the equality $H = W$.

Example 1.3.1. *Let* $N = 2$, $\Omega = \{x = (x_1, x_2) \in \mathbb{R}^2 \;:\; |x| < 1\}$ *and let*

$$\omega(x) = \begin{cases} \Big(\ln(2/|x|)\Big)^{\alpha}, & \text{for } x_1 x_2 > 0, \\ \Big(\ln(2/|x|)\Big)^{-\alpha}, & \text{for } x_1 x_2 < 0, \end{cases}$$

with $\alpha > 1$. *Then* $H \neq W$. *In fact, in the polar variables* $r = |x|$ *and* $\theta = \arccos(x_1/r)$, *we set*

$$u(x) = \begin{cases} 1, & \text{for } \; x_1 > 0 \text{ and } x_2 > 0, \\ \operatorname{sen}\theta, & \text{for } \; x_1 < 0 \text{ and } x_2 > 0, \\ 0, & \text{for } \; x_1 < 0 \text{ and } x_2 < 0, \\ \cos\theta, & \text{for } \; x_1 > 0 \text{ and } x_2 < 0. \end{cases}$$

We have that $u \in W(\Omega, \omega)$ *and* $u \notin H(\Omega, \omega)$ *(see counterexample 5.1 in [62]).*

If we have only the condition $\omega \in L^1_{\text{loc}}(\mathbb{R}^N)$ the H-solutions and W-solutions could be different when $N \geq 2$. See the example by F.S. Cassano (in [24]) where the H-solutions and the W-solutions with the same boundary data are different.

Definition 1.3.1. *If* $H = W$*, we call* ω *a regular weight.*

Sufficient conditions of the equality H=W

(1) F.S. Cassano has established an interesting fact: if the weighted Poincaré inequality

$$\int_B |\, u - u_B|^2 \omega \, dx \leq C(\text{diam}(B))^2 \int_B |\nabla u|^2 \omega \, dx$$

holds for each $u \in W(\mathbb{R}^N, \omega)$ then ω is a regular weight (see [24]).

Example 1.3.2. *An example of a weight function* ω *for which the weighted Poincaré inequality does not hold (see [23]).*

Let $\varphi : \; [0, \infty) \to [0, \infty)$ *be a function defined by*

$$\varphi(r) = r^\beta + \sum_{h=0}^{\infty} \varphi_h(r)$$

where $\varphi_h : [0,\infty)\to[0,\infty)$ *(*$h = 0,1,2,...$*) are the functions defined by*

$$\varphi_h(r) = \begin{cases} 0, \text{ if } 0\le r < 62^{-(h+3)} \text{ or } r\ge 2^{-h}, \\ 2^{-\alpha h}\left(1-\left(\dfrac{r-72^{-(h+3)}}{2^{-(h+3)}}\right)^2\right)^2, \\ \text{if } |r-72^{-(h+3)}| < 2^{-(h+3)}, \end{cases}$$

with α *and* β *are positive numbers such that* $0\le\alpha < \beta$.

Define now the function $\omega : \mathbb{R}^N \to [0,\infty)$ *as*

$$\omega(x) = \begin{cases} \dfrac{\varphi(|x|)}{|x|^{N-1}}, & \text{if } x\neq 0, \\ 0, & \text{if } x = 0. \end{cases}$$

Let $N\ge 2$, $p > 1$, α *and* β *be positive numbers verifying* $\alpha+p < \beta < pN-1$. *Then* ω *is a weight (with* ω *and* $\omega^{-1/(p-1)}\in L^1_{loc}(\mathbb{R}^N)$*) such that the Poincaré inequality does not hold.*

(2) V.V. Zhikov (see Theorem 4.1 in [62]) showed that: Let F be a closed subset of Ω and let $F_\varepsilon = \{x\in\Omega : \text{dist}(x,F)\le\varepsilon\}$. Assume that the weight ω has degeneracies only on F (i.e., $0 < c_1(\varepsilon)\le\omega(x)\le c_2(\varepsilon) < \infty$ in $\Omega\setminus F_\varepsilon$). If

$$\text{cap}(F) = 0 \quad \text{and} \quad \omega(x)\le \frac{C}{\text{cap}(F_\varepsilon)} \quad \text{on} \quad \Omega\setminus F_\varepsilon,$$

where

$$\text{cap}\,(F_\varepsilon) = \inf\left\{\int_\Omega |\nabla u|^2 dx : u\in C_0^\infty(\Omega),\ u = 1 \text{ in a neighborhood of } F_\varepsilon\right\},$$

then ω is a regular weight, that is, $H(\Omega,\omega) = W(\Omega,\omega)$ (here $\mathrm{cap}(A)$ is the Wiener capacity of the set A).

1.3.5 Weighted Sobolev spaces with variable exponent

Most materials can be modelled with sufficient accuracy using classical Lebesgue and Sobolev spaces, L^p and $W^{1,p}$, where p is a fixed constant. For some materials with inhomogeneities, for instance eletrorheological fluids, this is not adequate, but rather the exponent p should be able to vary. This leads us to study of variable exponent Lebesgue and Sobolev spaces, $L^{p(x)}$ and $W^{1,p(x)}$, where $p = p(x)$ is a real-valued function (see [28]).

The study of differential equations and variational problems involving $p(x)$-growth conditions is a consequence of their applications. Materials requiring such more advanced theory have been studied experimentally since the middle of the last century. The first major discovery in eletrorheological fluids was due to Willis Winslow in 1949. These fluids have the interesting property that their viscosity depends on the electric field in the fluid. Winslow noticed that in such fluids viscosity in an electrical field is inversely proportional to the strength of the field. The field induces string-like formations in the fluid, which are parallel to the field. They can raise the viscosity by as much as five orders of magnitude. This phenomenon is known as the Winslow effect. Electrorheological fluids have been used in robotics and space technology.

For a general account of the underlying physics consult [42] and [51].

Definition 1.3.2. *Let $\Omega \subset \mathbb{R}^N$ be a domain with nonempty boundary $\partial\Omega$. Denote*

$$L_+^\infty(\Omega) = \{p \in L^\infty(\Omega) \ : \ \operatorname{ess\,inf} p(x) > 1\}.$$

Definition 1.3.3. *Let ω be a measurable nonnegative and a.e. finite function defined in $\mathbb{R}^N$. For $p \in L_+^\infty(\Omega)$, define*

$$L^{p(x)}(\Omega, \omega)$$
$$= \{u \ : \ u \text{ is measurable on } \Omega \text{ and } \int_\Omega |u(x)|^{p(x)} \omega(x)\, dx < \infty\},$$

with the norm

$$\| u\|_{L^{p(x)}(\Omega,\omega)} = \inf\left\{\lambda > 0 \ : \ \int_\Omega \left|\frac{u(x)}{\lambda}\right|^{p(x)} \omega(x)\, dx \leq 1\right\}.$$

Definition 1.3.4. *Let ω and v are measurable nonnegative and a.e. finite functions defined in $\mathbb{R}^N$. For $p \in L_+^\infty(\Omega)$, define*

$$W^{1,p(x)}(\Omega, \omega, v) = \{u \in L^{p(x)}(\Omega, \omega) \ : \ |\nabla u| \in L^{p(x)}(\Omega, v)\},$$

with the norm

$$\|u\|_{W^{1,p(x)}(\Omega,\omega,v)} = \| u\|_{L^{p(x)}(\Omega,\omega)} + \|\nabla u\|_{L^{p(x)}(\Omega,v)}$$

or

$$\|u\|_{W^{1,p(x)}(\Omega,\omega,v)}$$
$$= \inf\left\{\lambda > 0 \ : \ \int_\Omega \left(\left|\frac{u(x)}{\lambda}\right|^{p(x)} \omega(x) + \left|\frac{\nabla u(x)}{\lambda}\right|^{p(x)} v(x)\right) dx \leq 1\right\}.$$

For more information properties see [29], [45] and [52].

1.4 Weighted Sobolev spaces with A_p-weights

The present section explores some properties of weighted Sobolev spaces $W^{k,p}(\Omega,\omega)$ and $W_0^{k,p}(\Omega,\omega)$ in particular case where $\omega\in A_p$.

The following results are proved exactly in the same way as in the non-weighted case (see [1] and [39]) using the completeness of $L^p(\Omega,\omega)$ and the fact that $L^p(\Omega,\omega)\subset L^1_{\mathrm{loc}}(\Omega)$ when $\omega\in A_p$.

Theorem 1.4.1. *Let $\Omega\subset\mathbb{R}^N$ be open, $1\le p<\infty$, k a nonnegative integer and $\omega\in A_p$, Then $W^{k,p}(\Omega,\omega)\subset W^{k,1}_{loc}(\Omega)$ and, if Ω is bounded, we have*

$$W^{k,p}(\Omega,\omega)\subset W^{k,1}(\Omega).$$

Theorem 1.4.2. *Suppose that $\Omega\subset\mathbb{R}^N$ is an open subset. Let $\varphi\in C_0^\infty(\mathbb{R}^N)$ be nonnegative such that $\operatorname{supp}\varphi\subset B(0;1)$ and $\int_{\mathbb{R}^N}\varphi(x)\,dx=1$, and define for $\varepsilon>0$, $\varphi_\varepsilon(x)=\frac{1}{\varepsilon^N}\varphi(x/\varepsilon)$, $x\in\mathbb{R}^N$. Suppose that $\omega\in A_p$, $1\le p<\infty$. If $u\in L^p(\Omega,\omega)$ we define $u_\varepsilon=u*\varphi_\varepsilon$ (the convolution of u and φ_ε). Then*

$$u_\varepsilon\in C^\infty(\Omega)\cap L^p(\Omega,\omega)$$

with $D^\alpha u_\varepsilon=u*D^\alpha\varphi_\varepsilon$, and $u_\varepsilon\to u$ in $L^p(\Omega,\omega)$ as $\varepsilon\to 0$.

Proof. We will prove the theorem in the case $p=1$.
Since $u\in L^1(\Omega,\omega)\subset L^1_{\mathrm{loc}}(\Omega)$, then $u_\varepsilon\in C^\infty(\Omega)$. Let $r>0$ be arbitrary and choose a function $\psi\in C_0^1(\Omega)$ such that $\|u-\psi\|_{L^1(\Omega,\omega)}<r$. We then get

$$\begin{aligned}\|u-u_\varepsilon\|_{L^1(\Omega,\omega)}&\le\|u-\psi\|_{L^1(\Omega,\omega)}+\|\psi-\psi_\varepsilon\|_{L^1(\Omega,\omega)}\\&+\|\psi_\varepsilon-u_\varepsilon\|_{L^1(\Omega,\omega)}\\&<r+\|\psi-\psi_\varepsilon\|_{L^1(\Omega,\omega)}+\|\psi_\varepsilon-u_\varepsilon\|_{L^1(\Omega,\omega)}.\end{aligned}\tag{1.3}$$

If $\operatorname{supp}\psi \subset B(0;R)$, then $\operatorname{supp}\psi_\varepsilon \subset B(0;2R)$, if $\varepsilon < R$. Also, for every $x \in \mathbb{R}^N$

$$\begin{aligned} |\psi(x) - \psi_\varepsilon(x)| &\leq \int_{|x-y|<\varepsilon} |\psi(x) - \psi(y)|\, \varphi_\varepsilon(x-y)\, dx \\ &\leq \sup_{|x-y|<\varepsilon} |\psi(x) - \psi(y)|. \end{aligned}$$

Thus, if we choose $\varepsilon < R$ so small that

$$|\psi(x) - \psi(y)| < \frac{r}{\omega(B(0;2r))}$$

for every x and y with $|x-y| < \varepsilon$, we see that the second term in (1.3) is $< r$. Let us now consider the last term in (1.3). By Fubini's theorem and $(M\omega)(x) \leq C_{1,\omega}\, \omega(x)$ a.e. (where M is the Hardy-Littlewood maximal operator, i.e., $(M\omega)(x) = \sup\limits_{r>0} \dfrac{1}{|B(x;r)|} \displaystyle\int_{B(x;r)} \omega(y)\, dy$), we obtain

$$\begin{aligned} &\int_{\mathbb{R}^N} |\psi_\varepsilon - u_\varepsilon|\, \omega\, dx \\ &\leq \int_{\mathbb{R}^N} \left(\int_{\mathbb{R}^N} \varphi_\varepsilon(x-y)\, \omega(x)\, dx \right) |\psi(x) - u(y)|\, dy \\ &\leq \int_{\mathbb{R}^N} (M\omega)\, |\psi - u|\, dy \\ &\leq C_{1,\omega} \int_{\mathbb{R}^N} |\psi - u|\, \omega\, dy \\ &< C_{1,\omega}\, r. \end{aligned}$$

This shows that $u_\varepsilon \to u$ as $\varepsilon \to 0$. Finally, if we replace $\psi - u$ by u in the last series of inequalities, we get $\|u_\varepsilon\|_{L^1(\Omega,\omega)} \leq C\, \|u\|_{L^1(\Omega,\omega)}$ and thus $u_\varepsilon \in L^1(\Omega,\omega)$, □

Theorem 1.4.2 has the following corollary, which is proved in the same way as when $\omega \equiv 1$.

Corollary 1.4.3. *Suppose that* $\Omega \subset \mathbb{R}^N$ *is an open subset and let* $\tilde{\Omega}$ *be an open set of* Ω *such that* $\operatorname{dist}(\tilde{\Omega}, \Omega^c) > 0$. *Suppose that* $u \in W^{k,p}(\Omega, \omega)$, $k \geq 1$, $1 \leq p < \infty$ *and* $\omega \in A_p$. *If we define* u_ε *as in Theorem 1.4.2, then as* $\varepsilon \to 0$, $u_\varepsilon \to u$ *in* $W^{k,p}(\tilde{\Omega}, \omega)$. *In the case* $\Omega = \mathbb{R}^N$, *we have convergence in* $W^{k,p}(\mathbb{R}^N, \omega)$.

Using Corollary 1.4.3, one can prove that $C^\infty(\Omega)$ is dense in $W^{k,p}(\Omega, \omega)$. In the non-weighted case, see Theorem 7.9 in [39].

Theorem 1.4.4. *If* $\Omega \subset \mathbb{R}^N$ *is open,* $1 \leq p < \infty$, $k \geq 1$ *a non-negative integer and* $\omega \in A_p$, *then* $C^\infty(\Omega)$ *is dense in* $W^{k,p}(\Omega, \omega)$.

Theorem 1.4.4 implies that functions in $W^{k,p}(\Omega, \omega)$ can be truncated. The proofs of the theorem and corollary below are identical the proofs in the case $\omega \equiv 1$ (see Lemma 7.6 and Lemma 7.7 in [39]).

Theorem 1.4.5. *Suppose that* $h \in C^1(\mathbb{R})$ *with* $h' \in L^\infty(\mathbb{R})$. *Let* $\Omega \subset \mathbb{R}^N$ *be open, and let* $u \in W^{1,p}(\Omega, \omega)$, $1 \leq p < \infty$ *and* $\omega \in A_p$. *Suppose that* $h \circ u \in L^p(\Omega, \omega)$. *Then* $h \circ u \in W^{1,p}(\Omega, \omega)$ *with* $D_j(h \circ u) = h'(u)\, D_j u$, $j = 1, ..., N$.

Corollary 1.4.6. *Suppose that* $\Omega \subset \mathbb{R}^N$ *is open and let* $u \in W^{k,p}(\Omega, \omega)$, $1 \leq p < \infty$ *and* $\omega \in A_p$. *Set* $u^+ = \max\{u, 0\}$ *and* $u^- = \min\{u, 0\}$. *Then* u^+, u^- *and* $|u| \in W^{1,p}(\Omega, \omega)$ *and, for* $j = 1, ..., N$

$$D_j\, u^+ = \begin{cases} D_j u, & \text{if } u > 0, \\ 0, & \text{if } u \leq 0, \end{cases}$$

$$D_j\, u^- = \begin{cases} 0, & \text{if } u \geq 0, \\ D_j u, & \text{if } u < 0, \end{cases}$$

$$D_j\, |u| = \begin{cases} D_j, & \text{if } u > 0, \\ 0, & \text{if } u = 0, \\ -D_j u, & \text{if } u < 0. \end{cases}$$

Theorem 1.4.7. *Let $\Omega \subset \mathbb{R}^N$ be a domain (i.e., open and connected set), $u \in W^{1,p}(\Omega,\omega)$ and $\omega \in A_p$ $(1 \leq p < \infty)$. If $\nabla u = 0$, then u is constant.*

Corollary 1.4.8. *If $u \in W_0^{1,p}(\Omega,\omega)$ with $\nabla u = 0$, then $u = 0$.*

Theorem 1.4.9. *Let $\Omega \subset \mathbb{R}^N$ be a open set, $\omega \in A_p$ and $1 < p < \infty$. Suppose that $\{u_m\}$ is a bounded sequence in $W^{1,p}(\Omega,\omega)$. Then there is a subsequence $\{u_{m_k}\}$ and a function $u \in W^{1,p}(\Omega,\omega)$ such that $u_{m_k} \to u$ weakly in $L^p(\Omega,\omega)$ and $\nabla u_{m_k} \to \nabla u$ weakly in $L^p(\Omega,\omega)$. Moreover, if $u \in W_0^{1,p}(\Omega,\omega)$, then $u \in W_0^{1,p}(\Omega,\omega)$.*

Proof. The existence of weakly convergent subsequence u_{m_k} and ∇u_{m_k} is a standard theorem of functional analysis (see [58] Theorem 1, Section 2, Chapter V). An appeal to Mazur's Lemma ([58], Theorem 2, Section 1, Chapter V) completes the proof. □

Theorem 1.4.10. *If $\omega \in A_p$, $1 < p < \infty$ and $\Omega \subset \mathbb{R}^N$ be a open set, then the dual space of $W_0^{1,p}(\Omega,\omega)$ is the space*

$$\begin{aligned}
&[W_0^{1,p}(\Omega,\omega)]^* \\
&= \big\{T = f_0 - \operatorname{div}(F) : F = (f_1, ..., f_N), \\
&\qquad \text{with } \frac{f_j}{\omega} \in L^{p'}(\Omega,\omega), j = 0, ..., N\big\}.
\end{aligned}$$

Proof. The proof of this theorem follows the line of Theorem 3.12 in [1]. □

Theorem 1.4.11. *Let $\omega \in A_p$, $1 < p < \infty$, and Ω be a bounded open set in $\mathbb{R}^N$. If $u_n \to u$ in $L^p(\Omega,\omega)$ then there exist a subsequence $\{u_{n_k}\}$ and a function $\Phi \in L^p(\Omega,\omega)$ such that*
(i) $u_{n_k}(x) \to u(x)$, $n_k \to \infty$, μ-a.e. on Ω;
(ii) $|u_{n_k}| \leq \Phi(x)$ μ-a.e. on Ω,
where $\mu(E) = \int_E \omega(x)\,dx$.

Proof. The proof of this theorem follows the lines of Theorem 2.8.1 in [34]. □

Theorem 1.4.12. *If u and v are in $W^{1,p}(\Omega,\omega)$, $\omega \in A_p$ ($1 < p < \infty$), then $\max\{u,v\}$ and $\min\{u,v\}$ are in $W^{1,p}(\Omega,\omega)$ with*

$$\nabla \max\{u,v\}(x) = \begin{cases} \nabla u(x) \text{ if } u(x) \geq v(x), \\ \nabla v(x), \text{ if } u(x) \leq v(x), \end{cases}$$

$$\nabla \min\{u,v\}(x) = \begin{cases} \nabla u(x), \text{ if } u(x) \leq v(x), \\ \nabla v(x), \text{ if } u(x) \geq v(x). \end{cases}$$

In particular, $|u| = u^+ - u^-$ belongs to $W^{1,p}(\Omega,\omega)$.

Proof. Since $\max\{u,v\} = (u-v)^+ + v$ and $\min\{u,v\} = u - (u - v)^+$, the assertions follow from Corollary 1.4.6. □

Corollary 1.4.13. *If $u \in W^{1,p}_{\text{loc}}(\Omega,\omega)$, $1 \leq p < \infty$, $\omega \in A_p$ and $\lambda \in \mathbb{R}$, then $\nabla u = 0$ a.e. on the set $\{x \in \Omega : u(x) = \lambda\}$.*

Remark 1.4.1. *(a) If $u, v \in W^{1,p}_0(\Omega,\omega)$, $\omega \in A_p$ ($1 < p < \infty$), then $\min\{u,v\}$ and $\max\{u,v\}$ are in $W^{1,p}_0(\Omega,\omega)$.*
(b) Note that the space $W^{1,2}_0(\Omega,\omega)$, $\omega \in A_2$, is a real Hilbert space when equipped with the inner product

$$\langle u,v \rangle = \int_\Omega u\, v\, \omega\, dx + \sum_{j=1}^N \int_\Omega D_j u\, D_j v\, \omega\, dx.$$

Definition 1.4.1. *Let $\Omega \subset \mathbb{R}^n$ be an open bounded set, $k \in \mathbb{N}$ and $1 \leq p < \infty$. Let ω be a given family of weight functions ω_α ($|\alpha| \leq k$): $\omega = \{\omega_\alpha(x) : x \in \Omega,\ |\alpha| \leq k\}$. We denote by $W^{k,p}(\Omega,\omega)$ the set of all functions $u \in L^p(\Omega,\omega_0)$ for which the weak derivatives $D^\alpha u$ (with $|\alpha| \leq k$) belong to $L^p(\Omega,\omega_\alpha)$. The weighted Sobolev space $W^{k,p}(\Omega,\omega)$ is a normed linear space if equipped with the norm*

$$\|u\|_{W^{k,p}(\Omega,\omega)} = \left(\sum_{|\alpha| \leq k} \|D^\alpha u\|^p_{L^p(\Omega,\omega_\alpha)} \right)^{1/p}. \tag{1.4}$$

Theorem 1.4.14. *(i) Let $1 < p < \infty$ and suppose that the weight functions ω_α satisfy $\omega_\alpha^{-1/(p-1)} \in L^1_{\text{loc}}(\Omega)$ $(|\alpha| \leq k)$. Then $W^{k,p}(\Omega,\omega)$ is a uniformly convex (and hence reflexive) Banach space.*

(ii) If we additionally suppose that also $\omega_\alpha \in L^1_{\text{loc}}(\Omega)$, $|\alpha| \leq k$, then $C_0^\infty(\Omega)$ is a subset of $W^{k,p}(\Omega,\omega)$ and we can introduce the space $W_0^{k,p}(\Omega,\omega)$ as the closure of $C_0^\infty(\Omega)$ with respect to the norm $\|.\|_{W^{k,p}(\Omega,\omega)}$.

Proof. See Theorem 1.3 in [32], [48]. □

Note that if $w_\alpha \in A_p$ ($1 < p < \infty$ and $|\alpha| \leq k$) then conditions (i) and (ii) of Theorem are satisfied.

For example, if $k = 1$ and $\omega = (\omega_0, \omega_1, ..., \omega_n)$, with $\omega_j \in L^1_{\text{loc}(\Omega)}$, then $W^{1,p}(\Omega,\omega)$ as the set of all functions $u \in L^p(\Omega, \omega_0$ such that $D_j u \in L^p(\Omega, \omega_j)$ with the norm

$$\|u\|_{W^{1,p}(\Omega,\omega)} = \left(\int_\Omega |u|^p \omega_0 + \sum_{j=1}^n \int_\Omega |D_j u|^p \omega_j \, dx \right)^{1/p}. \quad (1.5)$$

If we additionally suppose that $\omega_j \in L^1(\Omega)$ then

$$W_0^{1,p}(\Omega,\omega) = \overline{C_0^\infty(\Omega)}$$

the closure being taken with respect to the norm (1.5)

Chapter 2

$\mathbf{A_p}$ - weights

In this chapter we establish some basic properties of A_p-weights, that will be used throughout this book. Complete expositions can be found in [38], [43] and [55].

2.1 Basic results concerning A_p - weights

We begin by defining the class of A_p-weights.

Definition 2.1.1. *Let $1 \leq p < \infty$. A weight ω is said to be an A_p-weight, if there is a positive constant $C = C_{p,\omega}$ such that, for every ball $B \subset \mathbb{R}^N$,*

$$\left(\frac{1}{|B|}\int_B \omega\, dx\right)\left(\frac{1}{|B|}\int_B \omega^{-1/(p-1)}\, dx\right)^{p-1} \leq C, \quad \textit{if } p > 1,$$
$$\left(\frac{1}{|B|}\int_B \omega\, dx\right)\left(\operatorname*{ess\,sup}_{x \in B} \frac{1}{\omega(x)}\right) \leq C, \quad \textit{if } p = 1.$$

If $\omega \in A_p$, $1 \leq p < \infty$, then since $\omega^{-1/(p-1)}$ is locally integrable (when $p > 1$) and ω^{-1} is locally bounded (when $p = 1$) we have

$$L^p(\Omega, \omega) \subset L^1_{\mathrm{loc}}(\Omega)$$

for every open set $\Omega \subseteq \mathbb{R}^N$. Moreover, if $\omega \in A_p$, $p > 1$, then by writing $1 = \omega^{1/p}\omega^{-1/p}$, Hölder inequality implies that, for every ball B,

$$\begin{aligned} 1 &= \frac{1}{|B|}\int_B 1\,dx = \frac{1}{|B|}\int_B \omega^{1/p}\omega^{-1/p}\,dx \\ &\leq \frac{1}{|B|}\Big(\int_B \omega\,dx\Big)^{1/p}\Big(\int_B \omega^{-p'/p}\,dx\Big)^{1/p'} \\ &= \Big(\frac{1}{|B|}\int_B \omega\,dx\Big)\Big(\frac{1}{|B|}\int_B \omega^{-1/(p-1)}\,dx\Big)^{(p-1)/p} \\ &\leq C_{p,\omega}^{1/p}, \end{aligned}$$

i.e., the A_p constant of ω $C_{p,\omega} \geq 1$. Analogously, if $\omega \in A_1$ we also have $C_{1,\omega} \geq 1$.

Remark 2.1.1. *(i) The A_p condition is invariant under translations and dilations, i.e., if $\omega \in A_p$, then the weights $v_1(x) = \omega(x+a)$ and $v_2(x) = \omega(\delta\, x)$ ($a \in \mathbb{R}^N$ and $\delta > 0$ fixed), both belong to A_p with the same A_p constants as ω.*
(ii) As it sometimes is more convenient to work with cubes than balls, it is useful to notice that if one replaces the balls in the definitions of A_p with cubes, one gets the same class of weights and the different A_p constants are comparable. The proof of Lemma 2.2.1 is based on the Calderon-Zygmund Decomposition performed for dyadic cubes of $\mathbb{R}^N$. Throughout this book, however, we use balls instead of cubes in our estimates. This deviation from the standard proof causes only minor technical complications (see for example the Calderon-Zygmund decomposition for doubling weights in [43]).

Lemma 2.1.1. *If $1 \leq p < q < \infty$, then $A_1 \subset A_p \subset A_q$, and the A_q constant of a weight $\omega \in A_p$ equals the A_p constant of ω.*

Proof. We just need to observe that, if $1 < p < q$, since $(q-1)/(p-1) > 1$, by Hölder inequality we have that

$$\Big(\frac{1}{|B|}\int_B \omega^{-1/(q-1)}\,dx\Big)^{q-1} \leq \Big(\frac{1}{|B|}\int_B \omega^{-1/(p-1)}\,dx\Big)^{p-1}$$

$$\leq \operatorname*{ess\,sup}_{x\in B} \frac{1}{\omega(x)}.$$

□

Example 2.1.1. *There is a connection between A_p and BMO (the John-Nirenberg space of functions with bounded mean oscillation). In fact, if ω is a weight, then $\ln(\omega)\in BMO$ if and only if there is a constant $\eta>0$ such that $\omega^\eta\in A_2$ (see Proposition 6.1, Chapter IX in [55]).*

Lemma 2.1.2. *If $\omega\in A_p$, $1<p<\infty$.*
(a) Strong doubling property:

$$\left(\frac{|E|}{|B|}\right)^p \leq C_{p,\omega}\frac{\omega(E)}{\omega(B)}$$

whenever B is a ball in $\mathbb{R}^N$ and E is a measurable subset of B (where $\omega(E)=\displaystyle\int_E \omega(x)\,dx$).
(b) there exist $\delta>0$ and $C>0$ depending only on n,p and $C_{p,\omega}$ such that, every time we have a measurable set E contained in a cube K_0, the following inequality holds

$$\frac{\omega(E)}{\omega(K_0)}\leq C\left(\frac{|E|}{|K_0|}\right)^\delta.$$

Proof. (a) We have that, for $1<p<\infty$,

$$\begin{aligned}
|E| &= \int_E 1\,dx=\int_E \omega^{1/p}\,\omega^{-1/p}\,dx\\
&\leq \left(\int_E \omega\,dx\right)^{1/p}\left(\int_E \omega^{-p'/p}\,dx\right)^{1/p'}\\
&= \left(\int_E \omega\,dx\right)^{1/p}\left(\int_E \omega^{1/(p-1)}\,dx\right)^{(p-1)/p}
\end{aligned}$$

$$\begin{aligned}
&\leq [\omega(E)]^{1/p}\left(\frac{1}{|B|}\int_B \omega^{1/(p-1)}\,dx\right)^{(p-1)/p}|B|^{(p-1)/p}\\
&\leq C_{p,\omega}^{1/p}[\omega(E)]^{1/p}\left(\int_B \frac{1}{|B|}\int_B \omega\,dx\right)^{-1/p}|B|^{(p-1)/p}\\
&= C_{p,\omega}^{1/p}\,\frac{[\omega(E)]^{1/p}}{[\omega(B)]^{1/p}}\,\frac{|B|^{(p-1)/p}}{|B|^{-1/p}}\\
&= C_{p,\omega}^{1/p}\left(\frac{\omega(E)}{\omega(B)}\right)^{1/p}|B|.
\end{aligned}$$

Hence $\left(\dfrac{|E|}{|B|}\right)^p \leq C_{p,\omega}\,\dfrac{\omega(E)}{\omega(B)}$.

(b) See Theorem 2.9 in [38] or Lemma 15.8 in [43] . □

Therefore, by Lemma 2.1.2(a), $\omega(E) = 0$ if and only if $|E| = 0$; so there is no need to specify the measure when using the ubiquitous expression *almost everywhere* and *almost every*, both abbreviated a.e.. Thus, if $\{u_n\}$ is a sequence of functions defined in B and $u_n \to u$ μ-a.e. then $u_n \to u$ a.e. (where $\mu(E) = \displaystyle\int_E \omega(x)\,dx$).

Corollary 2.1.3. *If $\omega \in A_p$, then $\omega(B(x,2r)) \leq C\,\omega(B(x,r))$ for all balls $B = B(x,r)$ in $\mathbb{R}^N$, and $C = 2^{Np}\,C_{p,\omega}$.*

Proof. Since $B(x,r) \subset B(x,2r)$, then (by Lemma 2.1.2)

$$\left(\frac{|B(x,r)|}{|B(x,2r)|}\right)^p \leq C_{p,\omega}\frac{\omega(B(x,r))}{\omega(B(x,2r))},$$

and we obtain $\omega(B(x,2r)) \leq 2^{Np}\,C_{p,\omega}\,\omega(B(x,r)$. Hence, if $\omega \in A_p$, then ω is a doubling measure. □

Theorem 2.1.4. *(The Jones Factorization Theorem) For $1 < p < \infty$. Then $\omega \in A_p$ if and only if $\omega = \omega_0\,\omega_1^{1-p}$, where ω_0 and ω_1 are A_1 - weights.*

Proof. See Corollary 5.3, Chapter IV in [38]. □

Theorem 2.1.5. *(The weighted Sobolev inequality) Let Ω be an open bounded set in $\mathbb{R}^N$ and $\omega \in A_p$ $(1 < p < \infty)$. There exist positive constants C_Ω and δ such that for all $u \in W_0^{1,p}(\Omega,\omega)$ and all θ satisfying $1 \leq \theta \leq N/(N-1)+\delta$,*

$$\|u\|_{L^{p\theta}(\Omega,\omega)} \leq C_\Omega \| \, |\nabla u| \, \|_{L^p(\Omega,\omega)}, \tag{2.1}$$

where C_Ω depends only on N, p, the A_p-constant $C(p,\omega)$ of ω and the diameter of Ω, and δ depends on p.

Proof. Its suffices to prove the inequality for functions $u \in C_0^\infty(\Omega)$ (see Theorem 1.3 in [35]). To extend the estimates (2.1) to arbitrary function $u \in W_0^{1,p}(\Omega,\omega)$, we let $\{u_m\}$ be a sequence of $C_0^\infty(\Omega)$ functions tending to u in $W_0^{1,p}(\Omega,\omega)$. Applying the estimates (2.1) to differences $u_{m_1} - u_{m_2}$, we see that $\{u_m\}$ will be a Cauchy sequence in $L^p(\Omega,\omega)$. Consequently the limit function u will lie in the desired spaces and satisfy (2.1). □

Remark 2.1.2. *If $\omega \in A_p$, $1 < p < \infty$, then by Theorem 2.1.5 (with $\theta = 1$), the norms in $W_0^{1,p}(\Omega,\omega)$*

$$\|u\|_1 = \left(\int_\Omega |\nabla u|^p \, \omega \, dx \right)^{1/p}$$

$$\|u\|_2 = \left(\int_\Omega |u|^p \, \omega \, dx + \int_\Omega |\nabla u|^p \, \omega \, dx \right)^{1/p}$$

are equivalents $\|u\|_1 \leq \|u\|_2 \leq (C_\Omega^p + 1)^{1/p} \|u\|_1$. We will denote both by $\|.\|_{W_0^{1,p}(\Omega,\omega)}$.

Theorem 2.1.6. *(Reverse Hölder inequality) If $\omega \in A_p$, then there are numbers $r > 1$ and $C_r \geq 1$, depending on N, p and $C_{p,\omega}$, such that*

$$\left(\frac{1}{|B|} \int_B \omega^r \, dx \right)^{1/r} \leq C_r \left(\frac{1}{|B|} \int_B \omega \, dx \right),$$

for all balls $B \subset \mathbb{R}^N$.

Proof. See Theorem 15.3 in [43] or Theorem 3.5 in [55]. □

Theorem 2.1.7. *(Open-end property) If* $\omega \in A_p$, $1 < p < \infty$, *then* $\omega \in A_q$ *for some* $q < p$ *(or,* $\omega \in A_{p-\varepsilon}$ *for some* $\varepsilon > 0$*). Moreover,* q *and* $C_{q,\omega}$ *depend only on* n, p *and* $C_{p,\omega}$.

Proof. Scc Theorem 15.13 in [43] or Proposition 4.5 Chapter IX in [55]. □

Definition 2.1.2. *We say that the pair of weights* (v, ω) *satisfies the condition* A_p, $1 \le p < \infty$, *if there is a constant* C *such that*

$$(a)\ \left(\frac{1}{|B|}\int_B v\,dx\right)\left(\frac{1}{|B|}\int_B \omega^{-1/(p-1)}\,dx\right)^{p-1} \le C,$$

when $1 < p < \infty$,

$$(b)\ \frac{1}{|B|}\int_B v\,dx \le C\operatorname*{ess\,inf}_B \omega, \quad \text{when } p = 1,$$

for all balls B *in* $\mathbb{R}^N$. *The smallest constant* C *will be called the* A_p*-constant for the pair* (v, ω).

Lemma 2.1.8. *Let* $(v, \omega) \in A_p$, $1 < p < \infty$. *Then* $(v, \omega) \in A_q$ *for every* $p < q < \infty$.

Proof. By Hölder inequality with exponents $\theta = (q-1)/(q-p)$ and $\theta' = (q-1)/(p-1)$, we obtain

$$\begin{aligned}
\frac{1}{|B|}\int_B \omega^{-1/(q-1)}\,dx &= \int_B \left(\frac{1}{|B|}\right)^{1/\theta}\left(\frac{1}{|B|}\right)^{1/\theta'} \omega^{-1/(q-1)}\,dx \\
&\le \left(\int_B \frac{1}{|B|}\,dx\right)^{1/\theta}\left(\frac{1}{|B|}\int_B \omega^{-\theta'/(q-1)}\,dx\right)^{1/\theta'} \\
&= \left(\frac{1}{|B|}\int_B \omega^{-1/(p-1)}\,dx\right)^{(p-1)/(q-1)},
\end{aligned}$$

for every ball B. Then

$$\begin{aligned}&\left(\frac{1}{|B|}\int_B v\,dx\right)\left(\frac{1}{|B|}\int_B \omega^{-1/(q-1)}\,dx\right)^{q-1}\\ \leq&\left(\frac{1}{|B|}\int_B v\,dx\right)\left(\frac{1}{|B|}\int_B \omega^{-1/(p-1)}\,dx\right)^{p-1}\\ \leq& C,\end{aligned}$$

i.e., $(v,\omega)\in A_q$. □

Theorem 2.1.9. *(Hardy inequality) Let* $1 < r < p_1 < Nr$, $1/p_2 = 1/p_1 - 1/Nr$ *and* $(v,\omega)\in A_r$. *Then there exists a constant* $C_\Omega > 0$ *such that the inequality*

$$\left(\int_\Omega |u|^{p_2} v\,dx\right)^{1/p_2} \leq C_\Omega \left(\int_\Omega |\nabla u|^{p_1}\,\omega\,dx\right)^{1/p_1},$$

holds for every $u\in C_0^1(\Omega)$.

Proof. See Theorem 15.8 in [48]. □

Note that, by the same argument used in Theorem 2.1.5, Theorem 2.1.9 (with $v=\omega$) holds for $u\in W_0^{1,p}(\Omega,\omega)$.

2.2 Approximation of A_p-weights

For $k\in\mathbb{Z}$, we consider the lattice $\Gamma_k = 2^{-k}\mathbb{Z}^N$ formed by those points of $\mathbb{R}^N$ whose coordinate are integral multiples of 2^{-k}, i.e.,

$$\Gamma_k = \{x=(x_1,...,x_N)\in\mathbb{R}^N : x_j = p2^{-k},\, p\in\mathbb{Z}, j=1,...,N\}.$$

Let D_k be a collection of cubes determined by Γ_k, that is, those cubes with side length 2^{-k} and vertices in Γ_k. The cubes belonging to $D=\bigcup_{-\infty}^{\infty} D_k$ are called *dyadic cubes*. Note that if $Q_1, Q_2\in D$ and $|Q_1|\leq|Q_2|$, then either $Q_1\subset Q_2$ or else Q_1 and

Q_2 do not overlap (by which we mean that their interiors are disjoint).

Lemma 2.2.1. *Let $\alpha, \beta > 1$ be given and let $\omega \in A_p$ $(1 \leq p < \infty)$, with A_p-constant $C(\omega, p)$. Then there exist weights $\omega_{\alpha\beta} \geq 0$ a.e. such that the following conditions are met.*
(i) *$c_1(1/\beta) \leq \omega_{\alpha\beta} \leq c_2\,\alpha$ in Ω, where c_1 and c_2 depend only on ω and Ω.*
(ii) *There exist weights $\tilde{\omega}_1$ and $\tilde{\omega}_2$ such that $\tilde{\omega}_1 \leq \omega_{\alpha\beta} \leq \tilde{\omega}_2$, where $\tilde{\omega}_i \in A_p$ and $C(\tilde{\omega}_i, p)$ depends only on $C(\omega, p)$ $(i = 1, 2)$.*
(iii) *$\omega_{\alpha\beta} \in A_p$, with constant $C(\omega_{\alpha\beta}, p)$ depending only on $C(\omega, p)$ uniformly on α and β.*
(iv) *There exists a closed set $F_{\alpha\beta}$ such that $\omega_{\alpha\beta} \equiv \omega$ in $F_{\alpha\beta}$ and $\omega_{\alpha\beta} \approx \tilde{\omega}_1 \approx \tilde{\omega}_2$ in $F_{\alpha\beta}$ with equivalence constants depending on α and β (i.e., there are positive constants $c_{\alpha\beta}$ and $C_{\alpha\beta}$ such that*

$$c_{\alpha\beta}\,\tilde{\omega}_i \leq \omega_{\alpha\beta} \leq C_{\alpha\beta}\,\tilde{\omega}_i,$$

$i = 1, 2$). Moreover, $F_{\alpha\beta} \subset F_{\alpha'\beta'}$ if $\alpha \leq \alpha'$, $\beta \leq \beta'$, and the complement of $\displaystyle\bigcup_{\alpha,\beta \geq 1} F_{\alpha\beta}$ has zero measure.
(v) *$\omega_{\alpha\beta} \to \omega$ a.e. in $\mathbb{R}^N$ as $\alpha, \beta \to \infty$.*

Proof. **Case 1.**: Suppose first $\omega \in A_1$. Since we are interested to approximate in Ω, we may assume, without loss of generality, that $\omega \in L^1(\mathbb{R}^N)$. For each $\alpha > 1$, we define

$$U_\alpha^+ = \{x \in \mathbb{R}^N \ : \ M(\omega)(x) > \alpha\}$$

where $M(\omega)$ is the usual Hardy-Littlewood maximal operator for ω, i.e. $(M\omega)(x) = \displaystyle\sup_{x \in Q} \frac{1}{|Q|} \int_Q \omega(y)\, dy$, where the supremum is taken over all cubes Q containing x (cube will always mean a compact cube with sides parallel to the axes and nonempty interior). By Calderon-Zygmund decomposition (see Theorem 1.12,

Chapter II in [38]), there exists a family of non-overlapping cubes $\{Q_j^\alpha\}$ consisting of those maximal dyadic cubes over which the average of ω is greater than α, i.e.,

$$(CZ1) \qquad U_\alpha^+ = \bigcup_{j=1}^{\infty} Q_j^\alpha,$$

$$(CZ2) \qquad \alpha < \frac{1}{|Q_j^\alpha|}\int_{Q_j^\alpha} \omega(x)dx \leq 2^N\alpha,$$

$$(CZ3) \qquad \omega(x) \leq \alpha, \ \forall x \in F_\alpha^+ = (U_\alpha^+)^c,$$

$$(CZ4) \qquad |U_\alpha^+| \leq \frac{c}{\alpha}\int_{\mathbb{R}^N} \omega(x)dx.$$

Then,

$$\{x \in \mathbb{R}^N \ : \ M(\omega)(x) > 4^N\alpha\} \subset \bigcup_j 3Q_j^\alpha. \tag{2.2}$$

We explicitly note that, if $\alpha < \beta$, then $U_\beta^+ \subset U_\alpha^+$.
Define the weights ω_α by

$$\omega_\alpha(x) = \sum_{k=1}^{\infty}\left(\frac{1}{|Q_k^\alpha|}\int_{Q_k^\alpha} \omega(y)dy\right)\chi_{Q_k^\alpha}(x) + \omega(x)\chi_{F_\alpha^+}(x).$$

We will show that
(I) $\omega_\alpha \in A_1$ and $C(\omega_\alpha, 1)$ depends only on $C(\omega, 1)$.
(II) $\omega_\alpha \to \omega$ a.e. in $\mathbb{R}^N$ as $\alpha \to \infty$.
(III) $\min\{1, \omega\} \in A_1$ and $C(\min\{1, \omega\}, 1)$ depends only on $C(\omega, 1)$. Moreover, we also have $\min\{1, \omega\}(x) \leq \omega_\alpha(x) \leq C\,\omega(x)$, where C depends only on $C(\omega, 1)$.
(IV) $2^N\alpha \geq \omega_\alpha(x) \geq C = \dfrac{1}{|Q_0|}\displaystyle\int_{Q_0} \min\{1, \omega\}(y)dy$, $x \in \Omega$ and Q_0 is a fixed cube containing Ω ($\Omega \subset Q_0$).

First of all we note that, if $\omega \in L^1_{\rm loc}$ is a weight function such that for any dyadic cubes Q we have $\omega(3Q) \leq C_0\, \omega(Q)$ (C_0 depends only on N and $C(\omega,1)$), then we can restrict ourselves to test the A_1 condition only on dyadic cubes. In fact, denoting by $C^*(\omega,1)$ the *constant* A_1 *for dyadic cubes* , if $f \in L^1(\mathbb{R}^N, \omega)$ and $\{C_j^t\}$ is the Calderon-Zygmung decomposition for f, we have

$$\begin{aligned}
&\omega(\{x \,:\, M(f)(x) > 4^N t\}) \leq \sum_j \omega(3C_j^t) \\
&\leq C_0 \sum_j \omega(C_j^t) \leq C_0\, C^*(\omega,1) \sum_j |C_j^t| \operatorname{ess\,inf}_{C_j^t} \omega \\
&\leq C_0 C^*(\omega,1) \frac{1}{t} \sum_j \int_{C_j^t} |f(x)|\, \omega(x)\, dx \\
&\leq C_0\, C^*(\omega,1)\, \frac{1}{t} \int_\Omega |f(x)|\, \omega(x)\, dx,
\end{aligned}$$

and hence $\omega \in A_1$ by Theorem 2.1, Chapter IV in [38] and $C(\omega,1) = C_0\, C^*(\omega,1)$.
Thus, let us prove first that $\omega_\alpha(3Q) \leq C_0\, \omega_\alpha(Q)$ for any dyadic cube Q and for any $\alpha > 1$, where C_0 depends only on N and $C(\omega,1)$. Let Q be a dyadic cube. We have either $|Q \cap U_\alpha^+| < \frac{1}{2}|Q|$ or $|Q \cap U_\alpha^+| \geq \frac{1}{2}\,|Q|$. First, let us suppose that $|Q \cap U_\alpha^+| < \frac{1}{2}|Q|$. We will prove later, in (III), that $\omega_\alpha \leq \max\{1, C(\omega,1)\}\omega = C_1 \omega$; we have

$$\begin{aligned}
&\omega_\alpha(3Q) \leq C_1\, \omega(3Q) \\
&= C_1\, \frac{|3Q|}{|3Q|} \int_{3Q} \omega(x)\, dx = 3^N C_1 |Q| \frac{1}{|3Q|} \int_{3Q} \omega(x)\, dx \\
&\leq C_1\, C(\omega,1) 3^N |Q| \operatorname{ess\,inf}_{3Q} \omega \leq C_1\, C(\omega,1) 3^N |Q| \operatorname{ess\,inf}_{Q \cap F_\alpha^+} \omega
\end{aligned}$$

$$
\begin{aligned}
&= C_1\, C(\omega,1)3^N \frac{|Q|}{|Q\cap F_\alpha^+|}|Q\cap F_\alpha^+|\text{ess}\inf_{Q\cap F_\alpha^+}\omega \\
&\leq 2C_1\, C(\Omega,1)3^N|Q\cap F_\alpha^+|\text{ess}\inf_{Q\cap F_\alpha^+}\omega \\
&\leq 2C_1\, C(\omega,1)3^N \int_{Q\cap F_\alpha^+}\omega(x)\,dx \\
&= 2C_1\, C(\omega,1)3^N \int_{Q\cap F_\alpha^+}\omega_\alpha(x)\,dx,
\end{aligned}
$$

since $\omega_\alpha \equiv \omega$ on F_α^+, and therefore

$$
\begin{aligned}
\omega_\alpha(3Q) &\leq 2C_1\, C(\omega,1)3^N \int_{Q\cap F_\alpha^+}\omega_\alpha(x)\,dx \\
&\leq 2C_1\, C(\omega,1)3^N\omega(Q) = C_0\,\omega(Q).
\end{aligned}
$$

and hence, in this case we are done.

Suppose now $|Q\cap U_\alpha^+|\geq \frac{1}{2}|Q|$. Then either $Q\subset Q_{j_0}^\alpha$ for some j_0 (which in turn is unique) or $Q_j^\alpha \subset Q$ for $j\in\mathcal{J}$ (a set of indices). By definition of ω_α, it follows from (CZ2) e (CZ3) that $\omega_\alpha(x)\leq 2^N\alpha$ a.e.. Therefore, if $Q\subset Q_{j_0}^\alpha$ we get

$$
\begin{aligned}
\omega_\alpha(3Q) &= \int_{3Q}\omega_\alpha(x)dx \\
&\leq |3Q|2^N\alpha = 3^N\,2^N\,\alpha|Q| \\
&\leq 3^N\,2^n|Q|\frac{1}{|Q_{j_0}^\alpha|}\int_{Q_{j_0}^\alpha}\omega(x)\,dx \\
&= 6^N\int_Q\omega_\alpha(x)dx = 6^N\omega_\alpha(Q),
\end{aligned}
$$

since $\omega_\alpha \equiv \frac{1}{|Q_{j_0}^\alpha|}\int_{Q_{j_0}^\alpha}\omega(x)\,dx$ on Q. Otherwise, we obtain

$$
\omega_\alpha(3Q) = \int_{3Q}\omega_\alpha(x)\,dx
$$

$$
\begin{aligned}
&\leq 2^N \alpha |3Q| = 2^N\, 3^N \alpha |Q| \\
&\leq 6^N\, 2 |Q \cap U_\alpha^+| \alpha = 2\, 6^N \int_{Q\cap U_\alpha^+} \alpha \, dx \\
&= 2\, 6^N \sum_{j\in\mathcal{J}} \int_{Q_j^\alpha} \alpha \, dx \\
&\leq 2\, 6^N \sum_{j\in\mathcal{J}} \int_{Q_j^\alpha} \omega_\alpha(x)\, dx \\
&\leq 2\, 6^N \int_{\cup Q_j^\alpha} \omega_\alpha(x)\, dx \\
&\leq 2\, 6^N \int_{Q} \omega_\alpha(x)\, dx = 2\, 6^N \omega_\alpha(Q).
\end{aligned}
$$

Therefore $\omega_\alpha(3Q) \leq C_0\, \omega_\alpha(Q)$, $\forall Q$ dyadic cube.
Now we are ready to prove (I). Let Q be a fixed dyadic cube, then one of the three cases can happen:
(I1) $Q \cap Q_j^\alpha = \emptyset$, $\forall j$;
(I2) $Q \subset Q_j^\alpha$ for one and only one j;
(I3) $Q_j^\alpha \subset Q$ for some index $j \in \mathcal{J}$.

In case (I1), $Q \subset F_\alpha^+$ and hence $\omega_\alpha \equiv \omega$ in Q and we are done.
In case (I2), we have

$$
\omega_\alpha(Q) = \int_Q \omega_\alpha(x)\, dx = |Q| \left(\frac{1}{|Q_j^\alpha|} \int_{Q_j^\alpha} \omega \, dx \right) = |Q| \operatorname*{ess\,inf}_{x\in Q} \omega_\alpha(x),
$$

since $\omega_\alpha = \dfrac{1}{|Q_j^\alpha|} \displaystyle\int_{Q_j^\alpha} \omega(x)\, dx$ over Q_j^α.
Finally in case (I3), we have

$$
\omega_\alpha(Q) \;=\; \int_Q \omega_\alpha(x) dx = \int_{Q\cap U_\alpha^+} \omega_\alpha(x)\, dx + \int_{Q\cap F_\alpha^+} \omega_\alpha(x)\, dx
$$

$$
\begin{aligned}
&= \sum_{j\in\mathcal{J}} \int_{Q_j^\alpha} \omega_\alpha(x)\,dx + \omega_\alpha(Q\cap F_\alpha^+) \\
&= \sum_{j\in\mathcal{J}} \left(\frac{1}{|Q_j^\alpha|} \int_{Q_j^\alpha} \omega(x)dx \right) |Q_j^\alpha| + \omega(Q\cap F_\alpha^+) \\
&= \sum_{j\in\mathcal{J}} \int_{Q\cap Q_j^\alpha} \omega(x)\,dx + \omega(Q\cap F_\alpha^+) \\
&\leq \omega(Q) \\
&\leq C(\omega,1)\,|Q| \operatorname{ess}\inf_{y\in Q} \omega(y).
\end{aligned}
$$

On the other hand we note that if $y \in U_\alpha^+$, by definition we have $\omega_\alpha(y) > \alpha$. Thus, if $y \in Q\cap U_\alpha^+$ e Q_k^α and Q_k^α is any cube contained in Q we have

$$
\operatorname{ess}\inf_Q \omega \leq \operatorname{ess}\inf_{Q_k^\alpha} \omega \leq \frac{1}{|Q_k^\alpha|} \int_{Q_k^\alpha} \omega(x)\,dx \leq 2^N \alpha < 2^n \omega_\alpha(y).
$$

In addition, if $y \in Q\cap F_\alpha^+$, $\omega_\alpha(y) = \omega(y) \geq \operatorname{ess}\inf_Q \omega$.
Hence, $\operatorname{ess}\inf_Q \omega \leq 2^N \operatorname{ess}\inf_Q \omega_\alpha(y)$.
Therefore, in both cases,

$$
\omega_\alpha(Q) \leq C(\omega,1)|Q| \operatorname{ess}\inf_{y\in Q} \omega(y) \leq 2^N\, C(\omega,1)|Q| \operatorname{ess}\inf_{y\in Q} \omega_\alpha(y),
$$

that is,

$$
\frac{1}{|Q|} \int_Q \omega_\alpha(x)\,dx \leq 2^N\, C(\omega,1) \operatorname{ess}\inf_{y\in Q} \omega_\alpha(y).
$$

Consequently, in (I1), (I2) and (I3) we have

$$
\frac{1}{|Q|} \int_Q \omega_\alpha(x)\,dx \;\approx\; \operatorname{ess}\inf_Q \omega_\alpha
$$

i.e., $\omega_\alpha \in A_1$ and (I) is proved.

To prove (II), by definition we have

$$\omega_\alpha(x) = \sum_{j=1}^{\infty} \left(\frac{1}{|Q_j^\alpha|} \int_{Q_j^\alpha} \omega(x)\, dx \right) \chi_{Q_j^\alpha}(x) + \omega(x) \chi_{F_\alpha^+}(x).$$

Hence, $\omega_\alpha \equiv \omega$ in F_α^+ and that F_α^+ increases as α tends to infinity. Moreover, $|\cap (F_\alpha^+)^c| = 0$. Then $\omega_\alpha \to \omega$ a.e. as $\alpha \to \infty$.

Finally, to show (III) we know that, for any cube Q, either $\operatorname{ess\,inf}_Q \omega \geq 1$ or $\operatorname{ess\,inf}_Q \omega < 1$. In the first case,

$$\frac{1}{|Q|} \int_Q \min\{\omega, 1\}(x)\, dx = \frac{1}{|Q|} \int_Q 1 dx = 1 \leq \min\{\omega(y), 1\}$$

for any $y \in Q$, whereas if $\inf_Q \omega < 1$ then

$$\frac{1}{|Q|} \int_Q \min\{\omega, 1\}(x)\, dx \leq \frac{1}{|Q|} \int_Q \omega(x)\, dx \leq C(\omega, 1) \operatorname{ess\,inf}_Q \omega,$$

since $\omega \in A_1$. Put $\lambda = \operatorname{ess\,inf}_Q \omega < 1$ and assume by contradiction that $\inf_Q (\min\{\omega, 1\}) < \lambda$; then there exists $E \subset Q$, $|E| > 0$ such that $\min\{\omega, 1\} < \lambda' < \lambda$ in E and hence , since $\lambda < 1$, we obtain $\omega < \lambda'$ in E which is a contradiction. Thus we have proved the first part of (III), that is, $\min\{\omega, 1\} \in A_1$.

To prove the second part we note that if $x \in F_\alpha^+$ then

$$\omega_\alpha(x) = \omega(x) \geq \min\{\omega, 1\}(x).$$

If $x \in Q_j^\alpha$, for some j, then by definition and (CZ2), we get

$$\omega_\alpha(x) = \frac{1}{|Q_j^\alpha|} \int_{Q_j^\alpha} \omega(y) dy > \alpha \geq \min\{\omega, 1\}(x).$$

Analogously, if $x \in Q_j^\alpha$, for some j, then

$$\omega_\alpha(x) \leq C(\omega, 1) \operatorname{ess\,inf}_{Q_j^\alpha} \omega \leq C(\omega, 1)\omega(x).$$

Therefore, $\min\{\omega, 1\} \leq \omega_\alpha(x) \leq C(\omega, 1)\omega(x)$.

Finally, assertion (IV) follows straightforwardly from (III) by using (CZ1) and (CZ3).

Case 2. Suppose now $\omega \in A_p$, for $p > 1$. Then by Jones's Factorization Theorem (Theorem 2.1.4) there exist $\omega_0, \omega_1 \in A_1$ such that

$$\omega = \omega_0 \omega_1^{1-p},$$

and $C(\omega_i, 1)$ $(i = 0, 1)$ depends only on $C(\omega, p)$. Choose α, $\beta > 1$ and defines the weights $\omega_{\alpha\beta}$ by

$$\omega_{\alpha\beta} = (\omega_0)_\alpha \left[(\omega_1)_{\beta^{1/(p-1)}} \right]^{1-p}.$$

We need to show that $\omega_{\alpha\beta}$ satisfies properties (i) - (v).

(i) follows from (IV).
(ii) follows from (III) and Jones's Factorization Theorem, with

$$\tilde{\omega}_1 = \min\{\omega_0, 1\} \omega_1^{1-p},$$

$$\tilde{\omega}_2 = \omega_0 (\min\{\omega_1, 1\})^{1-p}.$$

(iii) from (I) and Jones's Factorization Theorem.
(v) from (II).
For (iv), we define

$$F_{\alpha\beta} = F_\alpha^0 \cap F_{\beta^{1/(p-1)}}^1$$

where $F_\alpha^0 = F_\alpha^+$ for the weight ω_0 and $F^1_{\beta^{1/(p-1)}} = \left(F_{\beta^{1/(p-1)}}\right)^+$ for the weight ω_1. By definition, we have $(\omega_0)_\alpha \equiv \omega_0$ and $(\omega_1)_{\beta^{1/(p-1)}} \equiv \omega_1$ on $F_{\alpha\beta}$.
Hence, to prove that $\omega_{\alpha\beta} \approx \tilde{\omega}_1$, we can replace $\omega_{\alpha\beta}$ by ω (on $F_{\alpha\beta}$). We have that $\min\{\omega_0, 1\} \approx \omega_0$ on F_α^0, since $\min\{\omega_0, 1\} \leq \omega_0$ (by (III)). Moreover, if $x \in F_\alpha^0$, we have $\min\{\omega_0(x), 1\} = 1$. Hence, by (CZ3) we obtain

$$\omega_0(x) \leq \alpha = \alpha \min\{\omega_0(x), 1\}.$$

Therefore, $\min\{\omega_0, 1\} \leq \omega_0 \leq \alpha \min\{\omega_0, 1\}$ on F_α^0 .
An analogous argument shows that $\omega_1 \approx \min\{\omega_1, 1\}$ on $F^1_{\beta^{1/(p-1)}}$.
Hence,

$$\omega_{\alpha\beta} \approx \min\left\{\omega_0, 1\right\}\left[(\omega_1)_{\beta^{1/(p-1)}}\right]^{1-p} = \tilde{\omega}_1 \quad \text{on } F_{\alpha\beta},$$

and analogously, we also have $\omega_{\alpha\beta} \approx \tilde{\omega}_2$. □

Remark 2.2.1. *If $(v, \omega) \in A_p$ and $\omega \leq v$, we have $\omega \in A_p$ and $v \in A_p$. Then by the Jones Factorization Theorem there exits $v_0, v_1, \omega_0, \omega_1 \in A_1$ such that $v = v_0\, v_1^{1-p}$ and $\omega = \omega_0\, \omega_1^{1-p}$. For the next lemma we will assume the following hypothesis:*
(a) $\omega_0 \leq v_0$ and $(v_0, \omega_0) \in A_1$;
(b) there is a constant $C > 0$ such that $v_1(x) \leq \omega_1(x) \leq C\, v_1(x)$ a.e. in $\mathbb{R}^N$.

Lemma 2.2.2. *Let $\alpha, \beta > 1$ be given and let there a pair of weights $(v, \omega) \in A_p$, $1 \leq p < \infty$. Then exist weights $\omega_{\alpha\beta} \geq 0$, $v_{\alpha\beta} \geq 0$ such that:*
***(i)** $C_{11}\,(1/\beta) \leq \omega_{\alpha\beta} \leq C_{12}\,\alpha$, $C_{21}(1/\beta) \leq v_{\alpha\beta} \leq C_{22}\,\alpha$ in Ω, with C_{11} and C_{12} depending only on ω and Ω, C_{21} and C_{22} depending only on v and Ω;*
***(ii)** There exist weights $\tilde{\omega}_1, \tilde{\omega}_2, \tilde{v}_1, \tilde{v}_2 \in A_p$, such that $\tilde{\omega}_1 \leq \omega_{\alpha\beta} \leq \tilde{\omega}_2$ and $\tilde{v}_1 \leq v_{\alpha\beta} \leq \tilde{v}_2$;*

(iii) *$\omega_{\alpha\beta}$ and $v_{\alpha\beta}$ are A_p - weights;*
(iv) *There exists a closed set $G_{\alpha\beta}$ such that $\omega_{\alpha\beta}(x) = \omega(x)$ and $v_{\alpha\beta}(x) = v(x)$ in $G_{\alpha\beta}$. Moreover, $G_{\alpha\beta} \subset G_{\alpha'\beta'}$, if $\alpha \leq \alpha'$ and $\beta \leq \beta'$, and the complement of $\bigcup_{\alpha,\beta>1} G_{\alpha,\beta}$ has zero measure.*
(v) *$\omega_{\alpha\beta} \to \omega$ and $v_{\alpha\beta} \to v$ a.e. in $\mathbb{R}^N$, as α, β tend to infinity.*

Proof. **Case 1.** In this case we denote $\omega_{\alpha\beta} = \omega_\alpha$, $v_{\alpha\beta} = v_\alpha$, and $G_{\alpha\beta} = G_\alpha$ (in (i)-(v)). Since we are interested to approximating within Ω, we can suppose, without loss of generality that $\omega, v \in L^1(\mathbb{R}^N)$.

If $(v, \omega) \in A_1$, that is,

$$\frac{1}{|Q|} \int_Q v(x)\, dx \leq C \operatorname{ess\,inf}_Q \omega$$

by Remark 2.2.1, we have $v \in A_1$ and $\omega \in A_1$. For each $\alpha > 1$, we define

$$\begin{aligned} U_\alpha^+ &= \{x \in \mathbb{R}^N \ : \ M(\omega)(x) > \alpha\} \quad \text{and} \\ V_\alpha^+ &= \{x \in \mathbb{R}^N \ : \ M(v)(x) > \alpha\} \end{aligned}$$

(U_α^+ and V_α^+ are open sets because $M(\omega)$ and $M(v)$ are lower semicontinuous functions) where M is the usual Hardy-Littlewood maximal operator , i.e.

$$M(\omega)(x) = \sup_{Q \ni x} \frac{1}{|Q|} \int_Q \omega(y)\, dy.$$

Using the fact that $0 < \omega(x) \leq v(x)$, we have $M(\omega)(x) \leq M(v)(x)$. Consequently $U_\alpha^+ \subset V_\alpha^+$.
By using Calderon-Zygmund decomposition (see Theorem 1.12, Chapter II in [38]) there are two families of non-overlapping cubes $\{Q_l^\alpha\}$ and $\{I_k^\alpha\}$, where Q_l^α and I_k^α are maximal dyadic cubes, such that

(CZ 1) $U_\alpha^+ = \bigcup_{l=1}^{\infty} Q_l^\alpha$ and $V_\alpha^+ = \bigcup_{k=1}^{\infty} I_k^\alpha$,

(CZ 2) $\alpha < \frac{1}{|Q_l^\alpha|}\int_{Q_l^\alpha} \omega(x)\,dx \leq 2^N\alpha$ and

$$\alpha < \frac{1}{|I_k^\alpha|}\int_{I_k^\alpha} v(x)\,dx \leq 2^N\alpha,$$

(CZ 3) $\omega(x) \leq \alpha$ in $F_\alpha^+ = (U_\alpha^+)^c$ and $v(x) \leq \alpha$ in $G_\alpha^+ = (V_\alpha^+)^c$,
(where E^c denotes the complement of a set E)

(CZ 4) $|U_\alpha^+| \leq \frac{c}{\alpha}\int_{\mathbb{R}^N} \omega(x)\,dx$ and $|V_\alpha^+| \leq \frac{c}{\alpha}\int_{\mathbb{R}^N} v(x)\,dx$.

Since $U_\alpha^+ \subset V_\alpha^+$, for each Q_l^α there exists only one I_k^α $(k = k(l))$ such that $Q_l^\alpha \subset I_k^\alpha$ (recall that Q_l^α and I_k^α are maximal dyadic cubes). We define the weights ω_α and v_α by

$$\omega_\alpha(x) = \sum_{l=1}^{\infty}\left(\frac{1}{|Q_l^\alpha|}\int_{Q_l^\alpha} \omega(y)\,dy\right)\chi_{Q_l^\alpha}(x) + \omega(x)\chi_{F_\alpha^+}(x)$$

$$v_\alpha(x) = \sum_{k=1}^{\infty}\left(\frac{1}{|I_k^\alpha|}\int_{I_k^\alpha} v(y)\,dy\right)\chi_{I_k^\alpha}(x) + v(x)\chi_{G_\alpha^+}(x),$$

where χ_E denotes the characteristic function of a set E.

We will show that $\omega_\alpha(x) \leq 2^N\, v_\alpha(x)$.

(1) If $x \in G_\alpha^+ \subset F_\alpha^+$, then $v_\alpha(x) = v(x)$ and $\omega_\alpha(x) = \omega(x)$. Consequently $\omega_\alpha(x) \leq v_\alpha(x)$.

(2) If $x \notin G_\alpha^+$ then $x \in I_k^\alpha$ (for a unique cube). Hence,

$$v_\alpha(x) = \frac{1}{|I_k^\alpha|}\int_{I_k^\alpha} v(y)\,dy.$$

For the weight ω_α, there are two possible cases as follows.

(a) If $x \in Q_l^\alpha \subset I_k^\alpha$. In this case, by property (CZ2) of Calderon-Zygmund decomposition, we have

$$\omega_\alpha(x) = \frac{1}{|Q_l^\alpha|}\int_{Q_l^\alpha} \omega(y)\,dy \leq 2^N\alpha \leq 2^N\frac{1}{|I_k^\alpha|}\int_{I_k^\alpha} v(y)\,dy = 2^N v_\alpha(x).$$

(b) If $x \in I_k^\alpha - \mathbf{Q}_k$, where $\mathbf{Q}_k = \cup Q_l^\alpha$, with $Q_l^\alpha \subset I_k^\alpha$, that is, $x \in F_\alpha^+$. Hence, using the property (CZ3) of Calderon-Zygmund decomposition, we obtain

$$\omega_\alpha(x) = \omega(x) \leq \alpha < \frac{1}{|I_k^\alpha|}\int_{I_k^\alpha} v(y)\,dy = v_\alpha(x).$$

Therefore $\omega_\alpha(x) \leq 2^N v_\alpha(x)$ in Ω.

Since $\omega \in A_1$ and $v \in A_1$ we have that
(I1) $\omega_\alpha \in A_1$ and $v_\alpha \in A_1$, with $C(\omega_\alpha, 1)$ depending only on $C(\omega, 1)$ and $C(v_\alpha, 1)$ depending only on $C(v, 1)$.
(II1) $\omega_\alpha \to \omega$ and $v_\alpha \to v$ a.e.in $\mathbb{R}^N$ when $\alpha \to \infty$.
(III1) $\min\{1, \omega\} \in A_1$, $\min\{1, v\} \in A_1$ and

$$\min\{1, \omega(x)\} \leq \omega_\alpha(x) \leq C\,\omega(x),$$
$$\min\{1, v(x)\} \leq v_\alpha(x) \leq \tilde{C}\,v(x),$$

where C depends only on $C(\omega, 1)$ and $\tilde{C}$ depends only on $C(v, 1)$;
(IV1) We also have

$$C_1 \leq \omega_\alpha(x) \leq 2^N\alpha, \quad \text{with } C_1 = \frac{1}{|Q_0|}\int_{Q_0} \min\{1, \omega\}(y)\,dy,$$

$$C_2 \le v_\alpha(x) \le 2^N\alpha, \quad \text{with } C_2 = \frac{1}{|Q_0|}\int_{Q_0} \min\{1, v\}(y)\,dy$$

where Q_0 is a fixed cube such that $\bar{\Omega} \subset Q_0$.

We need to show that ω_α and v_α satisfy properties (i)-(v). We have the following
▷ (i) Follows from (IV1),

$$2^N\alpha \ge \omega_\alpha(x) \ge C_1 \ge \frac{C_1}{\alpha} \quad \text{and} \quad 2^N\alpha \ge v_\alpha(x) \ge C_2 \ge \frac{C_2}{\alpha}.$$

▷ (ii) We define the weights $\tilde{\omega}_1$, $\tilde{\omega}_2$, $\tilde{v}_1$ and $\tilde{v}_2$ by

$$\tilde{\omega}_1(x) = \min\{1, \omega(x)\} \quad \text{and} \quad \tilde{\omega}_2(x) = \omega(x),$$

$$\tilde{v}_1(x) = \min\{1, v(x)\} \quad \text{and} \quad \tilde{v}_2(x) = v(x),$$

(C and $\tilde{C}$ are as in (III1)). We have that $\tilde{\omega}_1, \tilde{\omega}_2, \tilde{v}_1$ and $\tilde{v}_2 \in A_1$ (by (III1)) and
$\tilde{\omega}_1(x) \le \omega_\alpha(x) \le \tilde{\omega}_2(x)$, $\tilde{v}_1(x) \le v_\alpha(x) \le \tilde{v}_2(x)$.
▷ (iii) Follows from (I1).
▷ (iv) Define $G_\alpha = G_\alpha^+$. By definition $v_\alpha(x) = v(x)$ in G_α, and $\omega_\alpha(x) = \omega(x)$ in G_α (because $G_\alpha = G_\alpha^+ \subset F_\alpha^+$).
Moreover, if $\alpha \le \alpha'$ we have $U_{\alpha'}^+ \subset U_\alpha^+$. Then

$$G_\alpha = G_\alpha^+ = (U_\alpha^+)^c \subset (U_{\alpha'}^+)^c = G_{\alpha'}^+ = G_{\alpha'}.$$

Furthermore, by $(CZ4)$ we obtain $|(\bigcup_\alpha G_\alpha)^c| = 0$.
▷ (v) Follows straightforwardly from (II1).

Case 2. If $(v, \omega) \in A_p$, $p > 1$. By Remark 2.2.1 we have $\omega \in A_p$ and $v \in A_p$, and by Jones Factorization Theorem there are weights ω_0, ω_1, v_0 and v_1 in A_1 such that

$$v = v_0\, v_1^{1-p} \quad \text{and} \quad \omega = \omega_0\, \omega_1^{1-p}$$

and we assume the following hypotheses (see Remark 2.2.1):

(a) $\omega_0 \leq v_0$ and $(v_0, \omega_0) \in A_1$;
(b) $v_1(x) \leq \omega_1(x) \leq C_1 v_1(x)$.

Since we are interested to approximating within Ω, we may assume, without loss of generality that ω_0, ω_1, v_0 and $v_1 \in L^1(\mathbb{R}^N)$. For each $\alpha > 1$, we define

$$U_{0,\alpha}^+ = \{x \in \mathbb{R}^N \ : \ M(\omega_0)(x) > \alpha\},$$
$$V_{0,\alpha}^+ = \{x \in \mathbb{R}^N \ : \ M(v_0)(x) > \alpha,$$

($U_{0,\alpha}^+$ and $V_{0,\alpha}^+$ are open sets). By Calderon-Zygmund decomposition there exist two families of maximal dyadic cubes $\{Q_{0,l}^{\alpha}\}$ and $\{I_{0,k}^{\alpha}\}$, for ω_0 and v_0 respectively, satisfying the analogous conditions of Case (1) ((CZ1)-(CZ4)). Using the hypotheses that $\omega_0 \leq v_0$, we have
(a_o) $Q_{0,l}^{\alpha} \subset I_{0,k}^{\alpha}$ (with $k = k(l)$);
(b_o) $G_{0,\alpha}^+ = (V_{0,\alpha}^+)^c \subset (U_{0,\alpha}^+)^c = F_{0,\alpha}^+$.

For each $\beta > 1$, we define

$$U_{1,\beta}^+ = \{x \in \mathbb{R}^N \ : \ M(\omega_1)(x) > \beta^{1/(p-1)}\},$$
$$V_{1,\beta}^+ = \{x \in \mathbb{R}^N \ : \ M(v_1)(x) > \beta^{1/(p-1)}\}.$$

Since $v_1 \leq \omega_1$, we have $V_{1,\beta}^+ \subset U_{1\beta}^+$. By Calderon-Zygmund decomposition, there exist two families of maximal dyadic cubes $\{Q_{1,l}^{\beta}\}$ and $\{I_{1,k}^{\beta}\}$, for ω_1 and v_1 respectively, satisfying the conditions (CZ1)-(CZ4). We have
(a_1) $I_{1,k}^{\beta} \subset Q_{1,l}^{\beta}$ (with $l = l(k)$);

(b_1) $F^+_{1,\beta} = (U^+_{1,\beta})^c \subset (V^+_{1,\beta})^c = G^+_{1,\beta}$.

We define the weights

$$(\omega_0)_\alpha(x) = \sum_{l=1}^{\infty} \left(\frac{1}{|Q^\alpha_{0,l}|} \int_{Q^\alpha_{0,l}} \omega_0(y)\, dy \right) \chi_{Q^\alpha_{0,l}}(x) + \omega_0(x)\chi_{F^+_{0,\alpha}}(x),$$

$$(v_0)_\alpha(x) = \sum_{k=1}^{\infty} \left(\frac{1}{|I^\alpha_{0,k}|} \int_{I^\alpha_{0,k}} v_0(y)\, dy \right) \chi_{I^\alpha_{0,k}}(x) + v_0(x)\chi_{G^+_{0,\alpha}}(x),$$

$$(\omega_1)_\beta(x) = \sum_{l=1}^{\infty} \left(\frac{1}{|Q^\beta_{1,l}|} \int_{Q^\beta_{1,l}} \omega_1(y)\, dy \right) \chi_{Q^\beta_{1,l}}(x) + \omega_1(x)\chi_{F^+_{1,\beta}}(x),$$

and $(v_1)_\beta(x) = \sum_{k=1}^{\infty} \left(\frac{1}{|I^\beta_{1,k}|} \int_{I^\beta_{1,k}} v_1(y)\, dy \right) \chi_{I^\beta_{1,k}}(x) + v_1(x)\chi_{G^+_{1,\beta}}(x).$

We have, as in Case 1, $(\omega_0)_\alpha \le 2^N (v_0)_\alpha$ and $(v_1)_\beta \le 2^N (\omega_1)_\beta$.

Now we define the weights $\omega_{\alpha\beta}$ and $v_{\alpha\beta}$ by

$$\omega_{\alpha\beta}(x) = (\omega_0)_\alpha(x)[(\omega_1)_\beta]^{1-p}(x),$$

$$v_{\alpha\beta}(x) = (v_0)_\alpha(x)[(v_1)_\beta]^{1-p}(x).$$

We have that $(\omega_0)_\alpha$, $(v_0)_\alpha$, $(\omega_1)_\beta$, and $(v_1)_\beta$ satisfy the following conditions.
(I2) $(\omega_0)_\alpha$, $(v_0)_\alpha$, $(\omega_1)_\beta$ and $(v_1)_\beta \in A_1$.
(II2) $(\omega_0)_\alpha \to \omega_0$, $(v_0)_\alpha \to v_0$, $(\omega_1)_\beta \to \omega_1$ and $(v_1)_\beta \to v_1$ a.e. $\mathbb{R}^N$ when $\alpha \to \infty$, $\beta \to \infty$.

(III2) $\min\{1,\omega_0\}$, $\min\{1,\omega_1\}$, $\min\{1,v_0\}$ and $\min\{1,v_1\}$ are weights in A_1 and we have

$$\begin{aligned}
\min\{1,\omega_0(x)\} &\le (\omega_0)_\alpha(x) \le C_1\,\omega_0(x),\\
\min\{1,\omega_1(x)\} &\le (\omega_1)_\beta(x) \le C_2\,\omega_1(x),\\
\min\{1,v_0(x)\} &\le (v_0)_\alpha(x) \le C_3\,v_0(x),\\
\min\{1,v_1(x)\} &\le (v_1)_\beta(x) \le C_4\,v_1(x).
\end{aligned}$$

(IV2) We also have

$$\begin{aligned}
&\tilde{C}_1 \le (\omega_0)_\alpha(x) \le 2^N\alpha, \quad \text{with } \tilde{C}_1 = \frac{1}{|Q_0|}\int_{Q_0} \min\{1,\omega_0\}\,dy,\\
&\tilde{C}_2 \le (v_0)_\alpha(x) \le 2^N\alpha, \quad \text{with } \tilde{C}_2 = \frac{1}{|Q_0|}\int_{Q_0} \min\{1,v_0\}\,dy,\\
&\tilde{C}_3 \le (\omega_1)_\beta(x) \le 2^N\beta^{1/(p-1)}, \quad \text{with } \tilde{C}_3 = \frac{1}{|Q_0|}\int_{Q_0} \min\{1,\omega_1\}\,dy,\\
&\tilde{C}_4 \le (v_1)_\beta(x) \le 2^N\beta^{1/(p-1)}, \quad \text{with } \tilde{C}_4 = \frac{1}{|Q_0|}\int_{Q_0} \min\{1,v_1\}\,dy,
\end{aligned}$$

where Q_0 is a fixed cube such that $\bar{\Omega} \subset Q_0$.

We need to show that weights $\omega_{\alpha\beta}$ and $v_{\alpha\beta}$ satisfy properties (i)-(v).

▷ (i) Follows from (IV2). In fact,

$$\begin{aligned}
\omega_{\alpha\beta}(x) &= (\omega_0)_\alpha(x)[(\omega_1)_\beta]^{1-p}(x)\\
&\ge \tilde{C}_1(2^N\beta^{1/(p-1)})^{1-p}\\
&= \tilde{C}_1 2^{N(1-p)}\beta^{-1},
\end{aligned}$$

$$\begin{aligned}
\omega_{\alpha\beta}(x) &= (\omega_0)_\alpha(x)[(\omega_1)_\beta]^{1-p}(x)\\
&\le 2^N\alpha(\tilde{C}_3)^{1-p}
\end{aligned}$$

$$= 2^N(\tilde{C}_3)^{1-p}\alpha.$$

Analogously, $\tilde{C}_2\, 2^{N(1-p)}\, \beta^{-1} \leq v_{\alpha\beta}(x) \leq 2^N(\tilde{C}_4)^{1-p}\alpha$.

▷ (ii) We define the weights $\tilde{\omega}_1$, $\tilde{\omega}_2$, $\tilde{v}_1$ and $\tilde{v}_2$ by

$$\begin{aligned}
\tilde{\omega}_1(x) &= [\min\{1, \omega_0(x)\}]\omega_1^{1-p}(x),\\
\tilde{\omega}_2(x) &= [\min\{1, \omega_1(x)\}]^{1-p}\omega_0(x),\\
\tilde{v}_1(x) &= [\min\{1, v_0(x)\}]v_1^{1-p}(x),\\
\tilde{v}_2(x) &= [\min\{1, v_1(x)\}]^{1-p}v_0(x).
\end{aligned}$$

By (III2) and Jones Factorization Theorem, we have that $\tilde{\omega}_1$, $\tilde{\omega}_2$, $\tilde{v}_1$ and $\tilde{v}_2$ are A_p-weights.

▷ (iii) Follows from (I2) and Jones Factorization Theorem.

▷ (iv) We define the closed sets $G_{\alpha\beta}$ and $F_{\alpha\beta}$ by

$$G_{\alpha\beta} = G^+_{0,\alpha} \bigcap F^+_{1,\beta} \quad \text{and} \quad F_{\alpha\beta} = F^+_{0,\alpha} \bigcap G^+_{1,\beta}.$$

If $\alpha \leq \alpha'$ and $\beta \leq \beta'$ we have

$$\begin{aligned}
U^+_{0,\alpha'} \subset U^+_{0,\alpha}, &\quad V^+_{0,\alpha'} \subset V^+_{0,\alpha},\\
U^+_{1,\beta'} \subset U^+_{1,\beta} &\quad \text{and} \quad V^+_{1,\beta'} \subset V^+_{1,\beta}.
\end{aligned}$$

Hence,

$$F^+_{1,\beta} = (U^+_{1,\beta})^c \subset (U^+_{1,\beta'})^c = F^+_{1,\beta'}$$

and

$$G^+_{0,\alpha} = (V^+_{0,\alpha})^c \subset (V^+_{0,\alpha'})^c = G^+_{0,\alpha'}.$$

Thus

$$G_{\alpha\beta} = G^+_{0,\alpha} \cap F^+_{1,\beta} \subset G^+_{0,\alpha'} \cap F^+_{1,\beta'} = G_{\alpha'\beta'}.$$

Moreover, if $x \in G_{\alpha\beta}$ we have

$$\begin{aligned}\omega_{\alpha\beta}(x) &= (\omega_0)_\alpha(x)[(\omega_1)_\beta]^{1-p}(x)\\ &= \omega_0(x)(\omega_1)^{1-p}(x)\\ &= \omega(x).\end{aligned}$$

And by (b_o) and (b_1) we obtain $G_{\alpha\beta} \subset F_{\alpha\beta}$. In this case, if $x \in G_{\alpha\beta} \subset F_{\alpha\beta}$ we have

$$\begin{aligned}v_{\alpha\beta}(x) &= (v_0)_\alpha(x)[(v_1)_\beta]^{1-p}(x)\\ &= v_0(x)(v_1)^{1-p}(x)\\ &= v(x).\end{aligned}$$

Furthermore, by (CZ4), (b_o) and (b_1) we obtain

$$\begin{aligned}\Big|(\bigcup_{\alpha,\beta>1} G_{\alpha\beta})^c\Big| &= \Big|\bigcap_{\alpha,\beta>1} G^c_{\alpha\beta}\Big| = \Big|\bigcap_{\alpha,\beta>1}(G^+_{0,\alpha}\bigcap F^+_{1,\beta})^c\Big|\\ &= \Big|\bigcap_{\alpha,\beta>1}[(V^+_{0,\alpha})\bigcup(U^+_{1,\beta})]\Big|\\ &\le |V^+_{0,\alpha}| + |U^+_{1,\beta}|\\ &\le \frac{c}{\alpha}\int_{\mathbb{R}^N} v_0(x)\,dx + \frac{c}{\beta}\int_{\mathbb{R}^N}\omega_1(x)\,dx,\end{aligned}$$

for all $\alpha, \beta > 1$.
Thus $\Big|(\bigcup_{\alpha,\beta>1} G_{\alpha\beta})^c\Big| = 0$.

▷ (v) Follows from (II2).

□

Chapter 3

Degenerate Linear Elliptic Equations

Let L be a degenerate elliptic operator in divergence form

$$Lu = -\sum_{i,j=1}^{N} D_j(a_{ij}(x)D_i u(x)), \quad D_j = \partial/\partial x_j \tag{3.1}$$

and the coefficients a_{ij} are measurable, real-valued functions defined on a bounded open set $\Omega \subset \mathbb{R}^N$, satisfying the degenerate ellipticity condition

$$\lambda\,|\xi|^2\omega(x) \leq \sum_{i,j=1}^{N} a_{ij}(x)\xi_i\xi_j \leq \Lambda\omega(x)|\xi|^2, \ \forall\xi\in\mathbb{R}^N, \ \text{a.e. } x\in\Omega \tag{3.2}$$

$$a_{ij}(x) = a_{ji}(x) \tag{3.3}$$

where $\omega \in A_p$, and ω may either vanish, or be infinite, or both. When $\omega \equiv 1$ the problem $(P1)$ was studied in [39].

In this chapter we will study some problems, e.g., existence and uniqueness of solutions for the Dirichlet problem

$$(P1)\begin{cases} Lu = g - \sum_{j=1}^{N} D_j f_j \text{ in } \Omega, \\ u - \psi \in W_0^{1,2}(\Omega, \omega), \end{cases}$$

approximation by a sequence of solutions of non-degenerate elliptic equations and entropy solutions.

3.1 Existence, uniqueness and approximation of solution

Definition 3.1.1. *We say that an element $u \in W^{1,2}(\Omega, \omega)$ is a weak solution of the equation*

$$Lu = g - \sum_{j=1}^{N} D_j f_j, \quad \text{with } \frac{g}{\omega}, \frac{f_j}{\omega} \in L^2(\Omega, \omega) \ (j = 1, ..., N), \text{ if}$$

$$\begin{aligned} \sum_{i,j=1}^{N} \int_\Omega a_{ij}(x) D_i u(x) D_j \varphi(x)\, dx &= \sum_{j=1}^{N} \int_\Omega f_j(x) D_j \varphi(x)\, dx \\ &+ \int_\Omega g(x) \varphi(x)\, dx \end{aligned}$$

for all $\varphi \in W_0^{1,2}(\Omega, \omega)$.

Theorem 3.1.1. *Let L be the operator (3.1) and with coefficients a_{ij} satisfying $a_{ij} = a_{ji}$ and the degenerate ellipticity condition*

$$\lambda |\xi|^2 \omega(x) \le \sum_{i,j=1}^{N} a_{ij}(x) \xi_i \xi_j \le \Lambda\, \omega(x) |\xi|^2, \tag{3.4}$$

for all $\xi \in \mathbb{R}^N$, *a.e.* $x \in \Omega$, *where* λ *and* Λ *are positive constants. If* $\psi \in W^{1,2}(\Omega,\omega)$, $\omega \in A_2$, $\dfrac{g}{\omega} \in L^2(\Omega,\omega)$ *and* $\dfrac{f_j}{\omega} \in L^2(\Omega,\omega)$, *then the Dirichlet problem*

$$(P1)\begin{cases} Lu = g - \sum_{j=1}^{N} D_j f_j \text{ in } \Omega, \\ u - \psi \in W_0^{1,2}(\Omega,\omega), \end{cases}$$

has a unique solution $u \in W^{1,2}(\Omega,\omega)$ *and*

$$\|u\|_{W^{1,2}(\Omega,\omega)} \leq C\left(\|g/\omega\|_{L^2(\Omega,\omega)} + \sum_{j=1}^{N} \|f_j/\omega\|_{L^2(\Omega,\omega)} + \|\psi\|_{W^{1,2}(\Omega,\omega)}\right),$$

where $C = (C_\Omega^2 + 1)/\lambda$ *(*C_Ω *is the constant in Theorem 2.1.5).*

Proof. Let us first reduce the Dirichlet problem to the case of zero boundary values. Setting $v = u - \psi$, we obtain

$$\begin{aligned} Lv &= Lu - L\psi \\ &= g - \sum_{j=1}^{N} D_j f_j - \sum_{i,j=1}^{N} D_j(a_{ij}\, D_i\psi) \\ &= g - \sum_{j=1}^{N} D_j(f_j - \sum_{i=1}^{N} a_{ij}\, D_i\psi) \\ &= g - \tilde{f}_j, \end{aligned}$$

where $\tilde{f}_j = f_j - \sum_{i=1}^{N} a_{ij}\, D_i\psi$. We clearly have $\tilde{f}_j/\omega \in L^2(\Omega,\omega)$ $(j = 1, ..., N)$. Therefore it suffices to prove Theorem 3.1.1. for the case $\psi \equiv 0$.

Step 1. Let $B : W_0^{1,2}(\Omega,\omega) \times W_0^{1,2}(\Omega,\omega) \rightarrow \mathbb{R}$ and

$G : W_0^{1,2}(\Omega\omega) \to \mathbb{R}$

$$B(u,\varphi) = \sum_{i,j=1}^{N} \int_{\Omega} a_{ij}\, D_i u\, D_j \varphi\, dx,$$

$$G(\varphi) = \int_{\Omega} g\,\varphi\, dx + \sum_{j=1}^{N} \int_{\Omega} f_j\, D_j \varphi\, dx.$$

Then $u \in W_0^{1,2}(\Omega,\omega)$ is a (weak) solutions of the problem

$$(P_0)\begin{cases} Lu = g - \displaystyle\sum_{j=1}^{N} D_j f_j \text{ in } \Omega, \\ u = 0 \text{ on } \partial\Omega, \end{cases}$$

if $B(u,\varphi) = G(\varphi)$ for all $\varphi \in W_0^{1,2}(\Omega,\omega)$. We have,
(a) for all $u, \varphi \in W_0^{1,2}(\Omega,\omega)$,

$$\begin{aligned} B(u,u) &= \sum_{i,j=1}^{N} \int_{\Omega} a_{ij} D_i u\, D_j u\, dx \\ &\geq \int_{\Omega} \lambda |\nabla u|^2\, \omega\, dx \\ &\geq \frac{\lambda}{C_\Omega^2 + 1} \|u\|^2_{W_0^{1,2}(\Omega,\omega)}, \end{aligned}$$

that is, B is coercive. And we also have, using that $\mathcal{A} = (a_{ij})$ is a symmetric matrix,

$$\begin{aligned} |B(u,\varphi)| &\leq \int_{\Omega} |\langle \mathcal{A}\nabla u, \nabla\varphi\rangle|\, dx \\ &\leq \left(\int_{\Omega} \Lambda |\nabla u|^2\, \omega\, dx\right)^{1/2} \left(\int_{\Omega} \Lambda |\nabla \varphi|^2\, \omega\, dx\right)^{1/2} \\ &\leq \Lambda\, \|u\|_{W_0^{1,2}(\Omega,\omega)} \|\varphi\|_{W_0^{1,2}(\Omega,\omega)}, \end{aligned}$$

that is, B is bounded. Moreover, $G\in[W_0^{1,2}(\Omega,\omega)]^*$, since

$$
\begin{aligned}
|G(\varphi)| &\leq \int_\Omega |g|\,|\varphi|\,dx + \sum_{j=1}^N |f_j|\,|D_j\varphi|\,dx \\
&\leq \left(\|g/\omega\|_{L^2(\Omega,\omega)} + \sum_{j=1}^N \|f_j/\omega\|_{L^2(\Omega,\omega)}\right)\|\varphi\|_{W_0^{1,2}(\Omega,\omega)}.
\end{aligned}
$$

Therefore by Lax-Milgram Theorem, there exists a unique $u\in W_0^{1,2}(\Omega,\omega)$ such that $B(u,\varphi)=G(\varphi)$ for all $\varphi\in W_0^{1,2}(\Omega,\omega)$, that is, u is a unique solution for (P_0).

Step 2. For $\varphi=u$ we have $B(u,u)=G(u)$. Then,

$$
\frac{\lambda}{C_\Omega^2+1}\,\|u\|^2_{W_0^{1,2}(\Omega,\omega)} \leq \left(\|g/\omega\|_{L^2(\Omega,\omega)}+\sum_{j=1}^N \|f_j/\omega\|_{L^2(\Omega,\omega)}\right)\|u\|_{W_0^{1,2}(\Omega,\omega)}.
$$

Therefore, we obtain

$$
\|u\|_{W_0^{1,2}(\Omega,\omega)} \leq C\left(\|g/\omega\|_{L^2(\Omega,\omega)} + \sum_{j=1}^N \|f_j/\omega\|_{L^2(\Omega,\omega)}\right),
$$

where $C=(C_\Omega^2+1)/\lambda$.

Step 3. We have that $v=u+\psi\in W^{1,2}(\Omega,\omega)$ is a unique solution of the problem $(P1)$ and

$$
\|v\|_{W^{1,2}(\Omega,\omega)} \leq C\left(\|\psi\|_{W^{1,2}(\Omega,\omega)}+\|g/\omega\|_{L^2(\Omega,\omega)}+\sum_{j=1}^N \|f_j/\omega\|_{L^2(\Omega,\omega)}\right).
$$

□

Lemma 3.1.2. *Let α, $\beta>1$ be given and let $\omega\in A_p$, $1\leq p<\infty$, with A_p-constant $C(\omega,p)$ and let $a_{ij}=a_{ji}$ be measurable, real-valued function satisfying*

$$
\lambda\,\omega(x)|\xi|^2 \leq \sum_{i,j=1}^N a_{ij}(x)\xi_i\xi_j \leq \Lambda\,\omega(x)|\xi|^2 \tag{3.5}
$$

for all $\xi \in \mathbb{R}^N$ *and a.e.* $x \in \Omega \subset \mathbb{R}^N$. *Then there exist weights* $\omega_{\alpha\beta} \geq 0$ *a.e. and measurable real-valued functions* $a_{ij}^{\alpha\beta}$ *such that*

$$\lambda\, \omega_{\alpha\beta}(x)\, |\xi|^2 \leq \sum_{i,j=1}^{N} a_{ij}^{\alpha\beta}(x)\xi_i\xi_j \leq \Lambda\, \omega_{\alpha\beta}(x)|\xi|^2,$$

$$\sum_{i,j=1}^{N} |a_{ij}^{\alpha\beta}(x)| \leq C\, \omega_{\alpha\beta}(x),$$

for every $\xi \in \mathbb{R}^N$ *and a.e.* $x \in \Omega$, *with constant* C *is independent of* α, β.

Proof. With the same notation in Lemma 2.2.1. The weights $\omega_{\alpha\beta}$ are defined as in Lemma 2.2.1.
Case 1. If $\omega \in A_1$. We define the coefficients a_{ij}^{α} by

$$a_{ij}^{\alpha}(x) = \sum_{l=1}^{\infty} \left(\frac{1}{|Q_l^{\alpha}|} \int_{Q_l^{\alpha}} a_{ij}(y)dy \right) \chi_{Q_l^{\alpha}}(x) + a_{ij}(x)\chi_{F_\alpha^+}(x).$$

We have the following

(a) If $x \in F_\alpha^+$, then $a_{ij}^{\alpha}(x) = a_{ij}(x)$ and $\omega_\alpha(x) = \omega(x)$. Therefore, the condition

$$\lambda\, \omega_\alpha\, |\xi|^2 \leq \sum_{i,j=1}^{N} a_{ij}^{\alpha}(x)\, \xi_i\, \xi_j \leq \Lambda\, \omega_\alpha\, |\xi|^2$$

is valid.
(b) If $x \notin F_\alpha^+$, then $x \in Q_l^{\alpha}$ (for a unique cube). In this case,

$$a_{ij}^{\alpha}(x) = \frac{1}{|Q_l^{\alpha}|} \int_{Q_l^{\alpha}} a_{ij}(y)\, dy.$$

Consequently, using condition (3.5), we get

$$a_{ij}^{\alpha}(x)\xi_i\xi_j \;=\; \frac{1}{|Q_l^{\alpha}|} \int_{Q_l^{\alpha}} a_{ij}(y)\xi_i\xi_j dy$$

$$\begin{aligned} &\leq \frac{1}{|Q_l^\alpha|}\int_{Q_l^\alpha} \Lambda|\xi|^2\omega(y)dy \\ &= \Lambda|\xi|^2\omega_\alpha(x). \end{aligned}$$

Analogously we also have

$$a_{ij}^\alpha(x)\xi_i\xi_j \geq \lambda|\xi|^2\omega_\alpha(x).$$

Therefore, we get

$$\lambda\omega_\alpha(x)|\xi|^2 \leq \sum_{i,j=1}^N a_{ij}^\alpha(x)\xi_i\xi_j \leq \Lambda|\xi|^2\omega_\alpha(x).$$

Let us now check the second condition: $\sum_{i,j=1}^N |a_{ij}^\alpha(x)| \leq C\,\omega_\alpha(x)$.

(a) If $x \in F_\alpha^+$, then $a_{ij}^\alpha(x) = a_{ij}(x)$ e $\omega_\alpha(x) = \omega(x)$. The condition is true

(b) If $x \notin F_\alpha^+$, then $x \in Q_l^\alpha$ (for a unique cube). In this case

$$a_{ij}^\alpha(x) = \frac{1}{|Q_l^\alpha|}\int_{Q_l^\alpha} a_{ij}(y)dy.$$

Hence, we obtain

$$\begin{aligned} |a_{ij}^\alpha(x)| &\leq \frac{1}{|Q_l^\alpha|}\int_{Q_l^\alpha} |a_{ij}(y)|dy \\ &\leq C\frac{1}{|Q_l^\alpha|}\int_{Q_l^\alpha} \omega(y)\,dy \\ &= C\omega_\alpha(x). \end{aligned}$$

Therefore, in both cases we have

$$\sum_{i,j=1}^N |a_{ij}^\alpha(x)| \leq C\,\omega_\alpha(x),$$

with constant C independent of α.

Case 2. If $\omega \in A_p$, $1 < p < \infty$. By the Jones Factorization Theorem we have $\omega = \omega_0 \omega_1^{1-p}$, where $\omega_0, \omega_1 \in A_1$. With the same notation in Lemma 2.2.1, we define the coefficients $a_{ij}^{\alpha\beta}$ by

$$\begin{aligned} a_{ij}^{\alpha\beta}(x) &= \left[(\omega_1)_{\beta^{1/(p-1)}}\right]^{1-p}(x) \\ &\times \left\{ \sum_{k=1}^{\infty} \left(\frac{1}{|Q_k^\alpha|} \int_{Q_k^\alpha} a_{ij}(y)\omega_1^{p-1}(y)dy \right) \chi_{Q_k^\alpha}(x) \right. \\ &+ \left. a_{ij}(x)\omega_1^{p-1}(x)\chi_{F_\alpha^+}(x) \right\} \end{aligned}$$

where $\{Q_k^\alpha\}$ is the Calderon-Zygmund decomposition of ω_0.

Let us check the first condition for the coefficients $a_{ij}^{\alpha\beta}$

(1) If $x \in F_{\alpha\beta}$, then $x \in F_\alpha^o = F_\alpha^+$. Hence,

$$\begin{aligned} a_{ij}^{\alpha\beta}(x) &= \left[(\omega_1)_{\beta^{1/(p-1)}}\right]^{1-p}(x) a_{ij}(x)\omega_1^{p-1}(x) \\ &= \omega_1^{1-p}(x) a_{ij}(x)\omega_1^{p-1}(x) \\ &= a_{ij}(x). \end{aligned}$$

Using $\omega_{\alpha\beta}(x) = \omega(x)$ in $F_{\alpha\beta}$, we obtain

$$\lambda|\xi|^2\omega_{\alpha\beta}(x) \le \sum_{i,j=1}^{N} a_{ij}^{\alpha\beta}(x)\xi_i\xi_j = \sum_{i,j=1}^{n} a_{ij}(x)\xi_i\xi_j \le \Lambda|\xi|^2\omega_{\alpha\beta}(x).$$

(2) If $x \notin F_{\alpha\beta}$, then either $x \notin F_\alpha^o$ or else $x \notin F^1_{\beta^{1/(p-1)}}$.

(i) If $x \notin F_\alpha^+ = F_\alpha^o$, then $x \in Q_k^\alpha$ (for a unique cube). In this case

$$a_{ij}^{\alpha\beta}(x) = \left[(\omega_1)_{\beta^{1/(p-1)}}\right]^{1-p}(x) \left(\frac{1}{|Q_k^\alpha|} \int_{Q_k^\alpha} a_{ij}(y)\omega_1^{p-1}(y)dy \right)$$

and this implies, using condition (3.4) and $\omega = \omega_0 \omega_1^{1-p}$, we obtain

$$\begin{aligned}
a_{ij}^{\alpha\beta}(x)\xi_i\xi_j &= \left[(\omega_1)_{\beta^{1/(p-1)}}\right]^{1-p}(x)\frac{1}{|Q_k^\alpha|}\int_{Q_k^\alpha} a_{ij}(y)\xi_i\xi_j\omega_1^{p-1}(y)\,dy \\
&\leq \left[(\omega_1)_{\beta^{1/(p-1)}}\right]^{1-p}(x)\frac{1}{|Q_k^\alpha|}\int_{Q_k^\alpha} \Lambda|\xi|^2\omega(y)\omega_1^{p-1}(y)dy \\
&= \Lambda|\xi|^2\left[(\omega_1)_{\beta^{1/(p-1)}}\right]^{1-p}(x)\frac{1}{|Q_k|^\alpha}\int_{Q_k^\alpha} \omega_0(y)\,dy \\
&= \Lambda|\xi|^2\left[(\omega_1)_{\beta^{1/(p-1)}}\right]^{1-p}(x)(\omega_0)_\alpha(x) \\
&= \Lambda|\xi|^2\omega_{\alpha\beta}(x).
\end{aligned}$$

Analogously, we also have

$$\sum_{i,j=1}^{N} a_{ij}^{\alpha\beta}(x)\xi_i\xi_j \geq \lambda|\xi|^2\omega_{\alpha\beta}(x).$$

(ii) If $x \notin F^1_{\beta^{1/(p-1)}}$, then either $x \in F_\alpha^o$ or else $x \in (F_\alpha^o)^c$, which have already been analyzed.

Therefore, in both cases, we obtain

$$\lambda|\xi|^2\omega_{\alpha\beta}(x) \leq \sum_{i,j=1}^{N} a_{ij}^{\alpha\beta}(x)\xi_i\xi_j \leq \Lambda\,|\xi|^2\omega_{\alpha\beta}(x).$$

We need to show the second condition, that is,

$$\sum_{i,j=1}^{N} |a_{ij}^{\alpha\beta}(x)| \leq C\omega_{\alpha\beta}(x).$$

(a) If $x \in F_{\alpha\beta}$, then $a_{ij}^{\alpha\beta}(x) = a_{ij}(x)$ and $\omega_{\alpha\beta}(x) = \omega(x)$. Hence the condition is valid.

(b) If $x \notin F_{\alpha\beta}$, then $x \in Q_l^\alpha$ (for a unique cube). In this case, since $\omega = \omega_0\omega_1^{1-p}$, we have

$$\begin{aligned}
|a_{ij}^{\alpha\beta}(x)| &\leq \Big[(\omega_1)_{\beta^{1/(p-1)}}\Big]^{1-p}(x)\,\frac{1}{|Q_l^\alpha|}\int_{Q_l^\alpha}|a_{ij}(y)|\omega_1^{p-1}(y)dy \\
&\leq \Big[(\omega_1)_{\beta^{1/(p-1)}}\Big]^{1-p}(x)\,\frac{1}{|Q_l^\alpha|}\int_{Q_l^\alpha}C\omega(y)\omega_1^{p-1}(y)dy \\
&= C\Big[(\omega_1)_{\beta^{1/(p-1)}}\Big]^{1-p}(x)\,\frac{1}{|Q_l^\alpha|}\int_{Q_l^\alpha}\omega_0(y)dy \\
&= C\Big[(\omega_1)_{\beta^{1/(p-1)}}\Big]^{1-p}(x)(\omega_0)_\alpha(x) \\
&= C\omega_{\alpha\beta}(x).
\end{aligned}$$

Therefore, in both cases we obtain $\sum_{i,j=1}^{N}|a_{ij}^{\alpha\beta}(x)| \leq C\,\omega_{\alpha\beta}(x)$ with constant C independent of α and β. □

Theorem 3.1.3. *Assume* $\omega \in A_2$, $\dfrac{g}{\omega}$, $\dfrac{f_j}{\omega} \in L^2(\Omega,\omega)\cap L^2(\Omega,\omega^3)$.
(I) With the same notation and hypotheses as in Theorem 3.1.1. Then the solution $u \in W^{1,2}(\Omega,\omega)$ *of the problem*

$$(P1)\begin{cases} Lu = g - \displaystyle\sum_{j=1}^{N} D_j f_j \ \text{ in } \ \Omega, \\ u-\psi \in W_0^{1,2}(\Omega,\omega), \end{cases}$$

is the weak limit in $W^{1,2}(\Omega,\tilde{\omega}_1)$ *of a sequence of solutions* $\{u_m\}$, $u_m \in W^{1,2}(\Omega,\omega_m)$, *of the problems*

$$(P_m)\begin{cases} L_m u_m = g_m - \displaystyle\sum_{j=1}^{N} D_j f_{jm} \ \text{ in } \ \Omega, \\ u_m-\psi \in W_0^{1,2}(\Omega,\omega_m), \end{cases}$$

where $\omega_m = \omega_{mm}$, $L_m u_m = -\sum_{i,j=1}^{N} D_j(a_{ij}^{mm}(x) D_i u_m)$,
$f_{jm} = f_j(\omega/\omega_m)^{-1/2}$ *and* $g_m = g(\omega/\omega_m)^{-1/2}$ (ω_{mm}, a_{ij}^{mm} *and* $\tilde{\omega}_1$ *are as in Lemma 2.2.1 and Lemma 3.1.2).*
(II) Moreover, if $u \in W_0^{1,2}(\Omega, \omega)$ *is the solution of the equation*

$$Lu = -\sum_{j=1}^{N} D_j f_j$$

with $\dfrac{f_j}{\omega} \in L^p(\Omega, \omega)$ $(j = 1, ..., N)$, $p > 2N - \varepsilon'$ *(choosing* ε' *appropriately) then* u *is uniform limit of* $(u_m)_{m\in\mathbb{N}}$ *in any compact subset of* Ω, *where* $u_m \in W_0^{1,2}(\Omega, \omega_m)$ *is a solution of the problem*

$$L_m u_m = -\sum_{j=1}^{N} D_j f_{jm}, \quad \text{with} \quad f_{jm} = f_j(\omega/\omega_m)^{(1-p)/p}.$$

Proof. We shall denote $A^m(x) = (a_{ij}^{mm}(x))_{i,j=1,\dots,N}$.
(I) First, we note that

$$\|g_m/\omega_m\|_{L^2(\Omega,\omega_m)} = \|g/\omega\|_{L^2(\Omega,\omega)} \quad \text{and}$$
$$\|f_{jm}/\omega_m\|_{L^2(\Omega,\omega_m)} = \|f_j/\omega\|_{L^2(\Omega,\omega)}.$$

Consequently, we have

$$T_m = g_m - \sum_{j=1}^{N} D_j f_{jm} \in [W_0^{1,2}(\Omega, \omega)]^*.$$

Using the fact that u_m is a solution of the problem (P_m), by Theorem 3.1.1 we obtain (because $\omega_m \in A_2$)

$$\|u_m\|_{W^{1,2}(\Omega,\omega_m)}$$
$$\leq C\bigg(\|g_m/\omega_m\|_{L^2(\Omega,\omega_m)} + \sum_{j=1}^{N} \|f_{jm}/\omega_m\|_{L^2(\Omega,\omega_m)} + \|\psi\|_{W^{1,2}(\Omega,\omega)}\bigg)$$

$$= C\left(\|g/\omega\|_{L^2(\Omega,\omega)} + \sum_{j=1}^{N} \|f_j/\omega\|_{L^2(\Omega,\omega)} + \|\psi\|_{W^{1,2}(\Omega,\omega)} \right)$$
$$= \mathbf{C}_1 \tag{3.6}$$

where the constant C is independent of m. By Lemma 2.2.1(ii) $\tilde{\omega}_1 \leq \omega_m$, we obtain

$$\|u_m\|_{W^{1,2}(\Omega,\tilde{\omega}_1)} \leq \|u_m\|_{W^{1,2}(\Omega,\omega_m)}$$
$$\leq C\left(\|g/\omega\|_{L^2(\Omega,\omega)} + \sum_{j=1}^{N} \|f_j/\omega\|_{L^2(\Omega,\omega)} + \|\psi\|_{W^{1,2}(\Omega,\omega)} \right)$$
$$= \mathbf{C}_1. \tag{3.7}$$

Consequently $\{u_m\}$ is a bounded sequence in $W^{1,2}(\Omega, \tilde{\omega}_1)$. Therefore there exists a subsequence, again denoted by $\{u_m\}$, and $\tilde{u} \in W^{1,2}(\Omega, \tilde{\omega}_1)$ such that $u_m \to \tilde{u}$ weakly in $L^2(\Omega, \tilde{\omega}_1)$ and $\nabla u_m \to \nabla \tilde{u}$ weakly in $L^2(\Omega, \tilde{\omega}_1)$ (see Theorem 1.4.7). Therefore, we have that $\tilde{u} \in W^{1,2}(\Omega, \omega)$. We need to show that $\tilde{u}$ is a weak solution of the equation $L\tilde{u} = g - \sum_{j=1}^{N} D_j f_j$, that is,

$$\sum_{i,j=1}^{N} \int_\Omega a_{ij}(x) D_i \tilde{u}(x) D_j \varphi(x) dx = \int_\Omega g(x)\varphi(x) dx + \sum_{j=1}^{N} \int_\Omega f_j(x) D_j \varphi(x) dx,$$

for all $\varphi \in W_0^{1,2}(\Omega, \omega)$. Using the fact that u_m is a solution of (P_m), we have $L_m u_m = g_m - \sum_{j=1}^{N} D_j f_{jm}$, that is,

$$\sum_{i,j=1}^{N} \int_\Omega a_{ij}^{mm}(x) D_i u_m(x) D_j \varphi(x)\, dx$$

$$= \int_\Omega g_m(x)\varphi(x)\,dx + \sum_{j=1}^{N}\int_\Omega f_{jm}(x)D_j\varphi(x)\,dx,$$

for all $\varphi \in W_0^{1,2}(\Omega,\omega_m)$. Moreover, over F_k (for $m \geq k$) we have the properties:

(i) $\omega = \omega_m$,

(ii) $g = g_m$ and $f_j = f_{jm}$,

(iii) $a_{ij}^{mm}(x) = a_{ij}(x)$.

If $\varphi \in W_0^{1,2}(\Omega,\omega)$, we get

$$G : W^{1,2}(\Omega,\tilde{\omega}_1) \to \mathbb{R}$$
$$G(u) = \sum_{i,j=1}^{N}\int_\Omega a_{ij}(x)D_iu(x)D_j\varphi(x)\chi_{F_k}(x)dx$$

is a bounded linear functional. Using this fact and the properties (i)-(iii) we obtain

$$\sum_{i,j=1}^{N}\int_{F_k} a_{ij}(x)D_i\tilde{u}(x)D_j\varphi(x)\,dx$$
$$= \lim_{m\to\infty}\sum_{i,j=1}^{N}\int_{F_k} a_{ij}^{mm}(x)D_iu_m(x)D_j\varphi(x)\,dx$$
$$= \lim_{m\to\infty}\Bigg(\sum_{i,j=1}^{N}\int_\Omega a_{ij}^{mm}(x)D_iu_m(x)D_j\varphi(x)\,dx$$
$$-\sum_{i,j=1}^{N}\int_{\Omega\cap F_k^c} a_{ij}^{mm}(x)D_iu_m(x)D_j\varphi(x)\,dx\Bigg)$$
$$= \lim_{m\to\infty}\Bigg(\int_\Omega g_m(x)\varphi(x)\,dx + \sum_{j=1}^{N}\int_\Omega f_{jm}(x)D_j\varphi(x)\,dx$$

$$-\sum_{i,j=1}^{N}\int_{\Omega\cap F_k^c} a_{ij}^{mm}(x)D_iu_m(x)D_j\varphi(x)\,dx\Bigg), \tag{3.8}$$

(recall that u_m is a solution of (P_m)).
We have, by the Lebesgue Dominated Convergence Theorem and $\tilde{\omega}_2\in A_2$, and $\dfrac{g}{\omega}, \dfrac{f_j}{\omega}\in L^2(\Omega,\omega)\cap L^2(\Omega,\omega^3)$,

$$\int_\Omega g_m\varphi\,dx \to \int_\Omega g\varphi\,dx \;\text{ and }\; \int_\Omega f_{jm}D_j\varphi\,dx \to \int_\Omega f_jD_j\varphi dx. \tag{3.9}$$

In fact, since $g_m = g\dfrac{\omega_m^{1/2}}{\omega^{1/2}} \to g$ a.e. in Ω and $\omega_m=\omega$ in F_k $(m\geq k)$, we have

$$\begin{aligned}\int_\Omega g_m^2\,\omega\,dx &= \int_\Omega g^2\omega_m\,dx\\ &= \int_{F_k} g^2\omega_m\,dx + \int_{\Omega\cap F_k^c} g^2\omega_m\,dx\\ &\leq \int_\Omega g^2\omega\,dx + \int_{\Omega\cap F_k^c} g^2\omega_m\,dx\\ &= \int_\Omega\left(\frac{g}{\omega}\right)^2\omega^3\,dx + \int_{\Omega\cap F_k^c} g^2\omega_m\,dx.\end{aligned}$$

By Lemma 2.2.1 (iv) we know that $|\Omega\cap F_k^c|\to 0$ when $k\to\infty$. Then, for sufficiently large k we have $\displaystyle\int_{\Omega\cap F_k^c} g^2\omega_m\,dx\leq 1$. Therefore, for sufficiently large m we obtain

$$\int_\Omega g_m^2\omega\,dx \leq \int_\Omega\left(\frac{g}{\omega}\right)^2\omega^3\,dx + 1 < \infty.$$

noindent Hence, the sequence $\{g_m\}$ is bounded in $L^2(\Omega,\omega)$. Then there is a subsequence, still denoted by $\{g_m\}$, and a functions $\tilde{g}$

such that

$$g_m \rightharpoonup \tilde{g}, \text{ in } L^2(\Omega, \omega),$$
$$g_m \to \tilde{g}, \text{ a.e. in } \Omega.$$

Since $g_m \to \tilde{g}$ a.e. in Ω, then $\tilde{g} = g$ a.e. in Ω. Therefore, for all $\varphi \in W^{1,2}(\Omega, \omega)$ we have

$$\int_\Omega g_m \varphi \, dx \to \int_\Omega g \, \varphi \, dx.$$

Analogously, we also have $\int_\Omega f_{jm} \, D_j \varphi \, dx \to \int_\Omega f_j D_j \varphi \, dx$ for all $\varphi \in W^{1,2}(\Omega, \omega)$.

Using $\sum_{i,j=1}^{N} |a_{ij}^{mm}(x)| \leq C \, \omega_m(x)$, with C independent of m, we obtain

$$\begin{aligned}
&\sum_{i,j=1}^{N} \left| \int_{\Omega \cap F_k^c} a_{ij}^{mm} D_i u_m D_j \varphi \, dx \right| \\
&\leq \sum_{i,j=1}^{N} \int_{\Omega \cap F_k^c} |a_{ij}^{mm}| |D_i u_m| |D_j \varphi| \, dx \\
&\leq \sum_{i,j=1}^{N} \int_{\Omega \cap F_k^c} C \, \omega_m |D_i u_m| |D_j \varphi| \, dx \quad \text{(C is independent of } m) \\
&= C \sum_{i,j=1}^{N} \int_{\Omega \cap F_k^c} |D_i u_m| \omega_m^{1/2} |D_j \varphi| \omega_m^{1/2} \, dx \\
&\leq C \sum_{i,j=1}^{N} \left(\int_{\Omega \cap F_k^c} |D_i u_m|^2 \omega_m \, dx \right)^{1/2} \left(\int_{\Omega \cap F_k^c} |D_j \varphi|^2 \omega_m \, dx \right)^{1/2} \\
&\leq C \;\; \mathbf{C}_1 \, C_\varphi \left(\int_{F_k^c \cap \Omega} \omega_m \, dx \right)^{1/2} \quad \text{(by (3.5))}
\end{aligned}$$

$$= C\ \mathbf{C}_1\, C_\varphi[\omega_m(F_k^c \cap \Omega)]^{1/2}.$$

By Theorem 2.9, Chapter IV, of [38], there exist constants $\delta > 0$, $\mathbf{C} > 0$ such that, if $\overline{\Omega} \subset Q_0$ (Q_0 is a fixed cube) then

$$\omega_m(F_k^c \cap \Omega) \leq \tilde{\omega}_2(F_k^c \cap \Omega) \leq\ \mathbf{C}\,\tilde{\omega}_2(Q_0)\left(\frac{|F_k^c|}{|Q_0|}\right)^\delta.$$

Using Lemma 2.2.1, we know that $|F_k^c| \to 0$ when $k \to \infty$. Then

$$\omega_m(F_k^c \cap \Omega) \to 0 \quad \text{when} \quad k \to \infty$$

and we obtain

$$\lim_{k\to\infty} \sum_{i,j=1}^N \int_{\Omega \cap F_k^c} a_{ij}^{mm}(x) D_i u_m(x) D_j\varphi(x)\, dx = 0. \tag{3.10}$$

Therefore, by (3.9) and (3.10) we conclude

$$\begin{aligned}\sum_{i,j=1}^N \int_\Omega a_{ij}(x) D_i\tilde{u}(x) D_j\varphi(x)\, dx &= \int_\Omega g(x)\varphi(x)\, dx \\ &+ \sum_{j=1}^N \int_\Omega f_j(x) D_j\varphi(x)\, dx\end{aligned}$$

that is, $\tilde{u}$ is a solution of the equation $Lu = g - \sum_{j=1}^N D_j f_j$. Therefore $u = \tilde{u}$ (by the uniqueness of Theorem 3.1.1).

(II) Now we will prove the second part of the theorem.

If $\omega \in A_2$, then there exists $\varepsilon > 0$ such that $\omega \in A_{2-\varepsilon}$ (see Theorem 2.1.7 (open-end property). We choose $\varepsilon' = N\varepsilon$ and $p > 2N - \varepsilon' = 2N - \varepsilon N$ and $\omega \in A_{2-\varepsilon} \subset A_{p/N}$.

By Theorem 2.3.15 of [35], if u is a weak solution of the equation

$Lu = -\sum_{j=1}^{N} D_j f_j$, and $f_j/\omega \in L^p(\Omega,\omega)$, $\omega \in A_{p/N}$, then u is locally Hölder continuous in Ω, i.e., there exist constants $C > 0$ and λ $(0 < \lambda < 1)$, such that if $x_o \in \Omega$, $0 < \rho < R < \frac{1}{16}\operatorname{dist}(x_o, \partial\Omega)$, we have

$$\operatorname*{osc}_{B(x_o,\rho)} u \leq C\left[\left(\frac{1}{\omega(B_R)}\int_{B_R} u^2\omega\,dx\right)^{1/2} + \sum_{j=1}^{N}\|f_j/\omega\|_{L^p(B_R,\omega)}\right]\rho^{\lambda}, \tag{3.11}$$

with C and λ are independent of u, ρ and x_o, and $\operatorname{osc}_{B(x_o,\rho)} u$ is the oscillation over $B(x_o,\rho)$ of u. Applying this result for the solution of the equations

$$L_m u_m = -\sum_{j=1}^{N} D_j f_{jm}$$

$u_m \in W_0^{1,2}(\Omega,\omega_m)$, $\omega_m \in A_{p/n}$, with $f_{jm} = f_j(\omega/\omega_m)^{(1-p)/p}$, and using (3.6), we obtain

$$\begin{aligned}\operatorname*{osc}_{B(x_o,\rho)} u_m &\leq C\Bigg[\left(\frac{1}{\omega_m(B_R)}\int_{B_R} u_m^2\omega_m\,dx\right)^{1/2} \\ &+ \sum_{j=1}^{N}\|f_{jm}/\omega_m\|_{L^p(\Omega,\omega_m)}\Bigg]\rho^{\lambda} \\ &\leq C\left[\frac{\mathbf{C}_1}{\tilde{\omega}_1(B_R)} + \sum_{j=1}^{N}\|f_j/\omega\|_{L^p(\Omega,\omega)}\right]\rho^{\lambda}.\end{aligned}$$

Therefore, the sequence $\{u_m\}$ is locally equicontinuous. Moreover, by Lemma 2.3.14 of [35], we have

$$\operatorname*{ess\ sup}_{B(x_o,\rho)} |u_m| \leq C\left(\sum_{j=1}^{N}\left\|\frac{f_{jm}}{\omega_m}\right\|_{L^p(\Omega,\omega_m)}\right)\rho^{\lambda}.$$

Hence, using $\|f_{jm}/\omega_m\|_{L^p(\Omega,\omega_m)} = \|f_j/\omega\|_{L^p(\Omega,\omega)}$, we obtain

$$\operatorname*{ess\,sup}_{B(x_o,\rho)} |u_m| \leq C\left(\sum_{j=1}^{N} \left\|\frac{f_j}{\omega}\right\|_{L^p(\Omega,\omega)}\right)\rho^{\lambda},$$

that is, $\{u_m\}$ is a locally uniformly bounded sequence. We can apply Arzelà-Ascoli Theorem and conclude that $\{u_m\}$ converges to u uniformly in compact subsets of Ω. □

3.2 Entropy Solution

The main purpose of this section (see Theorem 3.2.8) is to establish the existence and uniqueness of entropy solutions (see Definition 3.2.2) for the Dirichlet problem

$$(P2)\begin{cases} Lu(x) & = f(x) \text{ in } \Omega, \\ u(x) & = 0 \text{ on } \partial\Omega, \end{cases}$$

where $f \in L^1(\Omega)$ and L is the degenerate elliptic operator in divergence form (3.1). In general we follow the approach of [2], where a theory of solvability of Dirichlet problem for non-degenerate elliptic equations with L^1-data was developed. Note that, in the proof of our main result, many ideas have been adapted from [2].

We propose to solve the problem (P2) by approximation with variational solutions: we take $f_n \in C_0^\infty(\Omega)$ such that $f_n \to f$ in $L^1(\Omega)$, we find a solution $u_n \in W_0^{1,2}(\Omega,\omega)$ for the problem with second member f_n and we will try to pass to the limit $n \to \infty$.

Now we introduce the function spaces we will need in this section.

Definition 3.2.1. *Let $X \subset \mathbb{R}^N$ be an open set.*
(I) We shall denote by $L_0(X,\omega)$ the set of all measurable functions $u : X \to \mathbb{R}$ such that $\mu(X_{u,\varepsilon}) < \infty$ for every $\varepsilon > 0$, where

$$X_{u,\varepsilon} = \{x \in X : |u(x)| > \varepsilon\},$$

and $\mu(E) = \int_E \omega(x)\,dx$. *We have that* $L^p(X,\omega)\subset L_0(X,\omega)$, $1\leq p<\infty$.

(II) For a given $k>0$, *we define the function* $T_k:\mathbb{R}\to\mathbb{R}$ *as*

$$T_k(s)=\begin{cases} s, & \text{if } |s|\leq k \\ k\ \text{sign}(s), & \text{if } |s|>k. \end{cases}$$

We define $\mathcal{T}^{1,1}_{\text{loc}}(X,\omega)$ *as the set of all measurable functions* $u:X\to\mathbb{R}$ *such that for every* $k>0$ *the function* $T_k(u)\in W^{1,1}_{\text{loc}}(X,\omega)$, *where*

$$T_k(u(x))=\begin{cases} u(x), & \text{if } |u(x)|\leq k \\ k\ \text{sign}(u(x)), & \text{if } |u(x)|>k. \end{cases}$$

(III) For $1<p<\infty$, *we define* $\mathcal{T}^{1,p}_{\text{loc}}(X,\omega)$ *as the subset of* $\mathcal{T}^{1,1}_{\text{loc}}(X,\omega)$ *consisting of the functions* $u:X\to\mathbb{R}$ *such that* $D_j(T_k(u))\in L^p_{\text{loc}}(X,\omega)$, $j=1,2,...,N$, *for every* $k>0$.

Likewise, $\mathcal{T}^{1,p}(X,\omega)$ *is the subset of* $\mathcal{T}^{1,1}_{\text{loc}}(X,\omega)$ *consisting of the functions* u *such that* $D_j(T_k(u))\in L^p(X,\omega)$, $j=1,2,...,N$, *for every* $k>0$.

(IV) We define $\mathcal{T}^{1,p}_0(X,\omega)$ *as the subset of* $\mathcal{T}^{1,p}(X,\omega)$ *consisting of the functions* u *that can be approximated by smooth functions* $C^\infty_0(X)$, *i.e.,* $u\in\mathcal{T}^{1,p}_0(X,\omega)$ *if* $u\in\mathcal{T}^{1,p}(X,\omega)$ *and for every* $k>0$ *there exists a sequence* $\{\varphi_n\}$, $\varphi_n\in C^\infty_0(X)$ *such that*

$$\varphi_n\to T_k(u) \quad \text{in} \quad L^1_{\text{loc}}(X,\omega);$$

$$D_j\varphi_n\to D_jT_k(u) \quad \text{in} \quad L^p(X,\omega).$$

Remark 3.2.1. *(a) If* $u\in W^{1,1}_{\text{loc}}(\Omega,\omega)$ *then we have*

$$\nabla T_k(u)=\chi_{\{|u|<k\}}\nabla u,$$

where χ_E *denotes the characteristic function of a measurable set* $E\subset\mathbb{R}^N$. *Hence, we have the inclusion* $W^{1,1}_{\text{loc}}(X,\omega)\subset\mathcal{T}^{1,1}_{\text{loc}}(X,\omega)$.

If $u\in\mathcal{T}^{1,1}_{\text{loc}}(X,\omega)$, *then there exists a unique measurable function* $v : X\to\mathbb{R}^N$ *such that*

$$\nabla T_k(u(x)) = v(x)\chi_{\{|u|<k\}}. \tag{3.12}$$

(b) If $u_1, u_2\in W^{1,p}_0(\Omega,\omega)$ *then* $\varphi = T_k(u_1+u_2)\in W^{1,p}_0(\Omega,\omega)\cap L^\infty(\Omega)$ *and we have*

$$\nabla\varphi = \nabla T_k(u_1+u_2) = \nabla(u_1+u_2)\,\chi_{\{|u_1+u_2|\le k\}}.$$

(c) Note that for given $h>0$ *and* $k>0$ *we have*

$$T_h(u-T_k(u)) = \begin{cases} 0, & \text{if } |u|\le k \\ (|u|-k)\operatorname{sign}(u) & \text{if } k<|u|\le k+h, \\ h\operatorname{sign}(u), & \text{if } |u|>k+h. \end{cases}$$

And if $\alpha\in\mathbb{R}$, $\alpha\neq 0$, *we have* $T_k(\alpha u) = T_{k/|\alpha|}(u)$.

If $u\in\mathcal{T}^{1,1}_{\text{loc}}(X,\omega)$ we define ∇u, the weak gradient of u, as the unique function v which satisfies (3.12). The definition of the weak gradient allows us to give the following definition of an entropy solution for problem (P2) (the concept of entropy solutions was introduced in [2]).

Definition 3.2.2. *Let L be the operator (3.1) and assume that (3.2) holds and* $f\in L^1(\Omega)$. *We define an entropy solution of problem*

$$(P2)\begin{cases} Lu(x) = f(x) & \text{in } \Omega, \\ u(x) = 0, & \text{on } \partial\Omega, \end{cases}$$

as a function $u\in\mathcal{T}^{1,2}_0(\Omega,\omega)$ *satisfying the inequality*

$$\sum_{i,j=1}^{N}\int_\Omega a_{ij}D_iuD_jT_k(u-\varphi)\,dx \le \int_\Omega fT_k(u-\varphi)\,dx, \tag{3.13}$$

for all $\varphi\in C^\infty_0(\Omega)$ *and for every* $k>0$.

Remark 3.2.2. *Notice that both integrals in (3.13) are well defined. In fact:*
(a) In the right-hand side, since $f \in L^1(\Omega)$*, we obtain*

$$\left| \int_\Omega f T_k(u-\varphi)\,dx \right| \leq k \int_\Omega |f|\,dx < \infty.$$

(b) In the left-hand side, using condition (3.2) we have

$$\begin{aligned}
&\sum_{i,j=1}^N \int_\Omega a_{ij} D_i u D_j T_k(u-\varphi)\,dx \\
&= \sum_{i,j=1}^N \int_{\{|u-\varphi|<k\}} a_{ij} D_i u [D_j u - D_j \varphi]\,dx \\
&= \sum_{i,j=1}^N \int_{\{|u-\varphi|<k\}} [a_{ij} D_i u D_j u - a_{ij} D_i u D_j \varphi]\,dx \\
&\geq \int_{\{|u-\varphi|<k\}} \lambda |\nabla u|^2 \omega\,dx - \sum_{i,j=1}^N \int_{\{|u-\varphi|<k\}} a_{ij} D_i u D_j \varphi\,dx \\
&\geq - \sum_{i,j=1}^N \int_{\{|u-\varphi|<k\}} a_{ij} D_i u D_j \varphi\,dx.
\end{aligned}$$

Considering $k^* = k + \|\varphi\|_\infty$*, we have*

$$\begin{aligned}
\Big(a_{ij} D_i u D_j \varphi \Big) \chi_{\{|u-\varphi|<k\}} &\leq |a_{ij}||D_i u||D_j \varphi| \chi_{\{|u(x)|<k^*\}} \\
&= |a_{ij}||D_i T_{k^*}(u)||D_j \varphi|.
\end{aligned}$$

Using condition (3.2) we have $\sum_{i,j=1}^N |a_{ij}(x)| \leq \Lambda\,\omega(x)$ *a.e.* $x \in \Omega$.

Hence,

$$\begin{aligned}
&\left|\sum_{i,j=1}^{N}\int_{\{|u-\varphi|<k\}} a_{ij}D_i u D_j\varphi\,dx\right|\\
&\leq \sum_{i,j=1}^{N}\int_{\{|u(x)|<k^*\}} |a_{ij}||D_i u||D_j\varphi|\,dx\\
&\leq \Lambda\sum_{i,j=1}^{N}\int_{\{|u(x)|<k^*\}} |D_i u||D_j\varphi|\,\omega\,dx\\
&= \Lambda\sum_{i,j=1}^{N}\int_{\Omega} |D_i T_{k^*}(u)||D_j\varphi|\omega\,dx\\
&= \Lambda\sum_{i,j=1}^{N}\int_{\operatorname{supp}(\varphi)} |D_i T_{k^*}(u)||D_j\varphi|\omega\,dx\\
&\leq \Lambda\|\nabla\varphi\|_{L^\infty(\Omega)}\left(\sum_{j=1}^{N}\int_{\operatorname{supp}(\varphi)} |D_i T_{k^*}(u)|\omega\,dx\right) < \infty,
\end{aligned}$$

because of $D_i T_{k^*}(u) \in L^1_{\text{loc}}(\Omega,\omega)$. *Therefore the integral in left - hand side of (3.13) is well defined.*

For the demonstration of the main result of this section, we need some preliminary results.

Lemma 3.2.1. *If* $u \in \mathcal{T}_0^{1,2}(\Omega,\omega)$ *is an entropy solution of problem* (P2), *then for every* $k > 0$ *we have*

$$\| |\nabla T_k(u)| \|^2_{L^2(\Omega,\omega)} \leq \frac{k}{\lambda}\|f\|_{L^1(\Omega)}.$$

Proof. Since u is an entropy solution of problem (P2), then we have

$$\sum_{i,j=1}^{N}\int_{\{|u-\varphi|<k\}} a_{ij}D_i u[D_j u - D_j\varphi]\,dx \leq \int_{\Omega} fT_k(u-\varphi)\,dx,$$

for all $\varphi \in C_0^\infty(\Omega)$. Then, for $\varphi = 0$, we have

$$\sum_{i,j=1}^{N} \int_{\{|u|<k\}} a_{ij} D_i u D_j u \, dx \le \int_\Omega f T_k(u) \, dx \le k \int_\Omega f \, dx \le k \|f\|_{L^1(\Omega)}.$$

In the left-hand side, using condition (3.2), we obtain

$$\begin{aligned} \sum_{i,j=1}^{N} \int_{\{|u|<k\}} a_{ij} D_i u D_j u \, dx &\ge \lambda \int_{\{|u|<k\}} |\nabla u|^2 \omega \, dx \\ &= \lambda \int_\Omega |\nabla T_k(u)|^2 \omega \, dx. \end{aligned}$$

Therefore, we have $\||\nabla T_k(u)|\|^2_{L^2(\Omega,\omega)} \le \frac{k}{\lambda} \|f\|_{L^1(\Omega)}$. □

Definition 3.2.3. *Let $\mathcal{F}$ be the class of functions $T \in C^2(\mathbb{R}) \cap L^\infty(\mathbb{R})$ satisfying*

$$\begin{aligned} T(0) &= 0, \quad T(-s) = -T(s), \\ T'(s) &\ge 0 \text{ and } T'(s) = 0 \text{ for sufficiently large s}, \\ T''(s) &\le 0, \text{ for } s \ge 0. \end{aligned}$$

Lemma 3.2.2. *The entropy condition (3.13) is equivalent to the statement that*

$$\sum_{i,j=1}^{N} \int_\Omega a_{ij} D_i u D_j T(u-\varphi) \, dx \le \int_\Omega f T(u-\varphi) \, dx \qquad (3.14)$$

holds for every $\varphi \in C_0^\infty(\Omega)$ and every function $T \in \mathcal{F}$.

Proof. The proof of this lemma follows the lines of Lemma 3.2 in [2]. □

Lemma 3.2.3. *If u is an entropy solution of problem* (P2). *Then (3.13) holds for every function* $\varphi \in \mathcal{T}_0^{1,2}(\Omega,\omega) \cap L^\infty(\Omega,\omega)$.

Proof. Let u be an entropy solution of problem (P2). Then by definition

$$\sum_{i,j=1}^{N} \int_{\{|u-\varphi|<k\}} a_{ij} D_i u [D_j u - D_j \varphi]\, dx \leq \int_\Omega f T_k(u-\varphi)\, dx,$$

for all $\varphi \in C_0^\infty(\Omega)$ and every $k > 0$.

Let $\Psi \in \mathcal{T}_0^{1,2}(\Omega,\omega) \cap L^\infty(\Omega,\omega)$. By definition of $\mathcal{T}_0^{1,2}(\Omega,\omega)$, for all $k > 0$ there exists a sequence $\{\varphi_n\}$, $\varphi_n \in C_0^\infty(\Omega)$ such that

$$\varphi_n \to T_k(\Psi) \text{ in } L^1_{\text{loc}}(\Omega,\omega), \tag{3.15}$$

$$D_j \varphi_n \to D_j T_k(\Psi) \text{ in } L^2(\Omega,\omega). \tag{3.16}$$

Replacing φ_n by $g(\varphi_n)$, with $g \in C^\infty(\mathbb{R}) \cap L^\infty(\mathbb{R},\omega)$, $g(s) = s$, for $s \leq \|\Psi\|_{L^\infty(\Omega,\omega)}$, we may assume that the functions φ_n are uniformly bounded in Ω. We may also assume that there exists a function $v \in L^2(\Omega,\omega)$ such that $|\nabla \varphi_n| \leq v$ μ-a.e. (see Theorem 2.8.1 in [47]) (and $|\nabla \varphi_n| \leq v$ a.e., by Lemma 2.1.2). We have $T_k(u-\varphi_n) \to T_k(u-\Psi)$ μ-a.e. and $|D_j T_k(u-\varphi_n)| \leq |D_j T_M(u)| + v$, with $M = k + \sup\|\varphi_n\|_{L^\infty(\Omega,\omega)}$.

Using (3.16), we have that $D_j T_k(u-\varphi_n) \to D_j T_k(u-\Psi)$ weakly in $L^2(\Omega,\omega)$.

Since u is an entropy solution, by definition we obtain

$$\sum_{i,j=1}^{N} \int_\Omega a_{ij} D_i u D_j T_k(u-\varphi_n)\, dx \leq \int_\Omega f T_k(u-\varphi_n)\, dx. \tag{3.17}$$

We need to show that condition (3.13) holds for Ψ.

(a) Since $f \in L^1(\Omega)$ and $T_k(u-\varphi_n) \to T_k(u-\Psi)$ μ-a.e. (and also a.e.), we obtain

$$\lim_{n\to\infty} \int_\Omega f T_k(u-\varphi_n)\,dx \le \int_\Omega f T_k(u-\Psi)\,dx.$$

(b) In the left-hand side of (3.17), by (3.2), (3.16) and with $M = k + \sup\|\varphi_n\|_{L^\infty(\Omega,\omega)}$, we have

$$\begin{aligned}
&\left|\sum_{i,j=1}^N \int_\Omega a_{ij} D_i u D_j T_k(u-\varphi_n) - a_{ij} D_i u D_j T_k(u-\Psi)\,dx\right| \\
&\le \sum_{i,j=1}^N \int_\Omega |a_{ij} D_i u [D_j T_k(u-\varphi_n) - D_j T_k(u-\Psi)]|\,dx \\
&\le \sum_{i,j=1}^N \int_\Omega |a_{ij}||D_i T_M(u)||D_j T_k(u-\varphi_n) - D_j T_k(u-\Psi)|\,dx \\
&\le \Lambda \sum_{i,j=1}^N \int_\Omega |D_i T_M(u)||D_j T_k(u-\varphi_n) - D_j T_k(u-\Psi)|\omega\,dx \\
&\le \Lambda \sum_{i,j=1}^N \left(\int_\Omega |D_i T_M(u)|^2 \omega\,dx\right)^{1/2} \\
&\times \left(\int_\Omega |D_j T_k(u-\varphi_n) - D_j T_k(u-\Psi)|^2 \omega\,dx\right)^{1/2} \\
&\to 0, \text{ as } n\to\infty.
\end{aligned}$$

Therefore the condition

$$\sum_{i,j=1}^N \int_\Omega a_{ij} D_i u D_j T_k(u-\Psi)\,dx \le \int_\Omega f T_k(u-\Psi)\,dx$$

holds for every $\Psi \in \mathcal{T}_0^{1,2}(\Omega,\omega) \cap L^\infty(\Omega,\omega)$. □

Remark 3.2.3. *For given $a > 0$ and $k > 0$ the function $T_{k,a}(s) = T_a(s - T_k(s))$ takes the values*

$$T_{k,a}(s) = \begin{cases} s - k\,\mathrm{sign}(s), & \text{for } k \leq |s| < k + a, \\ a, & \text{for } |s| \geq k + a, \\ 0, & \text{for } |s| \leq k, \end{cases}$$

and $|T_{k,a}(s)| \leq a$. If $\varphi \in \mathcal{T}_0^{1,2}(\Omega, \omega) \cap L^\infty(\Omega, \omega)$ we can write $T_{k,a}(u - \varphi)$ in the form

$$T_{k,a}(u - \varphi) = T_a(u - v),$$

with $v = \varphi + T_k(u - \varphi) \in \mathcal{T}_0^{1,2}(\Omega, \omega) \cap L^\infty(\Omega, \omega)$.

Lemma 3.2.4. *If u is an entropy solution of problem (P2) then*

$$\sum_{i,j=1}^{N} \int_{\{k<|u|<k+a\}} a_{ij} D_i u D_j u \, dx \leq \int_\Omega f T_{k,a}(u) \, dx \leq a \int_{\{|u|>k\}} |f| \, dx.$$

Proof. Let $\varphi \in \mathcal{T}_0^{1,2}(\Omega, \omega) \cap L^\infty(\Omega, \omega)$. Applying Lemma 3.2.3 for

$$v = \varphi + T_k(u - \varphi) \in \mathcal{T}_0^{1,2}(\Omega, \omega) \cap L^\infty(\Omega, \omega),$$

we obtain

$$\sum_{i,j=1}^{N} \int_\Omega a_{ij} D_i u D_j T_a(u - v) \, dx \leq \int_\Omega T_a(u - v) f \, dx.$$

Hence,

$$\sum_{i,j=1}^{N} \int_\Omega a_{ij} D_i u D_j T_{k,a}(u - \varphi) \, dx \leq \int_\Omega T_{k,a}(u - \varphi) f \, dx.$$

Setting $\varphi = 0$ we obtain

$$\sum_{i,j=1}^{N} \int_{\{k<|u|<k+a\}} a_{ij} D_i u D_j u \, dx \quad \leq \quad \int_\Omega f T_{k,a}(u) \, dx$$

$$\leq \quad a\int_{\{|u|>k\}}|f|\,dx.$$

Using the degenerate ellipticity condition (3.2) we also obtain

$$\lambda\int_{\{k<|u|<k+a\}}|\nabla u|^2\omega\,dx \leq a\int_{\{|u|>k\}}|f|\,dx.$$

□

A priori estimates

Now we prove that if $u\in\mathcal{T}_0^{1,2}(\Omega,\omega)$ is a entropy solution of problem (P2) then u is solution of equation $Lu=f$, i.e.,

$$\sum_{i,j=1}^{N}\int_\Omega a_{ij}(x)D_iu(x)D_j\varphi(x)\,dx=\int_\Omega f(x)\varphi(x)\,dx,$$

for every $\varphi\in C_0^\infty(\Omega)$ (see Theorem 3.2.7).

Definition 3.2.4. *Let $0<p<\infty$ and let ω be a weight function. We define the weighted Marcinkiewicz space $\mathcal{M}^p(\Omega,\omega)$ the set of measurable functions $f:\Omega\to\mathbb{R}$ such that the function*

$$\Gamma_f(k)=\mu(\{x\in\Omega\ :\ |f(x)|>k\}),\ k>0,$$

satisfies an estimate of the form $\Gamma_f(k)\leq Ck^{-p}$, $0<C<\infty$ (where $\mu(E)=\displaystyle\int_E\omega(x)\,dx$).

Remark 3.2.4. *If $1\leq q<p$, we have that*

$$L^p(\Omega,\omega)\subset\mathcal{M}^p(\Omega,\omega)\subset L_0(\Omega,\omega),$$
$$\mathcal{M}^p(\Omega,\omega)\subset L^q(\Omega,\omega),$$

(the proof follows the lines of Theorem 2.18.8 in [47]).

Lemma 3.2.5. *Let* $u \in \mathcal{T}_0^{1,p}(\Omega, \omega)$ *and* $\omega \in A_p$, $1 < p < \infty$, *be such that*

$$\frac{1}{k} \int_{\{|u|<k\}} |\nabla u|^p \omega \, dx \leq M, \tag{3.18}$$

for every $k > 0$. *Then* $u \in \mathcal{M}^{p_1}(\Omega, \omega)$, *where* $p_1 = \theta(p-1)$ *(where* θ *is the constant in Theorem 2.1.5). More precisely, there exists* $C > 0$ *such that* $\Gamma_k(u) \leq C M^\theta k^{-p_1}$.

Proof. For $k > 0$, by the weighted Sobolev inequality (Theorem 2.1.5) and (3.18), we obtain

$$\|T_k(u)\|_{L^{\theta p}(\Omega,\omega)} \leq C_\Omega \| |\nabla T_k(u)| \|_{L^p(\Omega,\omega)} \leq C_\Omega (Mk)^{1/p}.$$

For $0 < \varepsilon \leq k$, we have $\{x \in \Omega : |u(x)| \geq \varepsilon\} = \{x \in \Omega : |T_k(u(x))| \geq \varepsilon\}$. Hence,

$$\begin{aligned}
\mu(\{x \in \Omega : |u(x)| > \varepsilon\}) &= \int_{\{|u(x)|>\varepsilon\}} \omega(x) \, dx \\
&= \frac{1}{\varepsilon^{\theta p}} \int_{\{|T_k(u(x))|>\varepsilon\}} \varepsilon^{\theta p} \omega(x) \, dx \\
&\leq \frac{1}{\varepsilon^{\theta p}} \int_{\{|T_k(u(x))|>\varepsilon\}} |T_k(u(x))|^{\theta p} \omega(x) \, dx \\
&\leq \frac{1}{\varepsilon^{\theta p}} \|T_k(u)\|^{\theta p}_{L^{\theta p}(\Omega,\omega)} \\
&\leq \frac{1}{\varepsilon^{\theta p}} [C_\Omega (Mk)^{1/p}]^{\theta p} = \frac{1}{\varepsilon^{\theta p}} C (Mk)^\theta.
\end{aligned}$$

Setting $\varepsilon = k$, we obtain

$$\mu(\{x \in \Omega : |u(x)| > k\}) \leq C M^\theta k^{-\theta(p-1)} = C M^\theta k^{-p_1},$$

where $p_1 = \theta(p-1)$, that is, $\Gamma_u(k) \leq C M^\theta k^{-p_1}$.
Therefore $u \in \mathcal{M}^{p_1}(\Omega, \omega)$. □

Lemma 3.2.6. *Let $u\in\mathcal{T}_0^{1,p}(\Omega,\omega)$, where $\omega\in A_p$, $1<p<\infty$, be such that*

$$\frac{1}{k}\int_{\{|u|<k\}} |\nabla u|^p\omega\, dx \le M,$$

for every $k>0$. Then $|\nabla u|\in\mathcal{M}^{p_2}(\Omega,\omega)$, where $p_2 = p\,p_1/(p_1+1)$ (with θ as in Lemma 3.2.5) and $p_1=\theta(p-1)$). More precisely, there exists $C>0$ such that $\Gamma_k(|\nabla u|)\le CM^{(p_1+\theta)/(p_1+1)}k^{-p_2}$.

Proof. For $k>0$ and $\alpha>0$ we define the function

$$\Phi(k,\alpha)=\mu(\{x\in\Omega\ :\ |\nabla u(x)|^p>\alpha,\ |u(x)|>k\}).$$

From Lemma 3.2.5 we have

$$\Phi(k,0)\le CM^{\theta}k^{-p_1}.$$

Moreover, if $\alpha_1<\alpha_2$ then $\Phi(k,\alpha_2)\le\Phi(k,\alpha_1)$. Hence, for $k>0$ and $\alpha>0$ we have

$$\begin{aligned}\Phi(0,\alpha) &\le \frac{1}{\alpha}\int_0^{\alpha}\Phi(0,s)\,ds\\ &\le \Phi(k,0)+\frac{1}{\alpha}\int_0^{\alpha}[\Phi(0,s)-\Phi(k,s)]\,ds. \qquad (3.19)\end{aligned}$$

We get

$$\begin{aligned} &\Phi(0,s)-\Phi(k,s)\\ =\ &\mu(\{x\in\Omega\ :\ |\nabla u(x)|^p>s,\ |u(x)|>0\})\\ -\ &\mu(\{x\in\Omega\ :\ |\nabla u(x)|^p>s,\ |u(x)|>k\})\\ =\ &\mu(\{x\in\Omega\ :\ |\nabla u(x)|^p>s \text{ and } |u(x)|\le k\}).\end{aligned}$$

Hence, using Proposition 6.24 of [37]

$$\int_E |f|\,d\mu=\int_0^{\infty}\mu(\{x\in E\ :\ |f(x)|>\alpha\})\,d\alpha,$$

we obtain

$$\begin{aligned}
&\int_0^\infty [\Phi(0,s)-\Phi(k,s)]\,ds \\
&= \int_0^\infty \mu(\{x\in\Omega\ :\ |\nabla u(x)|^p > s \text{ and } |u(x)|\le k\})\,ds \\
&= \int_{\{|u(x)|<k\}} |\nabla u(x)|^p \omega(x)\,dx \le Mk.
\end{aligned}$$

Consequently in (3.19), we obtain

$$\begin{aligned}
\Phi(0,\alpha) &\le \Phi(k,0) + \frac{1}{\alpha}\int_0^\alpha [\Phi(0,s)-\Phi(k,s)]\,dx \\
&\le CM^\theta k^{-p_1} + \frac{1}{\alpha}Mk. \qquad (3.20)
\end{aligned}$$

Minimizing the right member of (3.20) with respect to k and setting $\alpha = h^p$, we obtain

$$\begin{aligned}
\Phi(0,h^p) &\le CM^{(p_1+\theta)/(p_1+1)} h^{-pp_1/(p_1+1)} \\
&= CM^{(p_1+\theta)/(p_1+1)} h^{-p_2}, \text{ where } p_2 = pp_1/(p_1+1).
\end{aligned}$$

Consequently,

$$\mu(\{x\in\Omega\ :\ |\nabla u(x)| > h\}) \le CM^{(p_1+\theta)/(p_1+1)} h^{-p_2},$$

that is, $|\nabla u|\in\mathcal{M}^{p_2}(\Omega,\omega)$. □

Theorem 3.2.7. *Let L be the operator (3.1) and assume that the condition (3.2) holds, $\omega\in A_2$ and $f\in L^1(\Omega)$. If $u\in\mathcal{T}_0^{1,2}(\Omega,\omega)$ is an entropy solution of problem* (P2) *then u is solution of equation $Lu=f$, that is,*

$$\sum_{i,j=1}^N \int_\Omega a_{ij}(x) D_i u(x) D_j \varphi(x)\,dx = \int_\Omega f(x)\varphi(x)\,dx,$$

for every $\varphi \in C_0^\infty(\Omega)$.

Proof. If $u\in\mathcal{T}_0^{1,2}(\Omega,\omega)$ is an entropy solution then, by Lemma 3.2.1, we have

$$\frac{1}{k}\int_{\{|u(x)|<k\}}|\nabla u(x)|^2\omega(x)\,dx \leq \frac{1}{\lambda}\int_\Omega |f(x)|\,dx = M.$$

Applying Lemma 3.2.6 (with $p=2$), for every $h>0$ we obtain

$$\mu(\{x\in\Omega\ :\ |\nabla u(x)|>h\})\leq CM^q h^{-p_2},$$

that is, $|\nabla u|\in\mathcal{M}^{p_2}(\Omega,\omega)$ where $p_2=2\theta/(\theta+1)$ and $q=(p_1+\theta)/(p_1+1)$. By Remark 3.2.4, we have $|\nabla u|\in L^1(\Omega,\omega)$.
For $\varphi\in C_0^\infty(\Omega)$, applying Lemma 3.2.3 with function $T_h(u)-\varphi$, we obtain

$$\begin{aligned}
&\sum_{i,j=1}^N\int_{\{|u-T_h(u)+\varphi|\leq k\}} a_{ij}D_iu[D_ju-D_j(T_h(u)-\varphi)]\,dx\\
&=\sum_{i,j=1}^N\int_{\{|u-T_h(u)+\varphi|\leq k\}} a_{ij}D_iu\Big(\chi_{\{|u|>h\}}D_ju+D_j\varphi\Big)\,dx\\
&\geq\sum_{i,j=1}^N\int_{\{|u-T_h(u)+\varphi|\leq k\}} a_{ij}D_iuD_j\varphi\,dx.
\end{aligned}$$

Therefore,

$$\sum_{i,j=1}^N\int_{\{|u-T_h(u)+\varphi|\leq k\}} a_{ij}D_iuD_j\varphi\,dx \leq \int_\Omega fT_k\Big(u-T_h(u)+\varphi\Big)\,dx.$$

Choosing $k>\|\varphi\|_{L^\infty(\Omega)}$ and letting $h\to\infty$ (we have $T_h(u)\to u$) we obtain

$$\sum_{i,j=1}^N\int_\Omega a_{ij}D_iuD_j\varphi\,dx \leq \int_\Omega f\varphi\,dx.$$

Replacing φ by $-\varphi$, we get the converse inequality. Therefore,

$$\sum_{i,j=1}^{N} \int_\Omega a_{ij} D_i u D_j \varphi \, dx = \int_\Omega f \varphi \, dx$$

for every $\varphi \in C_0^\infty(\Omega)$, that is, u is solution of $Lu = f$. □

The main result in this section is the following.

Theorem 3.2.8. *Let L be the operator (3.1) and assume that the condition (3.2) holds. If $\omega \in A_2$ and $f \in L^1(\Omega)$, then there exists a unique entropy solution in $\mathcal{T}_0^{1,2}(\Omega, \omega)$ for the Dirichlet problem* (P2).

Proof. **(I) Uniqueness.** The main obstacle to prove the uniqueness is that we cannot use the difference of two solutions, $u_1 - u_2$, as a test function. But we can apply Lemma 3.2.3

Let u_1 and u_2 be two functions in $\mathcal{T}_0^{1,2}(\Omega, \omega)$ which are entropy solutions of the equation $Lu = f$, that is, $Lu_1 = f$ and $Lu_2 = f$. If we write the entropy inequality corresponding to solution u_1 with test function $T_h u_2$ and to solution u_2 with test function $T_h u_1$ (using Lemma 3.2.3) we obtain

$$\sum_{i,j=1}^{N} \int_{\{|u_1 - T_h(u_2)| < k\}} a_{ij} D_i u_1 [D_j u_1 - D_j(T_h(u_2))] \, dx$$
$$\leq \int_\Omega f T_k(u_1 - T_h(u_2)) \, dx; \tag{3.21}$$

and

$$\sum_{i,j=1}^{N} \int_{\{|u_2 - T_h(u_1)| < k\}} a_{ij} D_i u_2 [D_j u_2 - D_j(T_h(u_1))] \, dx$$
$$\leq \int_\Omega f T_k(u_2 - T_h(u_1)) \, dx. \tag{3.22}$$

Adding up these inequalities we get

$$\sum_{i,j=1}^{N}\int_{\{|u_1-T_h(u_2)|<k\}} a_{ij}D_iu_1[D_ju_1-D_jT_h(u_2)]\,dx$$
$$+\sum_{i,j=1}^{N}\int_{\{|u_2-T_h(u_1)|<k\}} a_{ij}D_iu_2[D_ju_2-D_jT_h(u_1)]\,dx$$
$$\le\int_\Omega f[T_k(u_1-T_h(u_2))+T_k(u_2-T_h(u_1))]\,dx. \qquad (3.23)$$

We split the integrals in the first two terms of (3.23) into the contributions corresponding to different integration sets. Then the main contribution of the first member of (3.23) is located in the set

$$A_0=\{x\in\Omega\,:\,|u_1-u_2|<k,\ |u_1|<h,\ |u_2|<h\},$$

and amounts to

$$I_0=\sum_{i,j=1}^{N}\int_{A_0} a_{ij}[D_iu_1-D_iu_2][D_ju_1-D_ju_2]\,dx.$$

On the other hand, on the set $A_1=\{x\in\Omega\ :\ |u_1-T_h(u_2)|<k,\ |u_2|\ge h\}$ we have

$$\sum_{i,j=1}^{N}\int_{A_1} a_{ij}D_iu_1[D_ju_1-D_j(T_h(u_2))]\,dx$$
$$=\sum_{i,j=1}^{N}\int_{A_1} a_{ij}D_iu_1D_ju_1\,dx$$
$$\ge\lambda\int_{A_1}|\nabla u_1|^2\omega\,dx\ge 0,$$

while on the remaining set

$$A_2=\{x\in\Omega\,:\,|u_1-T_h(u_2)|<k,\ |u_2|<h,\ |u_1|\ge h\},$$

we get

$$\begin{aligned}
&\sum_{i,j=1}^{N}\int_{A_2} a_{ij}D_iu_1[D_ju_1 - D_j(T_h(u_2))]\,dx \\
&= \sum_{i,j=1}^{N}\int_{A_2} a_{ij}D_iu_1[D_ju_1 - D_ju_2]\,dx \\
&= \sum_{i,j=1}^{N}\int_{A_2} a_{ij}D_iu_1D_ju_1\,dx - \sum_{i,j=1}^{N}\int_{A_2} a_{ij}D_iu_1D_ju_2\,dx \\
&\geq \lambda\int_{A_2} |\nabla u_1|^2\omega dx - \sum_{i,j=1}^{N}\int_{A_2} a_{ij}D_iu_1D_ju_2\,dx \\
&\geq -\sum_{i,j=1}^{N}\int_{A_2} a_{ij}D_iu_1D_ju_2\,dx.
\end{aligned}$$

Analogously, we estimate the second integral of (3.23) in the sets

$$\begin{aligned}
A_1^* &= \{x\in\Omega : |u_2 - T_h(u_1)| < k,\ |u_1|\geq h\},\\
A_2^* &= \{x\in\Omega : |u_2 - T_h(u_1)| < k,\ |u_1| < h,\ |u_2|\geq h\}.
\end{aligned}$$

We obtain,

$$\begin{aligned}
&\sum_{i,j=1}^{N}\int_{A_1^*} a_{ij}D_iu_2[D_ju_2 - D_jT_h(u_1)]\,dx \\
&= \sum_{i,j=1}^{N}\int_{A_1^*} a_{ij}D_iu_2D_ju_2\,dx \geq \lambda\int_{A_1^*} |\nabla u_2|^2\omega\,dx \geq 0;
\end{aligned}$$

and

$$\sum_{i,j=1}^{N}\int_{A_2^*} a_{ij}D_iu_2[D_ju_2 - D_jT_h(u_1)]\,dx$$

$$= \sum_{i,j=1}^{N} \int_{A_2^*} a_{ij} D_i u_2 [D_j u_2 - D_j u_1] \, dx$$

$$\geq - \sum_{i,j=1}^{N} \int_{A_2^*} a_{ij} D_i u_2 D_j u_1 \, dx.$$

Summing up we estimate the first member of (3.23),

$$\begin{aligned} I &= \sum_{i,j=1}^{N} \int_{\{|u_1 - T_h u_2| < k\}} a_{ij} D_i u_1 [D_j u_1 - T_h u_2] \, dx \\ &+ \sum_{i,j=1}^{N} \int_{\{|u_2 - T_h u_1| < k\}} a_{ij} D_i u_2 [D_j u_2 - D_j T_h(u_1)] \, dx \\ &\geq \sum_{i,j=1}^{N} \int_{A_0} a_{ij} (D_i u_1 - D_i u_2)(D_j u_1 - D_j u_2) \, dx \\ &- \sum_{i,j=1}^{N} \int_{A_2} a_{ij} D_i u_1 D_j u_2 dx - \int_{A_2^*} a_{ij} D_i u_2 D_j u_1 dx \\ &= I_0 - I_2, \end{aligned}$$

where

$$I_2 = \sum_{i,j=1}^{N} \int_{A_2} a_{ij} D_i u_1 D_j u_2 \, dx + \sum_{i,j=1}^{N} \int_{A_2^*} a_{ij} D_i u_2 D_j u_1 dx.$$

Now, we have that $I_2 \to 0$ when $h \to \infty$.
Using the fact that $|a_{ij}(x)| \leq \Lambda \, \omega(x)$ a.e. in Ω, we get

$$\sum_{i,j=1}^{N} \left| \int_{A_2} a_{ij} D_i u_1 D_j u_2 \, dx \right| \leq \sum_{i,j=1}^{N} \int_{A_2} |a_{ij}||D_i u_1||D_j u_2| \, dx$$

$$\leq \Lambda \sum_{i,j=1}^{N} \int_{A_2} |D_i u_1||D_j u_2| \omega \, dx$$

$$\leq \Lambda \sum_{i,j=1}^{N} \left(\int_{\{h \leq |u_1| \leq h+k\}} |D_i u_1|^2 \omega \, dx \right)^{1/2} \times$$
$$\times \left(\int_{\{h-k \leq |u_2| \leq h\}} |D_j u_2|^2 \omega \, dx \right)^{1/2}$$
$$\to 0, \text{ when } h \to \infty.$$

In the same way, we obtain

$$\sum_{i,j=1}^{N} \int_{A_2^*} a_{ij} D_i u_2 D_j u_1 \, dx \to 0, \text{ when } h \to \infty.$$

- Now we shall estimate the right-hand side of equation (3.23),

$$\int_{\Omega} f[T_k(u_1 - T_h(u_2)) + T_k(u_2 - T_h(u_1))] \, dx.$$

(i) Let $B_0 = \{x \in \Omega \; : \; |u_1| < h \text{ and } |u_2| < h\}$. We have,

$$\int_{B_0} f[T_k(u_1 - T_h(u_2)) + T_k(u_2 - T_h(u_1))] \, dx$$
$$= \int_{B_0} f[T_k(u_1 - u_2) + T_k(u_2 - u_1)] \, dx = 0.$$

(ii) Let $B_1 = \{x \in \Omega \; : \; |u_1| \geq h\}$ and let $B_2 = \{x \in \Omega \; : \; |u_2| \geq h\}$. We get,

$$J_1 = \int_{B_1} f[T_k(u_1 - T_h(u_2)) + T_k(u_2 - T_h(u_1))] \, dx.$$

Hence,

$$|J_1| \leq 2\,k \int_{B_1} |f| \, dx.$$

Analogously,

$$J_2 = \int_{B_2} f[T_k(u_1 - T_h(u_2)) + T_k(u_2 - T_h(u_1))] \, dx$$

and

$$|J_2| \leq 2k \int_{B_2} |f|\, dx.$$

Since $|B_1|$ and $|B_2|$ go to zero when $h \to \infty$, we obtain $|J_1| + |J_2| \to 0$ when $h \to \infty$.

We now combine both estimates to get from (3.23),

$$I_0 = \sum_{i,j=1}^{N} \int_{A_0} a_{ij}(D_i u_1 - D_i u_2)(D_j u_1 - D_j u_2)\, dx \to 0,$$

when $h \to \infty$ and k fixed.

Since $A_0 = \{x \in \Omega \ : \ |u_1 - u_2| < k, \ |u_1| < h \text{ and } |u_2| < h\}$, we have

$$\sum_{i,j=1}^{N} \int_{A_0} a_{ij}(D_i u_1 - D_i u_2)(D_j u_1 - D_j u_2)\, dx$$
$$\to \sum_{i,j=1}^{N} \int_{\{x \in \Omega : |u_1 - u_2| < k\}} a_{ij}(D_i u_1 - D_i u_2)(D_j u_1 - D_j u_2)\, dx,$$

when $h \to \infty$. Thus,

$$\sum_{i,j=1}^{N} \int_{\{|u_1 - u_2| < k\}} a_{ij} D_i(u_1 - u_2) D_j(u_1 - u_2)\, dx \leq 0, \quad \text{for all } k.$$

Using the degenerate ellipticity condition (3.2), we get

$$\sum_{i,j=1}^{N} a_{ij} D_i(u_1 - u_2) D_j(u_1 - u_2) \geq \lambda |\nabla(u_1 - u_2)|^2 \omega \geq 0.$$

Therefore $\nabla u_1 = \nabla u_2$ μ-a.e. (and a.e.). Since,

$$\int_{\{|u_1 - u_2| < k\}} |\nabla(u_1 - u_2)|^2 \omega\, dx \leq 0,$$

by Lemma 3.2.5 (with $p = 2$), we obtain

$$u_1 - u_2 \in \mathcal{M}^{p_1}(\Omega, \omega) \subset L_0(\Omega, \omega).$$

Since u_1, $u_2 \in \mathcal{T}_0^{1,2}(\Omega, \omega)$, we conclude that $u_1 = u_2$ μ-a.e. (and a.e., by Lemma 2.1.2).

(II) Existence. We will splitting the proof in four steps.
Step 1. Considering a sequence $\{f_n\}$, $f_n \in C_0^\infty(\Omega)$, which $f_n \to f$ in $L^1(\Omega)$ and $\|f_n\|_1 \leq \|f\|_1$. For each n, there exist a solution $u_n \in W_0^{1,2}(\Omega, \omega)$ of the Dirichlet problem

$$(P_n) \begin{cases} Lu_n(x) = f_n(x) \text{ in } \Omega, \\ u_n(x) = 0 \text{ on } \partial\Omega, \end{cases}$$

(see Theorem 3.1.1) that is,

$$\sum_{i,j=1}^{N} \int_\Omega a_{ij} D_i u_n D_j \varphi \, dx = \int_\Omega f_n \varphi \, dx, \quad \text{for all } \varphi \in W_0^{1,2}(\Omega, \omega) \tag{3.24}$$

(in particular for all $\varphi \in C_0^\infty(\Omega)$). In (3.24), by convenient test functions, we obtain

$$\begin{aligned} \lambda \int_{\{|u_n|<k\}} |\nabla u_n|^2 \omega \, dx &\leq \sum_{i,j=1}^{N} \int_{\{|u_n|<k\}} a_{ij} D_i u_n D_j u_n \, dx \\ &\leq k \|f_n\|_{L^1(\Omega)} \\ &\leq k \|f\|_{L^1(\Omega)}, \end{aligned}$$

that is,

$$\int_{\{|u_n|<k\}} |\nabla u_n|^2 \omega \, dx \leq \frac{k}{\lambda} \|f\|_{L_1(\Omega)}. \tag{3.25}$$

Thus, using (3.25), we get that $\{|\nabla T_k(u_n)|\}$ is bounded sequence in $L^2_{\text{loc}}(\Omega, \omega)$ for all $k > 0$. Using equation (3.25) and Lemma

3.2.5, we obtain that $u_n \in \mathcal{M}^{p_1}(\Omega,\omega)$ and $\mu(\{x\in\Omega : |u_n(x)| > k\})$ is bounded uniformly in n for every k, that is,

$$\mu(\{x\in\Omega : |u_n(x)| > k\}) \leq C\, k^{-p_1}. \tag{3.26}$$

Step 2. In this step we prove that the sequences $\{u_n\}$ and $\{\nabla u_n\}$ are convergent in μ-measure.

• We have that $u_n \to u$ locally in μ - measure. In fact: for t and $\varepsilon > 0$ we can write

$$\{|u_n - u_m| > t\} \subset \{|u_n| > k\} \cup \{|u_m| > k\} \cup \{|T_k(u_n) - T_k(u_m)| > t\}.$$

Hence,

$$\begin{aligned} \mu(\{|u_n - u_m| > t\}) \;&\leq\; \mu(\{|u_n| > k\}) + \mu(\{|u_m| > k\}) \\ &+\; \mu(\{|T_k(u_n) - T_k(u_m)| > t\}). \end{aligned} \tag{3.27}$$

We get:

(i) For sufficiently large k, by equation (3.26),

$$\mu(\{|u_n| > k\}) < \varepsilon \;\text{ and }\; \mu(\{|u_m| > k\}) < \varepsilon.$$

(ii) Since $\{|\nabla T_k(u_n)|\}$ is bounded in $L^2_{\text{loc}}(\Omega,\omega)$ (by (3.26)) and we have $T_k(u_n) \in W^{1,2}_0(\Omega,\omega)$, we can assume that $\{T_k(u_n)\}$ is a Cauchy sequence in $L^q(\Omega\cap B_R,\omega)$ for any $0 < q < 2\theta = p^*$ (where θ as in Lemma 3.2.5) and any $R > 0$ and $T_k(u_n) \to T_k u$ in $L^q_{\text{loc}}(\Omega,\omega)$. Then,

$$\begin{aligned} &\mu(\{|T_k(u_n) - T_k(u_m)| > t\} \cap B_R) \\ &= \int_{\{|T_k(u_n) - T_k(u_m)| > t\} \cap B_R} \omega(x)\, dx \\ &\leq \frac{1}{t^q} \int_{\Omega\cap B_R} |T_k(u_n) - T_k(u_m)|^q \omega\, dx \\ &\leq \varepsilon, \end{aligned}$$

for all $n, m \geq n_0(k,t,R)$.

In equation (3.27), we obtain $\mu(\{x\in\Omega : |u_n(x) - u_m(x)| > t\}) \leq 3\varepsilon$, that is, $\{u_n\}$ is a Cauchy sequence in μ-measure. Therefore $u_n \to u$ locally in μ-measure.

• We have that $\{\nabla u_n\}$ converges in μ-measure (that is, $\{\nabla u_n\}$ is a Cauchy sequence in μ-measure) in any ball B_R. In fact: let t and $\varepsilon > 0$. We can write,

$$\begin{aligned}
&\{x\in\Omega : |\nabla u_n(x) - \nabla u_m(x)| > t\}\cap B_R \\
&\subset \{|\nabla u_n(x)| > b\}\cup\{|\nabla u_m(x)| > b\} \\
&\cup(\{|u_n(x) - u_m(x)| > k\}\cap B_R) \\
&\cup \{|u_n(x) - u_m(x)|\leq k,\ |\nabla u_n(x)|\leq b,\ |\nabla u_m(x)|\leq b \\
&\quad \text{and } |\nabla u_n(x) - \nabla u_m(x)| > t\}.
\end{aligned}$$

(i) For sufficiently large b, using Lemma 3.2.6, we obtain

$$\mu(\{x\in\Omega : |\nabla u_n(x)| > b\}) < \varepsilon, \quad \text{for all } n. \tag{3.28}$$

(ii) Since u_n is solution of problem (P_n), then

$$\sum_{i,j=1}^{N} \int_\Omega a_{ij} D_i u_n D_j \varphi \, dx = \int_\Omega f_n \varphi \, dx,$$

for all $\varphi\in W_0^{1,2}(\Omega,\omega)$. Thus,

$$\sum_{i,j=1}^{N} \int_\Omega a_{ij}[D_i u_n - D_i u_m] D_j \varphi \, dx = \int_\Omega [f_n - f_m]\varphi \, dx.$$

Choosing $\varphi = T_k(u_n - u_m)$, we get

$$\begin{aligned}
&\sum_{i,j=1}^{N} \int_{\{|u_n - u_m|\leq k\}} a_{ij} D_i(u_n - u_m) D_j(u_n - u_m)\, dx \\
&= \int_\Omega (f_n - f_m) T_k(u_n - u_m)\, dx
\end{aligned}$$

$$\leq 2k\|f\|_{L^1(\Omega)}.$$

Using the degenerate ellipticity condition (3.2), we obtain

$$\int_{\{|u_n-u_m|\leq k\}} |\nabla(u_n-u_m)|^2\omega\,dx \leq \frac{2k}{\lambda}\|f\|_{L^1(\Omega)}.$$

Consequently, if

$$E=\{|u_n-u_m|\leq k,\ |\nabla u_n|\leq b,\ |\nabla u_m|\leq b,\ |\nabla u_n-\nabla u_m|>t\},$$

we have

$$\begin{aligned}
\mu(E) &= \int_E \omega(x)\,dx \\
&= \frac{1}{t^2}\int_E t^2\omega(x)\,dx \\
&\leq \frac{1}{t^2}\int_E |\nabla(u_n-u_m)|^2\omega\ dx \\
&\leq \frac{1}{t^2}\int_{\{|u_n-u_m|\leq k\}} |\nabla(u_n-u_m)|^2\omega\,dx \\
&\leq \frac{1}{t^2}\frac{2k}{\lambda}\|f\|_{L^1(\Omega)} \leq \varepsilon, \qquad (3.29)
\end{aligned}$$

for $k\leq t^2\lambda\varepsilon/\|f\|_{L^1(\Omega)}$.

Since b and k have been already chosen and $\{u_n\}$ converges in μ - measure, for sufficiently n_0 and $n,m\geq n_0$ we get

$$\mu(\{x\in\Omega\ :\ |u_n(x)-u_m(x)|>k\}\cap B_R\})<\varepsilon. \qquad (3.30)$$

Therefore, by equations (3.28), (3.29) and (3.30), we have

$$\mu(\{x\in\Omega\ :\ |\nabla u_n(x)-\nabla u_m(x)|>t\}\cap B_R)\leq 4\varepsilon.$$

Thus, $\{\nabla u_n\}$ converges locally in μ-measure to some function v and $\nabla u_n\to v$ μ-a.e. (and also $\nabla u_n\to v$ a.e.).

• Since $\{\nabla T_k(u_n)\}$ is bounded in $L^2(\Omega,\omega)$, then $\{\nabla T_k(u_n)\}$ converges weakly in $L^2(\Omega,\omega)$, that is, for every $g\in L^2(\Omega,\omega)$

$$\int_\Omega g D_j T_k(u_n)\,\omega\,dx \to \int_\Omega g D_j T_k(u)\,\omega\,dx.$$

In particular, for $g=\chi_{\Omega\cap B_R}$ (characteristic function of $\Omega\cap B_R$), we get

$$\int_{\Omega\cap B_R} D_j T_k(u_n)\omega\,dx \to \int_{\Omega\cap B_R} D_j T_k(u)\omega\,dx,$$

that is, $\{D_j T_k(u_n)\}$ converges to $D_j T_k(u)$ in $L^1_{\rm loc}(\Omega,\omega)$.

Therefore $u\in\mathcal{T}^{1,1}_{\rm loc}(\Omega,\omega)$ and $\nabla u = v$ μ-a.e. (and also $\nabla u = v$ a.e.).

Step 3. We need to show that $u\in\mathcal{T}^{1,2}_0(\Omega,\omega)$ and $|\nabla u|\in L^1_{\rm loc}(\Omega,\omega)$.
• There exists a sequence $\varphi_n\in C_0^\infty(\Omega)$ such that

$$\|\nabla\varphi_n - \nabla(T_k u_n)\|_{L^2(\Omega,\omega)}\le\frac{1}{n};$$
$$\|\varphi_n - T_k(u_n)\|_{L^{p^*}(\Omega,\omega)}\le\frac{1}{n},$$

where $p^* = 2\theta$. We have,

$$\varphi_n\to T_k(u), \quad \text{in} \quad L^q_{\rm loc}(\Omega,\omega), \quad q<p^*,$$

$$D_j\varphi_n\rightharpoonup D_j T_k(u), \quad \text{in} \quad L^2(\Omega,\omega).$$

By Mazur's lemma (see Theorem 2, Section 1, Chapter V in [58]) there exists a sequence $\{\phi_n\}$ of convex combinations of φ_n such that

$$D_j\phi_n\to D_j T_k(u), \quad \text{in} \quad L^2(\Omega,\omega).$$

Therefore $u\in\mathcal{T}^{1,2}_0(\Omega,\omega)$.

• By (3.25) we have

$$\frac{1}{k}\int_{\{|u_n|<k\}} |\nabla u_n|^2 \omega\, dx \leq \frac{1}{\lambda}\|f\|_{L^1(\Omega)} = M. \tag{3.31}$$

Applying Lemma 3.2.4 (with $p = 2$), we obtain

$$\mu(\{x\in\Omega \ : \ |\nabla u_n(x)| > h\})\leq CM^q h^{-p_2},$$

where $p_2 = 2\theta/(\theta+1)$, that is, $|\nabla u_n|\in\mathcal{M}^{p_2}(\Omega,\omega)\subset L^q(\Omega,\omega)$, with $1\leq q < p_2$ (see Remark 3.2.4). Since $\{\nabla u_n\}$ converges locally in μ-measure to ∇u, we have $|\nabla u|\in L^q_{\text{loc}}(\Omega,\omega)$.

Therefore $|\nabla u|\in L^1_{\text{loc}}(\Omega,\omega)$.

Step 4. We need to show that u is an entropy solution of problem (P2). Since u_n is solution of (P_n), we have

$$\sum_{i,j=1}^N \int_\Omega a_{ij} D_i u_n D_j \varphi\, dx = \int_\Omega f_n \varphi\, dx, \tag{3.32}$$

for every $\varphi\in \mathrm{W}^{1,2}_0(\Omega,\omega)$ (in particular for every $\varphi\in C^\infty_0(\Omega)$).

We can choose $T\in\mathcal{F}$ (see Definition 3.2.3), bounded above by k, $k > 0$, and such that $T'(s) = 0$ for $|s|\geq k$. For $\varphi\in C^\infty_0(\Omega)$, we apply equation (3.32) to the function $T(u_n - \varphi)$ and we get

$$\sum_{i,j=1}^N \int_\Omega a_{ij} D_i u_n D_j T(u_n-\varphi)\, dx = \int_\Omega f_n T(u_n-\varphi)\, dx. \tag{3.33}$$

In the left-hand side of equation (3.33) we can write,

$$\begin{aligned}
&\sum_{i,j=1}^N \int_\Omega a_{ij} D_i u_n D_j T(u_n-\varphi)\, dx \\
&= \sum_{i,j=1}^N \int_\Omega a_{ij} D_i u_n T'(u_n-\varphi)[D_j u_n - D_j\varphi]\, dx
\end{aligned}$$

$$= \sum_{i,j=1}^{N} \int_\Omega a_{ij} D_i u_n D_j u_n T'(u_n - \varphi)\, dx$$
$$- \sum_{i,j=1}^{N} \int_\Omega a_{ij} D_i u_n D_j \varphi T'(u_n - \varphi)\, dx.$$

• Since $u_n \to u$ and $\nabla u_n \to \nabla u$ a.e., by Fatou's lemma we obtain

$$\sum_{i,j=1}^{N} \int_\Omega a_{ij} D_i u D_j u T'(u - \varphi)\, dx$$
$$\leq \liminf \sum_{i,j=1}^{N} \int_\Omega a_{ij} D_i u_n D_j u_n T'(u_n - \varphi)\, dx. \quad (3.34)$$

• We have that $a_{ij} D_i u_n D_j \varphi T'(u_n - \varphi) \to a_{ij} D_i u D_j \varphi T'(u - \varphi)$ a.e.. Since $\nabla u_n \to \nabla u$ in μ-measure, $\mu(\Omega) < \infty$ and that $\{\nabla u_n\}$ is uniformly bounded in $L^2_{\text{loc}}(\Omega, \omega)$, we have $\nabla u_n \to \nabla u$ in $L^1_{\text{loc}}(\Omega, \omega)$. Hence,

$$\sum_{i,j=1}^{N} \int_\Omega a_{ij} D_i u_n D_j \varphi T'(u_n - \varphi)\, dx$$
$$\to \sum_{i,j=1}^{N} \int_\Omega a_{ij} D_i u D_j \varphi T'(u - \varphi)\, dx. \quad (3.35)$$

Moreover,

$$\int_\Omega f_n T(u_n - \varphi)\, dx \to \int_\Omega f T(u - \varphi)\, dx. \quad (3.36)$$

Thus, using equations (3.34), (3.35) and (3.36) we get

$$\sum_{i,j=1}^{N} \int_\Omega a_{ij} D_i u D_j u T'(u - \varphi)\, dx$$

$$- \sum_{i,j=1}^{N} \int_{\Omega} a_{ij} D_i u D_j \varphi T'(u-\varphi)\, dx$$
$$\leq \int_{\Omega} f\, T(u-\varphi)\, dx,$$

that is,

$$\sum_{i,j=1}^{N} \int_{\Omega} a_{ij} D_i u D_j T(u-\varphi)\, dx \leq \int_{\Omega} f T(u-\varphi)\, dx.$$

Therefore u is an entropy solution of problem (P2) (by Lemma 3.2.2). □

Example 3.2.1. *Let $\Omega = \{(x,y)\in\mathbb{R}^2 \ : \ x^2+y^2<1\}$, and let $a\neq b$ positive numbers and $\omega(x,y) = |x|^{-1/2}$. By Theorem 3.2.8, the problem*

$$\begin{cases} -\dfrac{\partial}{\partial x}\left(a|x|^{-1/2}\dfrac{\partial u}{\partial x}\right) - \dfrac{\partial}{\partial y}\left(b|x|^{-1/2}\dfrac{\partial u}{\partial y}\right) = \dfrac{2}{\sqrt{x^2+y^2}} & \text{in } \ \Omega, \\ u(x,y)=0, \quad \text{on} \quad \partial\Omega, \end{cases}$$

has a unique entropy solution $u\in\mathcal{T}_0^{1,2}(\Omega,\omega)$.

3.3 Neumann Problem

In this section, we prove the existence and uniqueness of solutions in the space $\mathrm{H}(\Omega)$ (see Definition 3.3.1) for the Neumann problem

$$(P3)\begin{cases} Lu(x) = f(x) \ \text{ in } \ \Omega, \\ \langle \mathcal{A}(x)\nabla u(x), \vec{\eta}(x)\rangle = 0 \ \text{ on } \ \partial\Omega, \end{cases}$$

where L is a degenerate elliptic operator

$$\begin{aligned} Lu(x) &= -\sum_{i,j=1}^{N} D_j\big(a_{ij}(x)D_iu(x)\big) + \sum_{i=1}^{N} b_i(x)D_iu(x) \\ &+ g(x)u(x) + \alpha\, u(x)\, v(x), \end{aligned} \tag{3.37}$$

with $D_j = \dfrac{\partial}{\partial x_j}$, $(j = 1, ..., N)$, α is a constant, the coefficients a_{ij}, b_i and g are measurable, real-valued functions, the coefficient matrix $\mathcal{A}(x) = (a_{ij}(x))$ is symmetric and satisfies the *degenerate ellipticity condition*

$$|\xi|^2\omega(x) \leq \langle \mathcal{A}(x)\xi, \xi\rangle \leq |\xi|^2 v(x), \tag{3.38}$$

for all $\xi \in \mathbb{R}^N$ and almost every $x \in \Omega \subset \mathbb{R}^N$ a bounded open set with piecewise smooth boundary (i.e., $\partial\Omega \in C^{0,1}$), ω and v are weight functions ($\omega, v \in A_2$), $\vec{\eta}(x) = (\eta_1(x), ..., \eta_N(x))$ is the unit outward normal to $\partial\Omega$ at x, $\langle .,. \rangle$ denotes the usual inner product in $\mathbb{R}^N$ and the symbol ∇ indicates the gradient.

In the case $A(x) = I_d$ (where I_d denotes the identity matrix in $\mathbb{R}^N$) then the second equation of $(P3)$ is $\dfrac{\partial u}{\partial \vec{\eta}} = 0$ on $\partial\Omega$ namely the normal derivative of u vanishes on $\partial\Omega$. In the general case and with the summation convention the second equation of $(P3)$ can be written $a_{ij}(x)\dfrac{\partial u}{\partial x_j}\,\eta_i = 0$ on $\partial\Omega$. This expression is called *conormal derivative of* u.

Definition 3.3.1. *Let $\Omega \subset \mathbb{R}^N$ be a bounded and open set. We define the space* $\mathrm{H}(\Omega)$ *as the closure of $C^\infty(\overline{\Omega})$ with respect to the norm*

$$\|u\|_{\mathrm{H}(\Omega)} = \left(\int_\Omega |u|^2 v\,dx + \int_\Omega \langle \mathcal{A}\nabla u, \nabla u\rangle dx\right)^{1/2},$$

where $\mathcal{A} = (a_{ij})$ is the coefficient matrix of the operator L defined in (3.37), $\langle .,. \rangle$ denotes the usual inner product in $\mathbb{R}^N$ and the symbol ∇ indicates the gradient.

The space $\mathrm{H}(\Omega)$ is a Hilbert space with inner product

$$a(u,\varphi) = \left(\int_\Omega u\,\varphi\,v\,dx + \int_\Omega \langle \mathcal{A}\nabla u, \nabla\varphi \rangle\,dx \right)^{1/2}$$

Facts about $\mathrm{H}(\Omega)$ are given in [25].

Remark 3.3.1. *By the degeneracy condition (3.38), we have*

$$\int_\Omega |\nabla u|^2 \omega dx \le \int_\Omega \langle \mathcal{A}\nabla u, \nabla u \rangle dx \le \int_\Omega |\nabla u|^2 v dx.$$

Therefore, $W^{1,2}(\Omega, v) \subset \mathrm{H}(\Omega) \subset W^{1,2}(\Omega,\omega)$.
Note also that since $\mathcal{A}$ is symmetric, then

$$|\langle \mathcal{A}x, y \rangle| \le \langle \mathcal{A}x, x \rangle^{1/2} \langle \mathcal{A}y, y \rangle^{1/2}.$$

Remark 3.3.2. *Let $\Omega \subset \mathbb{R}^N$ be a bounded open set with boundary $\partial\Omega \in C^{0,1}$. Using integration by parts, with $u, \varphi \in \mathrm{H}(\Omega)$, if u satisfies the boundary condition in problem (P3), we have*

$$\begin{aligned} \int_\Omega \varphi L u\,dx &= \sum_{i,j=1}^N \int_\Omega a_{ij} D_i u D_j \varphi\,dx + \sum_{i=1}^N \int_\Omega b_i \varphi D_i u\,dx \\ &\quad + \int_\Omega g\,u\,\varphi\,dx \\ &+ \alpha \int_\Omega u\,\varphi\,v\,dx + \underbrace{\sum_{i,j=1}^N \int_{\partial\Omega} a_{ij} \frac{\partial u}{\partial x_j}\,\eta_i\,\varphi\,dx}_{=0} \\ &= B(u,\varphi) + \alpha \int_\Omega u\varphi v dx, \end{aligned}$$

where

$$B(u,\varphi)=\sum_{i,j=1}^{N}\int_\Omega a_{ij}D_iuD_j\varphi\,dx+\sum_{i=1}^{N}\int_\Omega b_i\varphi D_iu\,dx+\int_\Omega g\,u\,\varphi\,dx$$

is a bilinear form.

We introduce the following definition of solutions for the Neumann problem (P3).

Definition 3.3.2. *Let $\Omega\subset\mathbb{R}^N$ be a bounded open set with $\partial\Omega\in C^{0,1}$ and suppose that $f/v\in L^2(\Omega,v)$. A function $u\in \mathrm{H}(\Omega)$ is a (weak) solution of the Neumann problem (P3) if, for all $\varphi\in\mathrm{H}(\Omega)$, we have*

$$\int_\Omega\sum_{i,j=1}^{N}a_{ij}D_iuD_j\varphi+\int_\Omega\left[\sum_{i=1}^{N}b_iD_iu+gu\right]\varphi dx+\alpha\int_\Omega u\,\varphi\,v\,dx$$
$$=\int_\Omega f\,\varphi\,dx.$$

Lemma 3.3.1. *Suppose that $\omega\in A_2$, $v\in A_2$, $b_i/\omega\in L^\infty(\Omega)$ $(i=1,...,N)$ and $g/v\in L^\infty(\Omega)$. Then, there exists a constant $\mathbf{C}>0$ such that*
$B(u,u)+\mathbf{C}\|u\|^2_{L^2(\Omega,\omega)}\geq\dfrac{1}{2}\|u\|^2_{\mathrm{H}(\Omega)}$ *for all $u\in\mathrm{H}(\Omega)$.*

Proof. For all $u\in\mathrm{H}(\Omega)$, we have

$$B(u,u)=\sum_{i,j=1}^{N}\int_\Omega a_{ij}D_iuD_ju\,dx+\sum_{i=1}^{N}\int_\Omega b_iuD_iu\,dx+\int_\Omega g\,u^2\,dx$$
$$=\int_\Omega\langle\mathcal{A}\nabla u,\nabla u\rangle\,dx+\sum_{i=1}^{N}\int_\Omega\frac{b_i}{\omega}\,\omega\,u\,D_iu\,dx+\int_\Omega\frac{g}{v}\,u^2\,v\,dx$$
$$\geq\int_\Omega\langle\mathcal{A}\nabla u,\nabla u\rangle\,dx-\left(\max_{1\leq i\leq N}\left\|\frac{b_i}{\omega}\right\|_{L^\infty(\Omega)}\right)\sum_{i=1}^{N}\int_\Omega|u||D_iu|\omega\,dx$$

$$
\begin{aligned}
&-\left\|\frac{g}{v}\right\|_{L^\infty(\Omega)}\int_\Omega u^2\, v\, dx\\
&\geq \int_\Omega \langle \mathcal{A}\nabla u, \nabla u\rangle\, dx\\
&-C_1\sum_{i=1}^{N}\left(\int_\Omega u^2\omega\, dx\right)^{1/2}\left(\int_\Omega |D_i u|^2\omega\, dx\right)^{1/2}\\
&-C_2\int_\Omega u^2\, vdx\\
&\geq \int_\Omega \langle \mathcal{A}\nabla u, \nabla u\rangle\, dx\\
&-C_1\left(\int_\Omega u^2\, v\, dx\right)^{1/2}\left(\int_\Omega \langle \mathcal{A}\nabla u, \nabla u\rangle\, dx\right)^{1/2}\\
&-C_2\|u\|^2_{L^2(\Omega,v)}, \qquad (3.39)
\end{aligned}
$$

where $C_1 = \max\limits_{1\leq i\leq N}\left\|\dfrac{b_i}{\omega}\right\|_{L^\infty(\Omega)}$ and $C_2 = \left\|\dfrac{g}{v}\right\|_{L^\infty(\Omega)}$. Using the elementary inequality $a\,b\leq \varepsilon a^2 + \dfrac{1}{4\varepsilon}\, b^2$, for all $\varepsilon > 0$, we obtain in (3.39),

$$
\begin{aligned}
B(u,u) &\geq \int_\Omega \langle \mathcal{A}\nabla u, \nabla u\rangle\, dx\\
&- C_1\left(\varepsilon\|u\|^2_{L^2(\Omega,v)} + \frac{1}{4\varepsilon}\int_\Omega \langle \mathcal{A}\nabla u, \nabla u\rangle\, dx\right) - C_2\|u\|_{L^2(\Omega,v)}\\
&= \left(1-\frac{C_1}{4\varepsilon}\right)\int_\Omega \langle \mathcal{A}\nabla u, \nabla u\rangle\, dx\\
&-(C_1\varepsilon + C_2)\|u\|^2_{L^2(\Omega,v)}. \qquad (3.40)
\end{aligned}
$$

If $C_1 > 0$, we can choose $\varepsilon > 0$ such that $1 - \dfrac{C_1}{4\varepsilon} = \dfrac{1}{2}$, that is,

$\varepsilon = C_1/2$. Thus, in (3.40) we obtain

$$\begin{aligned} B(u,u) &\geq \frac{1}{2}\int_\Omega \langle \mathcal{A}\nabla u, \nabla u\rangle\, dx - \left(\frac{C_1^2}{2} + C_2\right)\|u\|^2_{L^2(\Omega)} \\ &= \frac{1}{2}\left(\int_\Omega u^2\, v\, dx + \int_\Omega \langle \mathcal{A}\nabla u, \nabla u\rangle\, dx\right) \\ &- \left(\frac{C_1^2}{2} + C_2 + \frac{1}{2}\right)\|u\|^2_{L^2(\Omega,v)} \\ &= \frac{1}{2}\|u\|^2_{\mathrm{H}(\Omega)} - \mathbf{C}\|u\|^2_{L^2(\Omega,v)}, \end{aligned} \tag{3.41}$$

where $\mathbf{C} = \dfrac{C_1^2}{2} + C_2 + \dfrac{1}{2} > 0$. Therefore,

$$B(u,u) + \mathbf{C}\|u\|^2_{L^2(\Omega,v)} \geq \frac{1}{2}\|u\|^2_{\mathrm{H}(\Omega)}.$$

If $C_1 = 0$ (that is, $b_i(x) \equiv 0$, $i = 1, ..., N$) then in (3.39) reduces to

$$\begin{aligned} B(u,u) &\geq \int_\Omega \langle \mathcal{A}\nabla u, \nabla u\rangle\, dx - C_2\|u\|^2_{L^2(\Omega,v)} \\ &\geq \frac{1}{2}\left(\int_\Omega |u|^2\, v\, dx + \int_\Omega \langle \mathcal{A}\nabla u, \nabla u\rangle\, dx\right) \\ &- (C_2 + \frac{1}{2})\|u\|^2_{L^2(\Omega,v)} \\ &= \frac{1}{2}\|u\|^2_{\mathrm{H}(\Omega)} - \mathbf{C}\|u\|^2_{L^2(\Omega,v)}. \end{aligned}$$

Therefore, we also have

$$B(u,u) + \mathbf{C}\|u\|^2_{L^2(\Omega,v)} \geq \frac{1}{2}\|u\|^2_{\mathrm{H}(\Omega)},$$

for all $u \in \mathrm{H}(\Omega)$, where $\mathbf{C} = \dfrac{C_1^2}{2} + C_2 + \dfrac{1}{2}$. □

Theorem 3.3.2. *Let $\Omega\subset\mathbb{R}^N$ be a bounded open set with boundary $\partial\Omega\in C^{0,1}$. Suppose that*
(H1) *$\omega\in A_2$, $v\in A_2$ and (3.38) holds;*
(H2) *$f/v\in L^2(\Omega, v)$;*
(H3) *$b_i/\omega\in L^\infty(\Omega)$ $(i=1,...,N)$ and $g/v\in L^\infty(\Omega)$.*
Then, there exists a constant $\mathbf{C}>0$ such that for all $\alpha\geq\mathbf{C}$ the Neumann problem (P3) has a unique solution $u\in\mathrm{H}(\Omega)$. Moreover, we have that

$$\|u\|_{\mathrm{H}(\Omega)}\leq 2\left\|\frac{f}{v}\right\|_{L^2(\Omega,v)}.$$

Proof. We define the bilinear form

$$\tilde{B}:\mathrm{H}(\Omega)\times\mathrm{H}(\Omega)\longrightarrow\mathbb{R},$$
$$\tilde{B}(u,\varphi)=B(u,\varphi)+\alpha\int_\Omega u\varphi\, v\,dx,$$

and the linear mapping $T:\mathrm{H}(\Omega)\to\mathbb{R}$, $T(\varphi)=\displaystyle\int_\Omega f\,\varphi\,dx$. Then $u\in\mathrm{H}(\Omega)$ is a solution of the Neumann problem (P3) if

$$\tilde{B}(u,\varphi)=T(\varphi),\text{for all }\varphi\in\mathrm{H}(\Omega).$$

Step 1. If $\alpha\geq\mathbf{C}$ then $\tilde{B}$ is coercive, that is, there exists a constant $C>0$ such that $\tilde{B}(u,u)\geq C\|u\|^2_{H(\Omega)}$ for all $u\in H(\Omega)$. In fact, by Lemma 3.3.1 there exists a constant $\mathbf{C}>0$ such that

$$B(u,u)+\mathbf{C}\,\|u\|^2_{L^2(\Omega,v)}\geq\frac{1}{2}\|u\|_{\mathrm{H}(\Omega)}.$$

Hence, if $\alpha\geq\mathbf{C}$, we have

$$\begin{aligned}\tilde{B}(u,u) &= B(u,u)+\alpha\int_\Omega u^2\,v\,dx\\ &= B(u,u)+\alpha\|u\|^2_{L^2(\Omega,v)}\end{aligned}$$

$$\begin{aligned} &\geq B(u,u) + \mathbf{C}\|u\|^2_{L^2(\Omega,v)} \\ &\geq \frac{1}{2}\|u\|^2_{\mathrm{H}(\Omega)}. \end{aligned} \tag{3.42}$$

Step 2. $\tilde{B}$ is bounded. In fact, using the fact the coefficient matrix $\mathcal{A} = (a_{ij})$ is symmetric, (H2) and (H3), we obtain

$$\begin{aligned} |\tilde{B}(u,\varphi)| \leq& |B(u,\varphi)| + \alpha\left|\int_\Omega u\varphi\, v\, dx\right| \\ \leq& \int_\Omega |\langle \mathcal{A}\nabla u, \nabla\varphi\rangle|\, dx + \sum_{i=1}^N \int_\Omega |b_i||\varphi||D_i u| dx \\ +& \int_\Omega |g||\varphi||u|\, dx + \alpha\int_\Omega |u||\varphi|\, v\, dx \\ \leq& \int_\Omega \langle \mathcal{A}\nabla u, \nabla u\rangle^{1/2} \langle \mathcal{A}\nabla\varphi, \nabla\varphi\rangle^{1/2}\, dx \\ +& \sum_{i=1}^N \int_\Omega \frac{|b_i|}{\omega}|\varphi||D_i u|\omega\, dx \\ +& \int_\Omega \frac{|g|}{v}|\varphi||u|\, v\, dx + \alpha\int_\Omega |u||\varphi|\, v\, dx \\ \leq& \left(\int_\Omega \langle \mathcal{A}\nabla u, \nabla u\rangle\, dx\right)^{1/2} \left(\int_\Omega \langle \mathcal{A}\nabla\varphi, \nabla\varphi\rangle\, dx\right)^{1/2} \\ +& \left(\max_{1\leq i\leq N}\left\|\frac{b_i}{\omega}\right\|_{L^\infty(\Omega)}\right) \sum_{i=1}^N \left(\int_\Omega |\varphi|^2\omega\, dx\right)^{1/2} \left(\int_\Omega |D_i u|^2\omega\, dx\right)^{1/2} \\ +& \left\|\frac{g}{v}\right\|_{L^\infty(\Omega)} \left(\int_\Omega |u|^2\, v\, dx\right)^{1/2} \left(\int_\Omega |\varphi|^2\, v\, dx\right)^{1/2} \\ +& \alpha\left(\int_\Omega |u|^2\, v\, dx\right)^{1/2} \left(\int_\Omega |\varphi|^2\, v\, dx\right)^{1/2} \end{aligned}$$

$$\leq \left(1+\left(\max_{1\leq i\leq N}\left\|\frac{b_i}{\omega}\right\|_{L^\infty(\Omega)}\right)+\left\|\frac{g}{v}\right\|_{L^\infty(\Omega)}+\alpha\right)\|u\|_{\mathrm{H}(\Omega)}\|\varphi\|_{\mathrm{H}(\Omega)}$$
$$= \tilde{C}\,\|u\|_{\mathrm{H}(\Omega)}\|\varphi\|_{\mathrm{H}(\Omega)},$$

for all $u, \varphi\in \mathrm{H}(\Omega)$.

Step 3. The linear mapping T is bounded (that is, $T\in[\mathrm{H}(\Omega)]^*$). In fact,

$$\begin{aligned}
|T(\varphi)| &\leq \int_\Omega |f||\varphi|\,dx \\
&= \int_\Omega \frac{|f|}{v}|\varphi|\,v\,dx \\
&\leq \left[\int_\Omega \left(\frac{|f|}{v}\right)^2 v\,dx\right]^{1/2}\left[\int_\Omega |\varphi|^2\,v\,dx\right]^{1/2} \\
&\leq \left\|\frac{f}{v}\right\|_{L^2(\Omega,v)}\|\varphi\|_{\mathrm{H}(\Omega)},
\end{aligned}$$

for all $\varphi\in \mathrm{H}(\Omega)$.
Therefore the bilinear form $\tilde{B}$ and the linear functional T satisfies the hypotheses of the Lax-Milgram Theorem. Thus, for every f, with $\frac{f}{v}\in L^2(\Omega, v)$, there is a unique solution $u\in\mathrm{H}(\Omega)$ such that

$$\tilde{B}(u,\varphi) = T(\varphi)$$

for all $\varphi\in \mathrm{H}(\Omega)$, that is, u is a unique solution of the Neumann problem (P3). In particular, by setting $\varphi = u$, we have

$$\tilde{B}(u,u) = \int_\Omega f\,u\,dx.$$

Using the definition of $\tilde{B}$, we obtain

$$\tilde{B}(u,u) \;=\; B(u,u)+\alpha\int_\Omega u^2\,v\,dx = \int_\Omega \frac{f}{v}\,u\,v\,dx$$

$$\leq \|u\|_{L^2(\Omega,v)}\left\|\frac{f}{v}\right\|_{L^2(\Omega,v)}$$
$$\leq \|u\|_{\mathrm{H}(\Omega)}\left\|\frac{f}{v}\right\|_{L^2(\Omega,v)}.$$

Using (3.42), we obtain

$$\frac{1}{2}\|u\|^2_{\mathrm{H}(\Omega)} \leq \tilde{B}(u,u) \leq \|u\|_{\mathrm{H}(\Omega)}\left\|\frac{f}{v}\right\|_{L^2(\Omega,v)}.$$

Therefore,

$$\|u\|_{\mathrm{H}(\Omega)} \leq 2\left\|\frac{f}{v}\right\|_{L^2(\Omega,v)}.$$

□

Example 3.3.1. *Consider the domain $\Omega = \{(x,y)\in\mathbb{R}^2 : x^2 + y^2 < 1\}$, the weight functions $\omega(x,y) = (x^2+y^2)^{-1/3}$ and $v(x,y) = (x^2+y^2)^{-1/2}$ and the coefficient matrix*

$$\mathcal{A}(x,y) = \begin{pmatrix} (x^2+y^2)^{-1/3} & 0 \\ 0 & (x^2+y^2)^{-1/2} \end{pmatrix}.$$

For all $\xi\in\mathbb{R}^2$ and almost every $(x,y)\in\Omega$, we have

$$\frac{1}{(x^2+y^2)^{1/3}}|\xi|^2 \leq \langle \mathcal{A}(x,y)\xi,\xi\rangle \leq \frac{1}{(x^2+y^2)^{1/2}}|\xi|^2.$$

If $(x,y)\in\partial\Omega = \{(x,y)\in\mathbb{R}^2 : x^2+y^2 = 1\}$, then $\vec{\eta}(x,y) = (x,y)$ is the unit outward normal to $\partial\Omega$. By Theorem 3.3.2 the Neumann problem

$$\begin{cases} Lu(x,y) = (x^2+y^2)^{-3/8}\cos(xy), & \text{in } \Omega, \\ \langle \mathcal{A}(x,y).\nabla u, \vec{\eta}\rangle = 0 & \text{on } \partial\Omega \end{cases}$$

where

$$\begin{aligned} Lu(x,y) &= -\left[\frac{\partial}{\partial x}\left(\frac{1}{(x^2+y^2)^{1/3}}\frac{\partial u}{\partial x}\right)+\frac{\partial}{\partial y}\left(\frac{1}{(x^2+y^2)^{1/2}}\frac{\partial u}{\partial y}\right)\right] \\ &+ \frac{\sin(xy)}{(x^2+y^2)^{1/3}}\frac{\partial u}{\partial x}+\frac{\cos(xy)}{(x^2+y^2)^{1/4}}\frac{\partial u}{\partial y}+\frac{u(x,y)\sin(xy)}{(x^2+y^2)^{1/3}} \\ &+ \alpha\frac{u(x,y)}{(x^2+y^2)^{1/2}} \end{aligned}$$

has a unique solution $u \in \mathrm{H}(\Omega)$ *(if* $\alpha \geq 2$*).*

In case $\omega = v$, the weak solution to the problem (P3) can be approximated by a sequence of solutions for non-degenerate elliptic equations, i.e.,

$$(P3)\begin{cases} Lu(x) = f(x) \text{ in } \Omega, \\ \langle A(x)\nabla u, \vec{\eta}(x)\rangle = 0 \text{ on } \partial\Omega, \end{cases}$$

with $D_j = \dfrac{\partial}{\partial x_j}$, $(j = 1, ..., n)$, θ is positive a constant, the coefficients a_{ij}, b_i and g are measurable, real-valued functions, the coefficient matrix $A(x) = (a_{ij}(x))$ is symmetric and satisfies the *degenerate ellipticity condition*

$$\lambda|\xi|^2\omega(x) \leq \langle A(x)\xi,\xi\rangle \leq \Lambda|\xi|^2\omega(x) \tag{3.43}$$

for all $\xi\in\mathbb{R}^n$ and almost every $x\in\Omega\subset\mathbb{R}^n$ a bounded open set with piecewise smooth boundary (i.e., $\partial\Omega \in C^{0,1}$), ω is a weight function (that is, locally integrable and nonnegative function on $\mathbb{R}^n$), λ and Λ are positive constants

Theorem 3.3.3. *Let* $\Omega \subset \mathbb{R}^n$ *be a bounded open set with boundary* $\partial\Omega \in C^{0,1}$*. Suppose that*
(H1) $\omega \in A_2$;
(H2*) $f/\omega \in L^2(\Omega,\omega) \cap L^2(\Omega,\omega^3)$;

(H3) *$b_i/\omega \in L^\infty(\Omega)$ (i=1,...,n) and $g/\omega \in L^\infty(\Omega)$.*
Then the unique solution $u\in W^{1,2}(\Omega,\omega)$ of problem (P) is the weak limit in $W^{1,2}(\Omega,\tilde{\omega}_1)$ of a sequence of solutions $u_m \in W^{1,2}(\Omega,\omega_m)$ of the problems

$$(P_m)\left\{\begin{array}{l} L_m u_m(x) = f_m(x), \text{ in } \Omega, \\ \langle A^m(x)\nabla u_m, \vec{\eta}(x)\rangle = 0, \text{ on } \partial\Omega, \end{array}\right.$$

with

$$L_m u_m = -\sum_{i,j=1}^{n} D_j(a_{ij}^{mm}\, D_i u_m) + \sum_{i=1}^{n} b_{mi}\, D_i u_m + g_m\, u_m + \theta\, u_m\, \omega_m,$$

$f_m = f\,(\omega_m/\omega)^{1/2}$, $g_m = g\,\omega_m/\omega$, $b_{mi} = b_i\,\omega_m/\omega$ and $\omega_m = \omega_{mm}$ (where ω_{mm}, a_{ij}^{mm} and $\tilde{\omega}_1$ are as Lemma 2.2.1 and $A^m(x) = \big(a_{ij}^{mm}(x)\big)$).

Proof. **Step 1.** First, if $f_m = f(\omega/\omega_m)^{-1/2}$, $g_m = g\,\omega_m/\omega$ and $b_{mi} = b_i\,\omega_m/\omega$, we note that

$$\begin{array}{l} \left\|\dfrac{f_m}{\omega_m}\right\|_{L^2(\Omega,\omega_m)} = \left\|\dfrac{f}{\omega}\right\|_{L^2(\Omega,\omega)}, \quad \left\|\dfrac{g_m}{\omega_m}\right\|_{L^\infty(\Omega)} = \left\|\dfrac{g}{\omega}\right\|_{L^\infty(\Omega)} \\ \left\|\dfrac{b_{mi}}{\omega_m}\right\|_{L^\infty(\Omega)} = \left\|\dfrac{b_i}{\omega}\right\|_{L^\infty(\Omega)}. \end{array} \tag{3.44}$$

Then, if $u_m \in W^{1,2}(\Omega,\omega_m)$ is a solution of problem (P_m) we have (by (3.44))

$$\begin{aligned} \|u_m\|_{W^{1,2}(\Omega,\omega_m)} &\leq \frac{2}{\lambda}\left\|\frac{f_m}{\omega_m}\right\|_{L^2(\Omega,\omega_m)} \\ &= \frac{2}{\lambda}\left\|\frac{f}{\omega}\right\|_{L^2(\Omega,\omega)} = C_3. \end{aligned}$$

Using Lemma 2.2.1, $\tilde{\omega}_1 \leq \omega_m$, we obtain

$$\|u_m\|_{W^{1,2}(\Omega,\tilde{\omega}_1)} \leq \|u_m\|_{W^{1,2}(\Omega,\omega_m)} \leq C_3. \tag{3.45}$$

Consequently, $\{u_m\}$ is a bounded sequence in $W^{1,2}(\Omega, \tilde{\omega}_1)$. Therefore, there is a subsequence, again denoted by $\{u_m\}$, and $\tilde{u} \in W^{1,2}(\Omega, \tilde{\omega}_1)$ such that

$$u_m \rightharpoonup \tilde{u} \text{ in } L^2(\Omega, \tilde{\omega}_1), \tag{3.46}$$

$$\frac{\partial u_m}{\partial x_j} \rightharpoonup \frac{\partial \tilde{u}}{\partial x_j} \text{ in } L^2(\Omega, \tilde{\omega}_1), \tag{3.47}$$

$$u_m \to \tilde{u} \text{ a.e. in } \Omega, \tag{3.48}$$

where the symbol "$\rightharpoonup$" denotes weak convergence (see Theorem 1.4.9).

Step 2. We have that $\tilde{u} \in W^{1,2}(\Omega, \omega)$. In fact, for $F_k = F_{kk}$ fixed (see Lemma 2.2.1), we have by (3.46) and (3.47), for all $\varphi \in W^{1,2}(\Omega, \tilde{\omega}_1)$, we obtain

$$\int_\Omega u_m \varphi\, \tilde{\omega}_1\, dx \to \int_\Omega \tilde{u}\, \varphi\, \tilde{\omega}_1\, dx,$$

$$\int_\Omega D_i u_m D_i \varphi\, \tilde{\omega}_1\, dx \to \int_\Omega D_i \tilde{u}\, D_i \varphi\, \tilde{\omega}_1\, dx.$$

If $\psi \in L^2(\Omega, \omega)$, then $\psi\, \chi_{F_k} \in L^2(\Omega, \tilde{\omega}_1)$ (since $\omega \sim \tilde{\omega}_1$ in F_k, i.e., there is a constant $c > 0$ such that $\tilde{\omega}_1 \leq c\,\omega$ in F_k, and χ_E denotes the characteristic function of a measurable set $E \subset \mathbb{R}^n$). Consequently,

$$\int_\Omega u_m \varphi \chi_{F_k}\, \tilde{\omega}_1\, dx \to \int_\Omega \tilde{u}\, \varphi\, \chi_{F_k}\, \tilde{\omega}_1\, dx,$$

$$\int_\Omega D_i u_m\, \varphi\, \chi_{F_k}\, \tilde{\omega}_1\, dx \to \int_\Omega D_i \tilde{u}\, \varphi\, \chi_{F_k}\, \tilde{\omega}_1\, dx,$$

for all $\varphi \in L^2(\Omega, \omega)$, that is, the sequence $\{\dfrac{\partial u_m}{\partial x_i}\, \chi_{F_k}\}$ is weakly convergent to a function in $L^2(\Omega, \omega)$, again since $\omega \sim \tilde{\omega}_1$ on F_k. Therefore, we have

$$\|\, |\nabla \tilde{u}|\, \|^2_{L^2(F_k, \omega)} \;=\; \int_{F_k} |\nabla \tilde{u}|^2 \omega\, dx$$

$$\leq \limsup_{m\to\infty} \int_{F_k} |\nabla u_m|^2 \, \omega \, dx,$$

and for $m \geq k$ we have $\omega = \omega_m$ in F_k. Hence, by (3.45), we obtain

$$\begin{aligned}
\| \, |\nabla \tilde{u}| \, \|^2_{L^2(F_k,\omega)} &\leq \limsup_{m\to\infty} \int_{F_k} |\nabla u_m|^2 \, \omega \, dx \\
&= \limsup_{m\to\infty} \int_{F_k} |\nabla u_m|^2 \, \omega_m \, dx \\
&\leq \limsup_{m\to\infty} \int_{\Omega} |\nabla u_m|^2 \omega_m \, dx \\
&\leq C_3^2.
\end{aligned}$$

By the Monotone Convergence Theorem we obtain

$$\| \, |\nabla \tilde{u}| \, \|_{L^2(\Omega,\omega)} \leq C_3.$$

Analogously, $\|\tilde{u}\|_{L^2(\Omega,\omega)} \leq C_3$. Therefore, we have $\tilde{u} \in W^{1,2}(\Omega,\omega)$.

Step 3. We need to show that $\tilde{u}$ is a solution of problem (P), i.e, for every $\varphi \in W^{1,2}(\Omega,\omega)$ we have

$$\sum_{i,j=1}^{n} \int_{\Omega} a_{ij} \, D_i\tilde{u} \, D_j\varphi \; dx + \sum_{i=1}^{n} \int_{\Omega} b_i \, \varphi \, D_i\tilde{u} \, dx + \int_{\Omega} g \, \tilde{u} \, \varphi \, dx$$

$$+\theta \int_{\Omega} \tilde{u} \, \varphi \, \omega \, dx = \int_{\Omega} f \, \varphi \, dx.$$

Using the fact that u_m is a solution of (P_m), we have

$$\sum_{i,j=1}^{n} \int_{\Omega} a_{ij}^{mm} \, D_i u_m \, D_j\varphi \, dx + \sum_{i=1}^{n} \int_{\Omega} b_{mi} \, \varphi \, D_i u_m \, dx$$

$$+ \int_{\Omega} g_m \, u_m \, \varphi \, dx + \theta \int_{\Omega} u_m \, \varphi \, \omega_m \, dx = \int_{\Omega} f_m \, \varphi \, dx,$$

for every $\varphi \in W^{1,2}(\Omega,\omega_m)$. Moreover, by Lemma 2.2.1 and (3.44), over $F_k = F_{kk}$ (for $m \geq k$) we have the following properties:

(i) $\omega = \omega_m$;
(ii) $f_m = f$, $g_m = g$ and $b_{mi} = b_i$;
(iii) $a_{ij}^{mm}(x) = a_{ij}(x)$.
For $\varphi \in W^{1,2}(\Omega, \omega)$ and $k > 0$(fixed), we define $G_1, G_2 : W^{1,2}(\Omega, \tilde{\omega}_1) \to \mathbb{R}$ by

$$\begin{aligned} G_1(u) &= \sum_{i,j=1}^{n} \int_\Omega a_{ij}\, D_i u\; D_j\varphi\, \chi_{F_k}\, dx, \\ G_2(u) &= \sum_{i=1}^{n} \int_\Omega \varphi\, b_i\, D_i u\, \chi_{F_k}\, dx + \int_\Omega g\, u\, \varphi\, \chi_{F_k}\, dx \\ &+ \theta \int_\Omega u\, \varphi\, \omega\, \chi_{F_k}\, dx. \end{aligned}$$

(a) We have that G_1 is linear and continuous functional. In fact, we have (by Lemma 2.2.1(iv)) $\omega \sim \tilde{\omega}_1$ in F_k (there is a constant $c > 0$ such that $\omega \le c\,\tilde{\omega}_1$ in F_k). And by (3.43) we obtain

$$\begin{aligned} |G_1(u)| &\le \int_{F_k} |\langle A\nabla u, \nabla\varphi\rangle|\, dx \\ &\le \int_{F_k} (\langle A\nabla u, \nabla u\rangle)^{1/2}\, (\langle A\nabla\varphi, \nabla\varphi\rangle)^{1/2}\, dx \\ &\le \left(\int_{F_k} \langle A\nabla u, \nabla u\rangle\, dx\right)^{1/2} \left(\int_{F_k} \langle A\nabla\varphi, \nabla\varphi\rangle\, dx\right)^{1/2} \\ &\le \Lambda \left(\int_{F_k} |\nabla u|^2\, \omega\, dx\right)^{1/2} \left(\int_{F_k} |\nabla\varphi|^2\, \omega\, dx\right)^{1/2} \\ &\le \Lambda \left(\int_{F_k} c\, |\nabla u|^2 \tilde{\omega}_1\, dx\right)^{1/2} \left(\int_\Omega |\nabla\varphi|^2\, \omega\, dx\right)^{1/2} \\ &\le \Lambda\, c^{1/2} \|\varphi\|_{W^{1,2}(\Omega,\omega)}\, \|u\|_{W^{1,2}(\Omega,\tilde{\omega}_1)}. \end{aligned}$$

(b) We have that G_2 is linear and continuous functional. In fact,

$$
\begin{aligned}
&|G_2(u)| \\
&\leq \sum_{i=1}^{n} \int_{F_k} |\varphi|\,|b_i|\,|D_i u|\,dx + \int_{F_k} |g|\,|u|\,|\varphi|\,dx + \theta \int_{F_k} |u|\,|\varphi|\,\omega\,dx \\
&\leq \sum_{i=1}^{n} \left\| \frac{b_i}{\omega} \right\|_{L^\infty(F_k)} \left(\int_{F_k} |D_i u|^2\,\omega\,dx \right)^{1/2} \left(\int_{F_k} |\varphi|^2\,\omega\,dx \right)^{1/2} \\
&+ \left\| \frac{g}{\omega} \right\|_{L^\infty(F_k)} \left(\int_{F_k} |u|^2\,\omega\,dx \right)^{1/2} \left(\int_{F_k} |\varphi|^2\,\omega\,dx \right)^{1/2} \\
&+ \theta \left(\int_{F_k} |u|^2\,\omega\,dx \right)^{1/2} \left(\int_{F_k} |\varphi|^2\,\omega\,dx \right)^{1/2} \\
&\leq \left(\max \left\| \frac{b_i}{\omega} \right\|_{L^\infty(\Omega)} \right) \left(c \int_{F_k} |D_i u|^2\,\tilde{\omega_1}\,dx \right)^{1/2} \left(\int_{\Omega} |\varphi|^2\,\omega\,dx \right)^{1/2} \\
&+ \left\| \frac{g}{\omega} \right\|_{L^\infty(\Omega)} \left(c \int_{F_k} |u|^2\,\tilde{\omega_1}\,dx \right)^{1/2} \left(\int_{\Omega} |\varphi|^2\,\omega\,dx \right)^{1/2} \\
&+ \theta \left(c \int_{F_k} |u|^2\,\tilde{\omega_1}\,dx \right)^{1/2} \left(\int_{\Omega} |\varphi|^2\,\omega\,dx \right)^{1/2} \\
&\leq \left(\max \left\| \frac{b_i}{\omega} \right\|_{L^\infty(\Omega)} + \left\| \frac{g}{\omega} \right\|_{L^\infty(\Omega)} + \theta \right) c^{1/2}\, \|u\|_{W^{1,2}(\Omega,\tilde{\omega}_1)} \|\varphi\|_{W^{1,2}(\Omega,\omega)}.
\end{aligned}
$$

Using (a), (b), properties (i),(ii) and (iii), and that u_m is solution of (P_m), we obtain

$$
\begin{aligned}
&\sum_{i,j=1}^{n} \int_{F_k} a_{ij}\, D_i\tilde{u}\, D_j\varphi\,dx + \sum_{i=1}^{n} \int_{F_k} \varphi\, b_i\, D_i\tilde{u}\,dx + \int_{F_k} g\,\tilde{u}\,\varphi\,dx \\
&+\theta \int_{F_k} \tilde{u}\,\varphi\,\omega\,dx \\
&= \lim_{m\to\infty} \left(G_1(u_m) + G_2(u_m) \right)
\end{aligned}
$$

$$
\begin{aligned}
&= \lim_{m\to\infty}\Bigg(\sum_{i,j=1}^{n}\int_{F_k} a_{ij}^{mm}\, D_i u_m\, D_j\varphi\, dx + \sum_{i=1}^{n}\int_{F_k}\varphi\, b_{mi}\, D_i u_m\, dx \\
&+ \int_{F_k} g_m\, u_m\, \varphi\, dx + \theta\int_{F_k} u_m\, \varphi\, \omega_m\, dx\Bigg) \\
&= \lim_{m\to\infty}\Bigg(\sum_{i,j=1}^{n}\int_{\Omega} a_{ij}^{mm}\, D_i u_m\, D_j\varphi\, dx + \sum_{i=1}^{n}\int_{\Omega}\varphi\, b_i\, D_i u_m\, dx \\
&+ \int_{\Omega} g_m\, u_m\, \varphi\, dx + \theta\int_{\Omega} u_m\, \varphi\, \omega_m\, dx \\
&- \sum_{i,j=1}^{n}\int_{\Omega\cap F_k^c} a_{ij}^{mm}\, D_i u_m\, D_j\varphi\, dx - \sum_{i=1}^{n}\int_{\Omega\cap F_k^c}\varphi\, b_i\, D_i u_m\, dx \\
&- \int_{\Omega\cap F_k^c} g_m\, u_m\, \varphi\, dx - \theta\int_{\Omega\cap F_k^c} u_m\, \varphi\, \omega_m\, dx\Bigg) \\
&= \lim_{m\to\infty}\Bigg(\int_{\Omega} f_m\, \varphi\, dx - \sum_{i,j=1}^{n}\int_{\Omega\cap F_k^c} a_{ij}^{mm}\, D_i u_m\, D_j\varphi\, dx \\
&- \sum_{i=1}^{n}\int_{\Omega\cap F_k^c}\varphi\, b_i\, D_i u_m\, dx - \int_{\Omega\cap F_k^c} g_m\, u_m\, \varphi\, dx \\
&- \theta\int_{\Omega\cap F_k^c} u_m\, \varphi\, \omega_m\, dx\Bigg), \qquad (3.49)
\end{aligned}
$$

where E^c denotes the complement of a set $E\subset\mathbb{R}^n$.

(I) By Lemma 2.2.1(v) we have $f_m = \dfrac{f\omega_m^{1/2}}{\omega^{1/2}} \to f$ a.e. in Ω. Since $\omega_m = \omega$ in F_k $(m\geq k)$ we also have

$$
\begin{aligned}
\int_{\Omega} f_m^2\, \omega\, dx &= \int_{\Omega} f^2\, \omega_m\, dx \\
&= \int_{F_k} f^2\, \omega_m\, dx + \int_{\Omega\cap F_k^c} f^2\, \omega_m\, dx
\end{aligned}
$$

$$\begin{aligned}
&= \int_{F_k} f^2\,\omega\,dx + \int_{\Omega\cap F_k^c} f^2\,\omega_m\,dx \\
&\le \int_{\Omega} f^2\,\omega\,dx + \int_{\Omega\cap F_k^c} f^2\,\omega_m\,dx \\
&= \int_{\Omega} \left(\frac{f}{\omega}\right)^2 \omega^3\,dx + \int_{\Omega\cap F_k^c} f^2\,\omega_m\,dx.
\end{aligned}$$

By Lemma 2(iv), we know that $|\Omega\cap F_k^c|\to 0$ when $k\to\infty$. Then, for sufficiently large k we have

$$\int_{\Omega\cap F_k^c} f^2\,\omega_m\,dx \le 1.$$

Therefore, for sufficiently large m and (H2*), we obtain

$$\int_{\Omega} f_m^2\,\omega\,dx \le \int_{\Omega}\left(\frac{f}{\omega}\right)^2 \omega^3\,dx + 1 < \infty.$$

Hence the sequence $\{f_m\}$ is bounded in $L^2(\Omega,\omega)$. Then there is a subsequence, still denoted by $\{f_m\}$, and a function $\tilde{f}$ such that

$$\begin{aligned}
&f_m \rightharpoonup \tilde{f} \text{ in } L^2(\Omega,\omega),\\
&f_m \to \tilde{f} \text{ a.e. in } \Omega.
\end{aligned}$$

Since $f_m\to f$ a.e. in Ω, then $\tilde{f}=f$ a.e. in Ω. Therefore, for all $\varphi\in W^{1,2}(\Omega,\omega)$, we have

$$\int_{\Omega} f_m\,\varphi\,dx \to \int_{\Omega} f\,\varphi\,dx.$$

(II) Since the matrix $A^m(x)=(a_{ij}^{mm})(x)$ is symmetric, we have

$$|\langle A^m\nabla u_m,\nabla\varphi\rangle| \le \langle A^m\nabla u_m,\nabla u_m\rangle^{1/2}\,\langle A^m\nabla\varphi,\nabla\varphi\rangle^{1/2}.$$

Then, by Lemma 2.2.1(vi) and (3.45), we obtain

$$\left|\sum_{i,j=1}^{n}\int_{\Omega\cap F_k^c} a_{ij}^{mm} D_i u_m D_j\varphi\,dx\right|$$

$$\leq \int_{\Omega\cap F_k^c} |\langle A^m \nabla u_m, \nabla\varphi\rangle|\, dx$$
$$\leq \Lambda\left(\int_{\Omega\cap F_k^c} |\nabla u_m|^2 \omega_m\, dx\right)^{1/2}\left(\int_{\Omega\cap F_k^c} |\nabla\varphi|^2 \omega_m\, dx\right)^{1/2}$$
$$\leq \Lambda \|u_m\|_{W^{1,2}(\Omega,\omega_m)}\left(\int_{\Omega\cap F_k^c} |\nabla\varphi|^2\, w_m\, dx\right)^{1/2}$$
$$\leq \Lambda\, C_3\left(\int_{\Omega\cap F_k^c} |\nabla\varphi|^2\, w_m\, dx\right)^{1/2}. \tag{3.50}$$

(III) By (H3), (3.44) and (3.45) we have

$$\left|\int_{\Omega\cap F_k^c} \varphi\, b_{mi}\, D_i u_m\, dx\right|$$
$$\leq \int_{\Omega\cap F_k^c} |\varphi|\,|b_{mi}|\,|D_i u_m|\, dx$$
$$\leq \left\|\frac{b_{mi}}{\omega_m}\right\|_{L^\infty(\Omega)}\left(\int_{\Omega\cap F_k^c} |D_i u_m|^2 \omega_m\, dx\right)^{1/2}\left(\int_{\Omega\cap F_k^c} |\varphi|^2\, \omega_m\, dx\right)^{1/2}$$
$$\leq \left\|\frac{b_i}{\omega}\right\|_{L^\infty(\Omega)} \|u_m\|_{W^{1,2}(\Omega,\omega_m)}\left(\int_{\Omega\cap F_k^c} |\varphi|^2 \omega_m\, dx\right)^{1/2}$$
$$\leq C_3\left\|\frac{b_i}{\omega}\right\|_{L^\infty(\Omega)}\left(\int_{\Omega\cap F_k^c} |\varphi|^2 \omega_m\, dx\right)^{1/2}, \tag{3.51}$$

and analogously

$$\left|\int_{\Omega\cap F_k^c} g_m\, u_m\, \varphi\, dx\right| \leq C_3\left\|\frac{g}{\omega}\right\|_{L^\infty(\Omega)}\left(\int_{\Omega\cap F_k^c} |\varphi|^2 \omega_m\, dx\right)^{1/2}, \tag{3.52}$$

and

$$\left|\int_{\Omega\cap F_k^c} u_m\, \varphi\, \omega_m\, dx\right| \leq C_3\left(\int_{\Omega\cap F_k^c} |\varphi|^2 \omega_m\, dx\right)^{1/2}. \tag{3.53}$$

Note now that $\omega_m \leq \tilde{\omega}_2$ and $\tilde{\omega}_2 \in A_2$ (by Lemma 2.2.1). Hence, by Lemma 2.1.2(b), there exist $\delta > 0$ and $C > 0$ such that, if K_0 is a cube containing $\bar{\Omega}$ then

$$\begin{aligned}\mu_m(\Omega \cap F_k^c) &= \int_{\Omega \cap F_k^c} \omega_m(x)\,dx \\ &\leq \int_{\Omega \cap F_k^c} \tilde{\omega}_2(x)\,dx \\ &= \tilde{\mu}_2(\Omega \cap F_k^c) \\ &\leq C\,\tilde{\mu}_2(K_0)\left(\frac{|F_k^c|}{|K_0|}\right)^{\delta},\end{aligned}$$

which is independent of m and tends to zero as $k \to \infty$ by Lemma 2.2.1(iv). Then

$$\lim_{k\to\infty}\left(\int_{\Omega\cap F_k^c} |\varphi|^2 \omega_m\,dx\right)^{1/2} = \lim_{k\to\infty}\left(\int_{\Omega\cap F_k^c} |\nabla\varphi|^2\,\omega_m\,dx\right)^{1/2} = 0,$$

and we obtain in (3.50), (3.51), (3.52) and (3.53)

$$\lim_{k\to\infty}\int_{\Omega\cap F_k^c} a_{ij}^{mm}(x) D_i u(x)\, D_j\varphi(x)\,dx = 0, \tag{3.54}$$

$$\lim_{k\to\infty}\int_{\Omega\cap F_k^c} \varphi\, b_{mi}\, D_i u_m\,dx = 0, \tag{3.55}$$

$$\lim_{k\to\infty}\int_{\Omega\cap F_k^c} g_m\, u_m\, \varphi\,dx = 0, \tag{3.56}$$

$$\lim_{k\to\infty}\int_{\Omega\cap F_k^c} u_m\, \varphi\, \omega_m\,dx = 0. \tag{3.57}$$

Therefore, by (3.49), (3.54), (3.55), (3.56) and (3.57) we conclude, when $k \to \infty$ (and $m \geq k$),

$$\sum_{i,j=1}^{n}\int_{\Omega} a_{ij} D_i\tilde{u}\, D_j\varphi\,dx + \sum_{i=1}^{n}\int_{\Omega} b_i\,\varphi\, D_i\tilde{u}\,dx + \int_{\Omega} g\,\tilde{u}\,\varphi\,dx$$

$$+\theta \int_\Omega \tilde{u}\,\varphi\,\omega\,dx$$
$$= \int_\Omega f\,\varphi\,dx,$$

for all $\varphi \in W^{1,2}(\Omega,\omega)$, that is, $\tilde{u}$ is a solution of problem (P3). Therefore, $u = \tilde{u}$ (by the uniqueness). □

Chapter 4

Degenerate Semilinear Elliptic Equations

4.1 Existence and uniqueness of solutions

In this section we study three semilinear Dirichlet problems. For this purpose we will use some results of the Theory of Monotone Operators. Complete expositions about the Theory of Monotone Operators can be found in [60] and [61].

We will begin this section demonstrating the existence of solutions in $W_0^{1,2}(\Omega, \omega)$ for the semilinear Dirichlet problems

$$(P4)\begin{cases} Lu(x) - \gamma\, u(x)\, g_1(x) + h(u(x))\, g_2(x) = f(x) \ \text{ in } \ \Omega, \\ u(x) = 0 \ \text{ on } \ \partial\Omega, \end{cases}$$

and

$$(P5)\begin{cases} Lu(x) - \gamma\, u(x) g(x) = -f(x, u(x)) \ \text{ in } \ \Omega, \\ u(x) = 0 \ \text{ on } \ \partial\Omega, \end{cases}$$

where L is an elliptic operator in divergence form

$$Lu = -\sum_{i,j=1}^{N} D_j(a_{ij}(x)\, D_i u(x)),\ D = \frac{\partial}{\partial x_j}, \tag{4.1}$$

where the coefficients a_{ij} are measurable, real-valued functions whose coefficient matrix $\mathcal{A} = (a_{ij})$ symmetric and satisfies the degenerate ellipticity condition

$$\lambda|\xi|^2\omega(x) \leq \sum_{i,j=1}^{N} a_{ij}(x)\,\xi_i\xi_j \leq \Lambda\,|\xi|^2\omega(x), \tag{4.2}$$

for all $\xi \in \mathbb{R}^N$ and almost every $\Omega \subset \mathbb{R}^N$ a bounded open set, ω is a weight function, $\gamma \in \mathbb{R}$, λ and Λ are positive constants.

Definition 4.1.1. *We say that an element $u \in W_0^{1,2}(\Omega,\omega)$ is the (weak) solution of problem (P4) if*

$$\sum_{i,j=1}^{N} \int_\Omega a_{ij}\, D_i u\, D_j\varphi\, dx - \gamma \int_\Omega u\,\varphi\, g_1\, dx + \int_\Omega h(u)\,\varphi g_2\, dx$$
$$= \int_\Omega f\,\varphi\, dx,$$

for every $\varphi \in W_0^{1,2}(\Omega,\omega)$.

Definition 4.1.2. *Let $B, N : X \to X^*$ be operators on the real separable reflexive Banach space X.*
(i) The operator N is said to be demicontinuous if

$$u_n \to u \text{ implies } Nu_n \rightharpoonup Nu, \text{ as } n \to \infty,$$

where the symbol "$\rightharpoonup$" denotes weak convergence.
(ii) The operator N is said to be strongly continuous if

$$u_n \rightharpoonup u \text{ implies } Nu_n \to Nu, \text{ as } n \to \infty.$$

(iii) $B + N$ is asymptotically linear if B is linear and

$$\frac{\|Nu\|}{\|u\|} \to 0 \text{ as } \|u\| \to \infty.$$

(iv) The operator B satisfies condition (S) *if*

$$u_n \rightharpoonup u \text{ and } \lim_{n\to\infty} (Bu_n - Bu|u_n - u) = 0 \text{ implies } u_n \to u,$$

where $(f|x)$ denotes the value of linear functional f at the point x.

The basic idea is to reduce $(P4)$ to an operator equation

$$Bu + Nu = T$$

and apply the following theorem.

Theorem 4.1.1. *Let $B, N : X \to X^*$ be operators on the real separable reflexive Banach space X. Assume*
(a) the operator $B : X \to X^*$ is linear and continuous;
(b) the operator $N : X \to X^*$ is demicontinuous and bounded;
(c) $B + N$ is asymptotically linear;
(d) for each $T \in X^*$ and for each $t \in [0, 1]$, the operator

$$A_t(u) = Bu + t(Nu - T)$$

satisfies condition (S) in X.

If $Bu = 0$ implies $u = 0$, then for each $T \in X^*$, the equation $Bu + Nu = T$ has a solution in X.

Proof. See Theorem 29.C in [61] □

Lemma 4.1.2. *If $\omega \in A_2$ then $W_0^{1,2}(\Omega, \omega) \hookrightarrow L^2(\Omega, \omega)$ is compact and*

$$\|u\|_{L^2(\Omega,\omega)} \le C_2 \|u\|_{W_0^{1,2}(\Omega,\omega)}.$$

Proof. The proof of this lemma follows the line of Theorem 4.6 in [36]. □

Definition 4.1.3. *Let X and Y be Hilbert spaces over $\mathbb{R}$, with the continuous embedding $X \subseteq Y$. Then $g : X \times X \to \mathbb{R}$ is called a*

Gårding form iff g is bilinear and bounded (i.e., $|g(u,v)| \leq C\,\|u\|\,\|v\|$, and there is a constant $C_1 > 0$ and a constant C_2 such that

$$g(u,u) \geq C_1 \|u\|_X^2 - C_2 \|u\|_Y^2$$

for all $u \in X$. The Gårding form g is called regular iff the embedding $X \subseteq Y$ is compact.

Lemma 4.1.3. *Let $g : X \times X \to \mathbb{R}$ be a regular Gårding form. Then there exists a decomposition of the form $g = T_1 + T_2$ where $T_1, T_2 : X \times X \to \mathbb{R}$ are bilinear and bounded, $T_1(.,.)$ is strongly positive (i.e., $T_1(u,u) \geq C\,\|u\|^2$) and $T_2(.,.)$ is compact (i.e., iff $u_n \rightharpoonup u$ and $v_n \rightharpoonup v$ implies $T_2(u_n, v_n) \to T_2(u,v)$ as $n \to \infty$) .*

Proof. See Lemma 22.38 in [60]. □

Theorem 4.1.4. *Suppose that*
(**H1**) *the function $h : \mathbb{R} \to \mathbb{R}$ is continuous and bounded ($|h(t)| \leq M$ for all $t \in \mathbb{R}$);*
(**H2**) $\omega \in A_2$;
(**H3**) $g_1/\omega \in L^\infty(\Omega)$, $g_2/\omega \in L^2(\Omega,\omega)$ *and* $f/\omega \in L^2(\Omega,\omega)$;
(**H4**) $\gamma > 0$ *is not an eigenvalue of the linearized problem*

$$(LP)\begin{cases} Lu(x) - \gamma\, u(x)\, \omega(x) = 0 & \text{in } \Omega, \\ u(x) = 0 & \text{on } \partial\Omega, \end{cases}$$

Then problem $(P4)$ has a solution $u \in W_0^{1,2}(\Omega,\omega)$.

Proof. We define the operators $B_1, B_2 : W_0^{1,2}(\Omega,\omega) \times W_0^{1,2}(\Omega,\omega) \to \mathbb{R}$ through

$$B_1(u,\varphi) = \sum_{i,j=1}^{N} \int_\Omega a_{ij}\, D_i u\, D_j \varphi\, dx - \gamma \int_\Omega u\, \varphi\, g_1\, dx,$$

$$B_2(u,\varphi) = \int_\Omega h(u)\, \varphi\, g_2\, dx,$$

and $T : W_0^{1,2}(\Omega,\omega) \to \mathbb{R}$ through $T(\varphi) = \displaystyle\int_\Omega f\,\varphi\,dx$. We have that $u \in W_0^{1,2}(\Omega,\omega)$ is a solution of problem $(P4)$ if

$$B_1(u,\varphi) + B_2(u,\varphi) = T(\varphi), \text{ for every } \varphi \in W_0^{1,2}(\Omega,\omega).$$

Step 1. By condition (4.2), we obtain $|a_{ij}(x)| \le C_3\,\omega(x)$. Using Lemma 4.1.2 and Hölder inequality, we get

$$\begin{aligned}
|B_1(u,\varphi)| &\le \sum_{i,j=1}^N \int_\Omega |a_{ij}|\,|D_i u|\,|D_j\varphi|\,dx + |\gamma| \int_\Omega |u|\,|\varphi|\,|g_1|\,dx \\
&\le C_3 \sum_{i,j=1}^N \int_\Omega |D_i u|\,|D_j\varphi|\,\omega\,dx + |\gamma| \int_\Omega |u||\varphi|\,\frac{|g_1|}{\omega}\,\omega\,dx \\
&\le C_3 \sum_{i,j=1}^N \left(\int_\Omega |D_i u|^2\omega\,dx\right)^{1/2} \left(\int_\Omega |D_j\varphi|^2\omega\,dx\right)^{1/2} \\
&\quad + |\gamma| \left\|\frac{g_1}{\omega}\right\|_{L^\infty(\Omega)} \left(\int_\Omega |u|^2\omega\,dx\right)^{1/2} \left(\int_\Omega |\varphi|^2\omega\,dx\right)^{1/2} \\
&\le \left(C_3 + C_2|\gamma| \left\|\frac{g_1}{\omega}\right\|_{L^\infty(\Omega)}\right) \|u\|_{W_0^{1,2}(\Omega,\omega)} \|\varphi\|_{W_0^{1,2}(\Omega,\omega)}.
\end{aligned}$$

By condition (H1) and Lemma 4.1.2, we obtain

$$\begin{aligned}
|B_2(u,\varphi)| &\le \int_\Omega |h(u)|\,|\varphi|\,|g_2|\,dx \\
&\le M \int_\Omega |\varphi|\,\frac{|g_2|}{\omega}\,\omega\,dx \\
&\le M \|\varphi\|_{L^2(\Omega,\omega)} \left\|\frac{g_2}{\omega}\right\|_{L^2(\Omega,\omega)} \\
&\le M\,C_2 \left\|\frac{g_2}{\omega}\right\|_{L^2(\Omega,\omega)} \|\varphi\|_{W_0^{1,2}(\Omega,\omega)}. \qquad (4.3)
\end{aligned}$$

Moreover, we also have

$$\begin{aligned} |T(\varphi)| &\leq \int_\Omega |f|\,|\varphi|\,dx \\ &= \int_\Omega \frac{|f|}{\omega}\,\omega^{1/2}\,|\varphi|\,\omega^{1/2}\,dx \\ &\leq \left\|\frac{f}{\omega}\right\|_{L^2(\Omega,\omega)} \|\varphi\|_{L^2(\Omega,\omega)} \\ &\leq C_2\left\|\frac{f}{\omega}\right\|_{L^2(\Omega,\omega)} \|\varphi\|_{W_0^{1,2}(\Omega,\omega)}. \end{aligned}$$

Step 2. Since $W_0^{1,2}(\Omega,\omega)$ is a real Hilbert space, using the Identification Principle (see Theorem 21.18 in [60]) we set $W_0^{1,2}(\Omega,\omega) = [W_0^{1,2}(\Omega,\omega)]^*$ and $\langle u, v\rangle = (u|v)$ (where $\langle .,.\rangle$ denotes the inner product in a Hilbert space).

We define the operators $B, N : W_0^{1,2}(\Omega,\omega) \to W_0^{1,2}(\Omega,\omega)$ through

$$\begin{aligned} (Bu|\varphi) &= B_1(u,\varphi), \\ (Nu|\varphi) &= B_2(u,\varphi), \end{aligned}$$

for every $u\,\varphi \in W_0^{1,2}(\Omega,\omega)$. Since $T \in [W_0^{1,2}(\Omega,\omega)]^*$, problem $(P4)$ is equivalent to operator equation

$$Bu + Nu = T, \quad u \in W_0^{1,2}(\Omega,\omega).$$

Step 3. Using that $W_0^{1,2}(\Omega,\omega) \hookrightarrow L^2(\Omega,\omega)$ is compact, we have that $B_1(.,.)$ is a regular Gårding form. In fact, by condition (4.2), we obtain

$$\begin{aligned} B_1(u,u) &= \sum_{i,j=1}^{N} \int_\Omega a_{ij}\,D_i u\,D_j u\,dx - \gamma \int_\Omega u^2\,g_1\,dx \\ &\geq \lambda \int_\Omega |\nabla u|^2 \omega\,dx - \gamma \int_\Omega u^2\,\frac{g_1}{\omega}\,\omega\,dx \end{aligned}$$

$$\geq \quad \lambda \|u\|^2_{W_0^{1,2}(\Omega,\omega)} - \gamma \left\| \frac{g_1}{\omega} \right\|_{L^\infty(\Omega)} \|u\|_{L^2(\Omega,\omega)}.$$

Hence, there exist a decomposition of the form $B = T_1 + T_2$, where T_1 and T_2 are bilinear and bounded, $T_1(.,.)$ is strongly positive and $T_2(.,.)$ is compact (see Lemma 4.1.3). Thus, B is Fredholm of index zero (see Definition 8.13 in [59] and Theorem 21.F in [60] and B satisfies condition (S) (see Proposition 27.12 in [61]).

Step 4 By (4.3), we have

$$|(Nu, \varphi)| = |B_2(u, \varphi)| \leq C_2 M \left\| \frac{g_2}{\omega} \right\|_{L^2(\Omega,\omega)} \|\varphi\|_{W_0^{1,2}(\Omega,\omega)}.$$

Hence, $\|Nu\| \leq C$ for all $u \in W_0^{1,2}(\Omega, \omega)$. Therefore,

$$\frac{\|Nu\|}{\|u\|} \to 0, \quad \text{as} \quad \|u\|_{W_0^{1,2}(\Omega,\omega)} \to \infty,$$

i.e., $B+N$ is asymptotically linear and the operator N is strongly continuous (see Corollary 26.14 in [61]).

Step 5. For each $t \in [0,1]$ the operator $A_t(u) = Bu + t(Nu - T)$ is a strongly continuous perturbation of the operator B. Hence, the operator A_t satisfies condition (S) (see Proposition 27.12 in [61]).

If γ is not an eigenvalue of the linearized problem (LP), $Bu = 0$ implies $u = 0$. Therefore, by Theorem 4.1.1, the operator equation $Bu + Nu = T$ has a solution $u \in W_0^{1,2}(\Omega, \omega)$, and u is solution of problem $(P4)$. □

Example 4.1.1. *Let* $\Omega = \{(x,y) \in \mathbb{R}^2 : |x| < 1 \text{ and } |y| < 1\}$ *and* $0 < a < b$. *By Theorem 4.1.4, with* $h(t) = t\,\mathrm{e}^{-t^2}$, $f(x,y) = x\,|y|$, $\omega(x,y) = (x^2+y^2)^{-1/2}$, $g_1(x,y) = (x^2+y^2)^{-1/2}\cos(xy)$, $g_2(x,y) = (x^2+y^2)^{-1/4}\sin(xy)$ *and*

$$\mathcal{A}(x,y) = \begin{pmatrix} a(x^2+y^2)^{-1/2} & 0 \\ 0 & b(x^2+y^2)^{-1/2} \end{pmatrix},$$

the problem

$$\begin{cases} Lu(x,y) - \gamma\, u(x,y)\, g_1(x,y) + h(u(x,y))\, g_2(x,y) = f(x,y) \text{ in } \Omega, \\ u(x,y) = 0 \quad \text{on} \quad \partial\Omega, \end{cases}$$

where

$$Lu(x) = -\frac{\partial}{\partial x}\left(a(x^2+y^2)^{-1/2}\frac{\partial u}{\partial x}\right) - \frac{\partial y}{\partial x}\left(b(x^2+y^2)^{-1/2}\,\frac{\partial u}{\partial y}\right),$$

has a solution $u \in W_0^{1,2}(\Omega,\omega)$ *if* $\gamma > 0$ *is not an eigenvalue of linearized problem (LP).*

Definition 4.1.4. *We say that an element* $u \in W_0^{1,2}(\Omega,\omega)$ *is a (weak) solution of problem (P5) if*

$$\sum_{i,j=1}^{N} \int_\Omega a_{ij}(x) D_i u(x) D_j \varphi(x)\, dx - \gamma \int_\Omega u(x) g(x) \varphi(x)\, dx$$
$$= - \int_\Omega f(x,u(x))\, \varphi(x)\, dx,$$

for all $\varphi \in W_0^{1,2}(\Omega,\omega)$.

The following definitions are basics.

Definition 4.1.5. *Let* X *be a real Banach space.*
(i) The operator $A : X \to X^*$ *is called monotone iff*

$$(Au_1 - Au_2 | u_1 - u_2) \geq 0$$

for all $u_1, u_2 \in X$ *(where* $(f|u)$ *means the value* $f(u)$ *of the linear functional* $f \in X^*$*).*
(ii) The operator $A : X \to X^*$ *is called strictly monotone iff*

$$(Au - Av | u - v) > 0$$

for all $u, v \in X$ *with* $u \neq v$.
(iii) The operator $A : X \to X^*$ *is called strongly monotone if there is a constant* $C > 0$ *such that*

$$(Au - Av|u - v) \geq C \|u - v\|^2$$

for all $u, v \in X$.
(iv) The operator $A : X \to X^*$ *is called coercive if*

$$\lim_{|u\| \to \infty} \frac{(Au|u)}{\|u\|} = \infty.$$

(v) The operator $A : X \to X^*$ *is called hemicontinuous iff the real function* $t \mapsto (A(u + tv_1)|v_2)$ *is continuous on* $[0, 1]$, *for all* $u, v_1, v_2 \in X$.
(vi) $A : X \to X^*$ *is called angle-bounded iff* A *is linear, monotone and there is a constant* $C \geq 0$ *such that*

$$|(Au|v) - (Av|u)|^2 \leq C(Au|u)\,(Av|v),$$

for all $u, v \in X$.
(vii) The operator $A : X \to X^*$ *is called pseudomonotone iff* $u_n \rightharpoonup u$ *as* $n \to \infty$ *and* $\overline{\lim\limits_{n \to \infty}} (Au_n|u_n - u) \leq 0$ *implies*

$$(Au|u - v) \leq \underline{\lim_{n \to \infty}} (Au_n|u_n - v)$$

for all $v \in X$.

The basic idea is to reduce the problem (P5) to an operator equation $u + KFu = 0$ and apply the theorem below.

Theorem 4.1.5. *Let* X *be a real separable Banach spaces.*
If $F : X^* \to X$ *and* $K : X \to X^*$ *are operators, with the following assumptions:*
(a) the operator K *is linear, monotone and angle-bounded;*
(b) the operator F *is monotone and hemicontinuous.*
Then the equation $u + KFu = 0$ *has exactly one solution* $u \in X^*$.

Proof. See Theorem 28.A in [60]. □

Theorem 4.1.6. *Let $\omega \in A_2$. Suppose that*
(**H1**) *The function $f : \Omega \times \mathbb{R} \to \mathbb{R}$ satisfies the Carathéodory condition, that is, $x \mapsto f(x,t)$ is measurable on Ω for all $t \in \mathbb{R}$ and $t \mapsto f(x,t)$ is continuous on $\mathbb{R}$ for almost all $x \in \Omega$.*
(**H2**) *There exist two nonnegative functions g_1 and g_2 with*

$$g_1 \in L^2(\Omega,\omega) \cap L^2(\Omega,\omega^{-1}),$$
$$g_2 \in L^\infty(\Omega) \text{ and } g_2/\omega \in L^\infty(\Omega),$$

such that $|f(x,t)| \leq g_1(x) + g_2(x)|t|$.
(**H3**) *The function $t \mapsto f(x,t)$ is monotone increasing on $\mathbb{R}$ for all $x \in \Omega$.*
(**H4**) $g/\omega \in L^\infty(\Omega)$.
Then there exist a number $C_1 > 0$ such that if $0 < \gamma < C_1$ the problem $(P5)$ has exactly one solution $u \in W_0^{1,2}(\Omega,\omega)$. Moreover, there exists a number $C_2 > 0$ such that if $0 < \gamma < C_2$ then

$$\|u\|_{W_0^{1,2}(\Omega,\omega)} \leq C_3 \|g_1\|_{L^2(\Omega,\omega^{-1})}.$$

Proof. We define $B : W_0^{1,2}(\Omega,\omega) \times W_0^{1,2}(\Omega,\omega) \to \mathbb{R}$ by

$$B(u,\varphi) = \int_\Omega \sum_{i,j=1}^N a_{ij}(x) D_i u(x) D_j \varphi(x)\, dx - \gamma \int_\Omega u(x) g(x) \varphi(x)\, dx.$$

Then $u \in W_0^{1,2}(\Omega,\omega)$ is a (weak) solution to problem (P5) if

$$B(u,\varphi) = -\int_\Omega f(x,u(x))\varphi(x)\, dx, \text{ for all } \varphi \in W_0^{1,2}(\Omega,\omega).$$

Step 1. We define the operator

$$F : L^2(\Omega,\omega) \to L^2(\Omega,\omega)$$
$$(Fu)(x) = f(x,u(x)).$$

We have that the operator F is bounded and continuous. In fact,
(i) Using (H3) and Theorem 2.1.5 (with $\theta = 1$) we obtain

$$\begin{aligned}\|Fu\|^2_{L^2(\Omega,\omega)} &= \int_\Omega |Fu(x)|^2 \omega\, dx = \int_\Omega |f(x,u(x))|^2 \omega\, dx \\ &\leq \int_\Omega (g_1(x) + g_2(x)|u(x)|)^2 \omega\, dx \\ &\leq 2\int_\Omega \left(g_1^2(x) + g_2^2(x)|u(x)|^2\right) \omega\, dx \\ &\leq 2(\|g_1\|^2_{L^2(\Omega,\omega)} + \|g_2\|^2_{L^\infty(\Omega)}\|u\|^2_{L^2(\Omega,\omega)}).\end{aligned}$$

(ii) Let $u_n \to u$ in $L^2(\Omega,\omega)$ as $n \to \infty$. We need to show that $Fu_n \to Fu$ in $L^2(\Omega,\omega)$. Using Theorem 1.4.9, there exist a subsequence $\{u_{n_k}\}$ and a function $\Phi \in L^2(\Omega,\omega)$ such that $u_{n_k}(x) \to u(x)$ μ-a.e. and $|u_{n_k}(x)| \leq \Phi(x)$ μ-a.e. in Ω (where $\mu(E) = \displaystyle\int_E \omega(x)\, dx$). Hence, we obtain

$$\begin{aligned}&\|Fu_{n_k} - Fu\|^2_{L^2(\Omega,\omega)} = \int_\Omega |Fu_{n_k}(x) - Fu(x)|^2 \omega\, dx \\ &= \int_\Omega |f(x,u_{n_k}) - f(x,u)|^2 \omega\, dx \\ &\leq 2\int_\Omega \left(|f(x,u_{n_k})|^2 + |f(x,u)|^2\right)\omega\, dx \\ &\leq 2\left[\int_\Omega (g_1(x) + g_2(x)|u_{n_k}|)^2 \omega\, dx + \int_\Omega (g_1(x) + g_2(x)|u|)^2 \omega\, dx\right] \\ &\leq C\left(\|g_1\|^2_{L^2(\Omega,\omega)} + \|g_2\|^2_{L^\infty(\Omega)}\|\Phi\|^2_{L^2(\Omega,\omega)}\right).\end{aligned}$$

By condition (H1) we have $Fu_n(x) = f(x,u_n(x)) \to f(x,u(x)) = Fu(x)$, as $n \to +\infty$. Therefore, by the Lebesgue Dominated Convergence Theorem, we obtain $\|Fu_{n_k} - Fu\|_{L^2(\Omega,\omega)} \to 0$, that is, $Fu_{n_k} \to Fu$ in $L^2(\Omega,\omega)$. By the Convergence Principle in Banach spaces (see Proposition 10.13 in [59]),we have $Fu_n \to Fu$ in $L^2(\Omega, \omega)$.

Step 2. The map $B : W_0^{1,2}(\Omega, \omega) \times W_0^{1,p}(\Omega, \omega) \to \mathbb{R}$ is bilinear, bounded and strongly positive. In fact:
(i) By (4.2) we have $|a_{ij}(x)| \leq \Lambda\, \omega(x)$ μ-a.e. $x \in \Omega$. Using Theorem 2.1.5 (with $\theta = 1$) and (H4) we obtain

$$
\begin{aligned}
|B(u,\varphi)| &\leq \sum_{i,j=1}^{N} \int_\Omega |a_{ij}||D_i u||D_j\varphi|\, dx + |\gamma| \int_\Omega |u||g||\varphi|\, dx \\
&\leq \sum_{i,j=1}^{N} \int_\Omega \Lambda |D_i u||D_j\varphi|\,\omega\, dx + |\gamma| \|g/\omega\|_{L^\infty(\Omega)} \int_\Omega |u||\varphi|\,\omega\, dx \\
&\leq \Lambda \sum_{i,j=1}^{N} \left(\int_\Omega |D_i u|^2 \omega\, dx \right)^{1/2} \left(\int_\Omega |D_j \varphi|^2 \omega\, dx \right)^{1/2} \\
&+ |\gamma| \|g/\omega\|_{L^\infty(\Omega)} \left(\int_\Omega |u|^2 \omega\, dx \right)^{1/2} \left(\int_\Omega |\varphi|^2 \omega\, dx \right)^{1/2} \\
&\leq \Lambda \|u\|_{W_0^{1,2}(\Omega,\omega)} \|\varphi\|_{W_0^{1,2}(\Omega,\omega)} \\
&+ |\gamma| \|g\omega\|_{L^\infty(\Omega)}\, C_\Omega^2\, \|u\|_{W_0^{1,2}(\Omega,\omega)} \|\varphi\|_{W_0^{1,2}(\Omega,\omega)} \\
&= C \|u\|_{W_0^{1,2}(\Omega,\omega)} \|\varphi\|_{W_0^{1,2}(\Omega,\omega)},
\end{aligned}
$$

where $C = \Lambda + |\gamma| C_\Omega^2 \|g/\omega\|_{L^\infty(\Omega)}$.
(ii) We also have, by condition (4.2), (H4) and Theorem 2.1.5,

$$
\begin{aligned}
B(u,u) &= \sum_{i,j=1}^{N} \int_\Omega a_{ij} D_i u D_j u\, dx - \gamma \int_\Omega u^2 g\, dx \\
&\geq \lambda \int_\Omega |Du|^2 \omega\, dx - \gamma \|g/\omega\|_{L^\infty(\Omega)} \int_\Omega u^2 \omega\, dx \\
&\geq \lambda \int_\Omega |Du|^2 \omega\, dx - \gamma\, \|g/\omega\|_{L^\infty(\Omega)}\, C_\Omega \int_\Omega |Du|^2 \omega\, dx
\end{aligned}
$$

$$= (\lambda - \gamma \, \|g/\omega\|_{L^\infty(\Omega)} \, C_\Omega) \|u\|^2_{W_0^{1,2}(\Omega,\omega)}$$
$$= C_4 \, \|u\|^2_{W_0^{1,2}(\Omega,\omega)}$$

with $C_4 = \lambda - \gamma \, \|g/\omega\|_{L^\infty(\Omega)} C_\Omega > 0$
(i.e., with $0 < \gamma < \lambda / \|g/\omega\|_{L^\infty(\Omega)} C_\Omega = C_1$).Therefore the bilinear map B is strongly positive if $0 < \gamma < C_1$.
Step 3 We consider the linear boundary value problem

$$\begin{cases} Lu(x) - \gamma u(x) g(x) = h(x) \text{ in } \Omega, \\ u(x) = 0 \text{ on } \partial\Omega, \end{cases}$$

That is, we set $-f(x, u(x)) \equiv h(x)$. By (4.1) and Theorem 22.C in [60] the corresponding problem

$$B(u, \varphi) = -\int_\Omega h(x)\, \varphi(x)\, dx, \quad \text{for all } \varphi \in W_0^{1,2}(\Omega, \omega)$$

has exactly one solution $u \in W_0^{1,2}(\Omega, \omega) \subset L^2(\Omega, \omega)$.
We set $u = Kh$. By Corollary 22.20 in [60], the solution operator $K : L^2(\Omega, \omega) \to L^2(\Omega, \omega)$ is linear, monotone, compact and angle-bounded. Consequently, the problem $(P5)$ is equivalent to the operator equation

$$u + KFu = 0, \quad u \in L^2(\Omega, \omega). \tag{4.4}$$

Therefore, by Theorem 4.1.5 the equation (4.4) has exactly one solution.
Step 4. In particular, for $\varphi = u \in W_0^{1,p}(\Omega, \omega)$ we have

$$B(u, u) = -\int_\Omega f(x, u(x))\, u(x)\, dx.$$

By Step 2, $B(u, u) \geq C_4 \|u\|^2_{W_0^{1,2}(\Omega,\omega)}$. And we also have by (H2) and Theorem 2.1.5 (with $\theta = 1$),

$$\left| \int_\Omega f(x, u)\, u\, dx \right| \leq \int_\Omega |f(x, u)|\, |u|\, dx$$

$$\begin{aligned}
&\leq \int_\Omega (g_1 + g_2\,|u|)\,|u|\,dx \\
&= \int_\Omega g_1\,\omega^{-1/2}|u|\,\omega^{1/2}\,dx + \int_\Omega g_2\,\omega^{-1}\,|u|^2\,\omega\,dx \\
&\leq \left(\int_\Omega g_1^2\,\omega^{-1}\,dx\right)^{1/2}\left(\int_\Omega |u|^2\omega\,dx\right)^{1/2} + \|g_2/\omega\|_{L^\infty(\Omega)}\int_\Omega u^2\,\omega\,dx \\
&\leq C_\Omega\|g_1\|_{L^2(\Omega,\omega^{-1})}\|u\|_{W_0^{1,2}(\Omega,\omega)} + C_\Omega^2\,\|g_2/\omega\|_{L^\infty(\Omega)}\|u\|_{W_0^{1,2}(\Omega,\omega)}.
\end{aligned}$$

Hence, we obtain

$$\begin{aligned}
C_4\|u\|^2_{W_0^{1,2}(\Omega,\omega)} &\leq C_\Omega\|g_1\|_{L^2(\Omega,\omega^{-1})}\|u\|_{W_0^{1,2}(\Omega,\omega)} \\
&+ C_\Omega^2\|g_2/\omega\|_{L^\infty(\Omega)}\|u\|^2_{W_0^{1,2}(\Omega,\omega)}.
\end{aligned}$$

Therefore,

$$\|u\|_{W_0^{1,2}(\Omega,\omega)} \leq C_3\|g_1\|_{L^2(\Omega,\omega^{-1})},$$

where
$C_3 = C_\Omega/(C_4 - C_\Omega^2\|g_2/\omega\|_{L^\infty(\Omega)}) > 0$ if $C_4 - C_\Omega^2\|g_2/\omega\|_{L^\infty(\Omega)} > 0$ (i.e., if $0 < \gamma < \|g/\omega\|_{L^\infty(\Omega)}C_\Omega/(\lambda - C_\Omega^2\|g_2/\omega\|_{L^\infty(\Omega)}) = C_2$). □

Example 4.1.2. *Let* $\Omega = \{(x,y)\in\mathbb{R}^2\ : x^2+y^2<1\}$, *and consider the weight* $\omega(x,y) = (x^2+y^2)^{-1/2}$ $(\omega\in A_2)$, *and the function* $f\ :\ \Omega\times\mathbb{R}\to\mathbb{R}$, $f((x,y),t) = \dfrac{\cos(xy)}{(x^2+y^2)^{1/2}} + t\mathrm{e}^{-(x^2+y^2)}$. *Let us consider the partial differential operator*

$$Lu(x,y) = -\frac{\partial}{\partial x}\left(a(x^2+y^2)^{-1/2}\frac{\partial u}{\partial x}\right) - \frac{\partial}{\partial y}\left(b(x^2+y^2)^{-1/2}\frac{\partial u}{\partial y}\right)$$

with $0<a<b$, $g(x,y) = \sin(xy)/(x^2+y^2)^{1/2}$, $g_1(x,y) = (x^2+y^2)^{-1/2}$ *and* $g_2(x,y) = \mathrm{e}^{-(x^2+y^2)}$. *By Theorem 4.1.6 the problem*

$$\begin{cases} Lu(x,y) - \gamma\,u(x,y)\,g(x,y) = -f((x,y),u(x,y)) \ \text{ in } \ \Omega, \\ u(x,y) = 0 \ \text{ on } \ \partial\Omega, \end{cases}$$

has a solution $u\in W_0^{1,2}(\Omega,\omega)$.

Now we will prove the existence and uniqueness of (weak) solutions in the weighted Sobolev spaces $W_0^{1,p}(\Omega,\omega)$ for the Dirichlet problem

$$(P6)\begin{cases} Lu(x)+\operatorname{div}(\Phi(u(x)))=g(x)-\sum_{j=1}^{N} D_j f_j \text{ in } \Omega, \\ u(x)=0 \text{ on } \partial\Omega, \end{cases}$$

where L is the partial differential operator

$$Lu(x)=-\sum_{i,j=1}^{N} D_j(a_{ij}(x)D_i u(x))+b(x)\,u(x) \tag{4.5}$$

with $D_j=\partial/\partial x_j$, where Ω is a bounded open set in $\mathbb{R}^N$ and we assume that Ω has a Lipschitz boundary $\partial\Omega$ (i.e., $\Omega\in C^{0,1}$) with outward unit normal $\vec{\eta}(x)=(\eta_1(x),...,\eta_N(x))$, the coefficients a_{ij} are measurable, real valued functions whose coefficient matrix $\mathcal{A}=(a_{ij})$ is symmetric and satisfies the degenerate elliptic condition

$$\lambda\,|\xi|^2\,\omega(x)\le\sum_{i,j=1}^{N} a_{ij}(x)\,\xi_i\xi_j\le\Lambda\,|\xi|^2\,\omega(x), \tag{4.6}$$

for all $\xi\in\mathbb{R}^n$ and a.e. $x\in\Omega$, ω is a weight function, λ and Λ are positive constants, and $\Phi:\mathbb{R}\to\mathbb{R}^N$.

Equations like (4.5) have been studied by many authors in the non-degenerate case (i.e. with $\omega(x)\equiv 1$) (see e.g. [39], [44] and the references therein).

The following lemma is due to Stampacchia (see Lemme 4.1 in [53]).

Lemma 4.1.7. *Let α, β, C, k_0 be real positive numbers, where $\beta>1$. Let $\varphi:\mathbb{R}_+\to\mathbb{R}_+$ be a decreasing function such that*

$$\varphi(l)\le\frac{C}{(l-k)^{\alpha}}[\varphi(k)]^{\beta}$$

for all $l > k \geq k_0$. *Then* $\varphi(k_0 + d) = 0$, *where* $d^\alpha = C\,[\varphi(k_0)]^{\beta-1} 2^{\alpha\,\beta/(\beta-1)}$.

Definition 4.1.6. *We say that an element* $u \in W_0^{1,2}(\Omega,\omega)$ *is a (weak) solution of problem (P6) if*

$$\sum_{i,j=1}^{N} \int_\Omega a_{ij}(x)\, D_i u(x)\, D_j\varphi(x)\, dx + \int_\Omega b(x)\, u(x)\, \varphi(x)\, dx$$
$$-\sum_{j=1}^{N} \int_\Omega \Phi_j(u(x))\, D_j\varphi(x)\, dx$$
$$= \int_\Omega g(x)\, \varphi(x)\, dx + \sum_{j=1}^{N} \int_\Omega f_j(x) D_j\varphi(x)\, dx \tag{4.7}$$

for all $\varphi \in W_0^{1,2}(\Omega,\omega)$.

Remark 4.1.1. *Since* $p > N \geq 2$ *and* $p' < 2$, *if* $\omega \in A_{p'}$ *then* $\omega \in A_p$ *and* $\omega \in A_2$ *(by Lemma 2.1.1) and we also have*

$$L^p(\Omega,\omega) \subset L^2(\Omega,\omega) \subset L^{p'}(\Omega,\omega)$$

(where $\mu(\Omega) = \int_\Omega \omega(x)\, dx < \infty$*).*

Theorem 4.1.8. *Let* Ω *be an open bounded set in* $\mathbb{R}^N$ *with a Lipschitz boundary* $\partial\Omega$. *Suppose that*
(H1) $f_j/\omega \in L^p(\Omega,\omega)$, $(j = 1, ..., N)$ *with* $p > Nr \geq 2$;
(H2) $g/\omega \in L^q(\Omega,\omega) \cap L^2(\Omega,\omega)$, *with* $1/q = 1/p + 1/Nr$;
(H3) $\omega \in A_r$, *with* $1 < r < p' < Nr$ *(where* $1/p + 1/p' = 1$*)*;
(H4) $b(x) \geq 0$ *for a.e.* $x \in \Omega$ *and* $b/\omega \in L^\infty(\Omega)$;
(H5) $\Phi : \mathbb{R} \to \mathbb{R}^N$ $(\Phi = (\Phi_1, ..., \Phi_N))$, *with* $|\Phi(u)|/\omega \in L^2(\Omega,\omega)$ *if* $u \in W_0^{1,2}(\Omega,\omega)$ *and the functions* Φ_j *are continuous* $(j = 1, ..., N)$.

Then the problem (P6) has a solution $u \in W_0^{1,2}(\Omega,\omega)$. *Moreover, we have that* $u \in L^\infty(\Omega)$ *with the estimate*

$$\|u\|_{L^\infty(\Omega)} \le C\left(\|g/\omega\|_{L^q(\Omega,\omega)} + \sum_{j=1}^{N}\|f_j/\omega\|_{L^p(\Omega,\omega)}\right) \qquad (4.8)$$

where C *is a constant independent of* u, g, f_j *and* Φ_j.
(H6) *If* $|\Phi_j(u(x)) - \Phi_j(v(x))| \le C_0\,|u(x) - v(x)|\,\omega(x)$ *a.e.* $x \in \Omega$, *for all* $u, v \in W_0^{1,2}(\Omega,\omega)$ *and* C_0 *is a positive constant, then the problem* $(P6)$ *has a unique solution.*

Remark 4.1.2. *The estimate (4.8) is an important ingredient in the proof of the existence of a weak solution to problem* $(P6)$. *Under the assumption* $\Phi_j = 0$ $(j = 1, ..., N)$ *and* $\omega \equiv 1$ *(non-degenerate case) then (4.8) is the usual* L^∞*-estimate of Stampacchia (see [44]).*

Proof. **Step 1.** Proof of (4.8). Assuming that the problem (P6) has a solution $u \in W_0^{1,2}(\Omega,\omega)$.
Set $\Omega(k) = \{x \in \Omega : |u(x)| > k\}$ (where $k \ge 0$). We choose for $\tilde{\varphi}$ in (4.7) the function

$$\tilde{\varphi} = (u-k)^+ + (u+k)^- \qquad (4.9)$$

where $(u-k)^+ = \max\{u-k, 0\}$ and $(u+k)^- = \min\{u+k, 0\}$. We have $(u-k)^+$, $(u+k)^-$ and $\tilde{\varphi} \in W_0^{1,2}(\Omega,\omega)$, and

$$D_i(u-k)^+ = \chi_{\{u>k\}} D_i u \text{ and } D_i(u+k)^- = \chi_{\{u<-k\}} D_i u,$$

(see Corollary 1.4.4), where χ_E denotes the characteristic function of a measurable set $E \subset \mathbb{R}^n$. Moreover, if we set $\psi_j(s) = \int_0^s \Phi_j(t+k)\,dt$, we have by the Divergence Theorem,

$$\int_\Omega \Phi_j(u) D_j(u-k)^+\,dx = \int_\Omega D_j\psi_j((u-k)^+)\,dx$$

$$= \int_{\partial\Omega} \psi_j((u-k)^+)\,\eta_j\,d\sigma(x) = 0, \tag{4.10}$$

since $\psi_j(0) = 0$ and $(u-k)^+ = 0$ on $\partial\Omega$. Analogously,

$$\int_\Omega \Phi_j(u)\,D_j(u+k)^-\,dx = 0.$$

Moreover, on $u > k > 0$, u is positive and on $u < -k < 0$, u is negative. So we have

$$\begin{aligned}\int_\Omega b(x)\,u(x)\,\tilde{\varphi}(x)\,dx &= \int_\Omega b(x)\,u(x)\,(u(x)-k)^+\,dx \\ &- \int_\Omega b(x)\,u(x)\,(u(x)+k)^-\,dx \geq 0.\end{aligned} \tag{4.11}$$

Using (4.9), (4.10), (4.11),(4.6) and (4.7), we obtain

$$\begin{aligned}\lambda\int_\Omega |\nabla\tilde{\varphi}|^2\omega\,dx &\leq \sum_{i,j=1}^N \int_\Omega a_{ij}(x)D_i\tilde{\varphi}\,D_j\tilde{\varphi}\,dx \\ &\leq \sum_{i,j=1}^N \int_\Omega a_{ij}(x)D_iu\,D_j\tilde{\varphi}\,dx + \int_\Omega b(x)\,u\,\tilde{\varphi}\,dx \\ &- \sum_{j=1}^N \int_\Omega \Phi_j(u)D_j\tilde{\varphi}\,dx \\ &= \int_\Omega g\,\tilde{\varphi}\,dx + \sum_{j=1}^N \int_\Omega f_jD_j\tilde{\varphi}\,dx.\end{aligned}$$

Hence, by the Hölder inequality, we obtain

$$\begin{aligned}\|\,|\nabla\tilde{\varphi}|\,\|^2_{L^2(\Omega,\omega)} &\leq \frac{1}{\lambda}\Bigg(\left\|\frac{g}{\omega}\right\|_{L^q(\Omega,\omega)} \|\tilde{\varphi}\|_{L^{q'}(\Omega,\omega)} \\ &+\sum_{j=1}^N \left\|\frac{f_j}{\omega}\right\|_{L^p(\Omega,\omega)} \|\,|\nabla\tilde{\varphi}|\,\|_{L^{p'}(\Omega,\omega)}\Bigg),\end{aligned} \tag{4.12}$$

(where $1/q + 1/q' = 1$). Since $p > Nr \geq 2$, then $p' < 2 \leq Nr$. If $\omega \in A_r$ and $r < p' < p$ then $\omega \in A_{p'}$ and $\omega \in A_p$ (by Remark 4.1.1). By Theorem 2.1.5 (with $\theta = 1$) we have

$$\|\tilde{\varphi}\|_{L^p(\Omega,\omega)} \leq C_1 \, \| \, |\nabla \tilde{\varphi}| \, \|_{L^p(\Omega,\omega)}, \tag{4.13}$$

$$\|\tilde{\varphi}\|_{L^{p'}(\Omega,\omega)} \leq C_2 \, \| \, |\nabla \tilde{\varphi} \, |\|_{L^{p'}(\Omega,\omega)}. \tag{4.14}$$

Since $1/q' = 1/p' - 1/nr$, using Theorem 2.1.9 (with $v = \omega$) we have

$$\|\tilde{\varphi}\|_{L^{q'}(\Omega,\omega)} \leq C_3 \, \|\nabla \tilde{\varphi}\|_{L^{p'}(\Omega,\omega)}. \tag{4.15}$$

Hence, by (4.12), we get

$$\begin{aligned} &\| \, |\nabla \tilde{\varphi}| \, \|^2_{L^2(\Omega,\omega)} \\ &\leq \frac{C_4}{\lambda} \left(\left\| \frac{g}{\omega} \right\|_{L^q(\Omega,\omega)} + \sum_{j=1}^{N} \left\| \frac{f_j}{\omega} \right\|_{L^p(\Omega,\omega)} \right) \| \, |\nabla \tilde{\varphi}| \, \|_{L^{p'}(\Omega,\omega)}. \end{aligned} \tag{4.16}$$

Now let us remark that $\tilde{\varphi} = 0$ outside $\Omega(k)$, by Hölder's inequality we obtain

$$\begin{aligned} \| \, |\nabla \tilde{\varphi}| \, \|^{p'}_{L^{p'}(\Omega,\omega)} &= \int_\Omega |\nabla \tilde{\varphi}|^{p'} \omega \, dx = \int_{\Omega(k)} |\nabla \tilde{\varphi}|^{p'} \omega \, dx \\ &\leq \left(\int_{\Omega(k)} |\nabla \tilde{\varphi}|^2 \omega \, dx \right)^{p'/2} \left(\int_{\Omega(k)} \omega \, dx \right)^{(2-p')/2} \\ &= \| \, |\nabla \tilde{\varphi}| \, \|^{p'}_{L^2(\Omega,\omega)} [\mu(\Omega(k))]^{(2-p')/2}. \end{aligned} \tag{4.17}$$

Hence, we obtain

$$\| \, |\nabla \tilde{\varphi}| \, \|^2_{L^{p'}(\Omega,\omega)} \leq \| \, |\nabla \tilde{\varphi}| \, \|^2_{L^2(\Omega,\omega)} \, [\mu(\Omega(k))]^{(2-p')/p'}. \tag{4.18}$$

From (4.16) and (4.18), we then deduce

$$\| \, |\nabla \tilde{\varphi}| \, \|_{L^{p'}(\Omega,\omega)}$$

$$\leq \frac{C_4}{\lambda}\left(\left\| \frac{g}{\omega} \right\|_{L^q(\Omega,\omega)} + \sum_{j=1}^{N} \left\| \frac{f_j}{\omega} \right\|_{L^p(\Omega,\omega)} \right) [\mu(\Omega(k))]^{(2-p')/p'}. \tag{4.19}$$

If $h > k$ then $\Omega(h) \subset \Omega(k)$, $\tilde{\varphi} = \pm(|u| - k)$ on $\Omega(k)$ and $|\tilde{\varphi}| \geq (h - k)$ on $\Omega(h)$ for $h > k$. We obtain

$$\begin{aligned} (h-k)[\mu(\Omega(h))]^{1/q'} &\leq \left(\int_{\Omega(h)} |\tilde{\varphi}|^{q'} \, \omega \, dx \right)^{1/q'} \\ &\leq \left(\int_{\Omega(k)} |\tilde{\varphi}|^{q'} \, \omega \, dx \right)^{1/q'} \\ &= \|\tilde{\varphi}\|_{L^{q'}(\Omega,\omega)}. \end{aligned} \tag{4.20}$$

Using (4.15), (4.19) and (4.20) we get

$$\begin{aligned} &(h-k)[\mu(\Omega(h))]^{1/q'} \leq \|\tilde{\varphi}\|_{L^{q'}(\Omega,\omega)} \\ &\leq C_3 \, \| \, |\nabla \tilde{\varphi}| \, \|_{L^{p'}(\Omega,\omega)} \\ &\leq \frac{C_5}{\lambda}\left(\left\| \frac{g}{\omega} \right\|_{L^q(\Omega,\omega)} + \sum_{j=1}^{N} \left\| \frac{f_j}{\omega} \right\|_{L^p(\Omega,\omega)} \right) [\mu(\Omega(k))]^{(2-p')/p'}. \end{aligned} \tag{4.21}$$

Hence

$$\begin{aligned} &\mu(\Omega(h)) \\ &\leq \left[\frac{C_5}{\lambda} \frac{\left(\|g/\omega\|_{L^q(\Omega,\omega)} + \sum_{j=1}^{N} \|f_j/\omega\|_{L^p(\Omega,\omega)} \right)}{(h-k)} \right]^{q'} [\mu(\Omega(k))]^{(2-p')q'/p'}. \end{aligned} \tag{4.22}$$

Since $p > N \geq 2$ and $1/q' = 1/p' - 1/N$, we have that $\beta = (2 - p')\, q'/p' = (p-2)\, N/(Np - N - p) > 1$. By Lemma 4.1.7

applied to $\varphi(h) = \mu(\Omega(h))$ we have that $\varphi(d) = \mu(\Omega(d)) = 0$ where

$$d = \frac{C_6}{\lambda}\left(\left\|\frac{g}{\omega}\right\|_{L^q(\Omega,\omega)} + \sum_{j=1}^{N}\left\|\frac{f_j}{\omega}\right\|_{L^p(\Omega,\omega)}\right)[\varphi(0)]^{(\beta-1)/q'}2^{\beta/(\beta-1)}.$$

By Lemma 2.1.2, if $\mu(\Omega(d)) = 0$ then $|\,\Omega(d)| = 0$. Therefore

$$\|u\|_{L^\infty(\Omega)} \leq \frac{C_7}{\lambda}\left(\left\|\frac{g}{\omega}\right\|_{L^q(\Omega,\omega)} + \sum_{j=1}^{N}\left\|\frac{f_j}{\omega}\right\|_{L^p(\Omega,\omega)}\right).$$

Step 2. *Proof of existence of a solution.*

Let us denote by $M = \dfrac{C_7}{\lambda}\left(\left\|\dfrac{g}{\omega}\right\|_{L^q(\Omega,\omega)} + \displaystyle\sum_{j=1}^{N}\left\|\dfrac{f_j}{\omega}\right\|_{L^p(\Omega,\omega)}\right)$, and we define, for all $j = 1, ..., N$ and $t \in \mathbb{R}$

$$\tilde{\Phi}_j(t) = \begin{cases} \Phi_j(-M), \ \text{if } t < -M, \\ \Phi_j(t), \ \text{if } |t| \leq M, \\ \Phi_j(M), \ \text{if } t > M. \end{cases}$$

We have, by (H5), that the $\tilde{\Phi}_j$ are bounded. For each $v \in L^2(\Omega,\omega)$, there exists a unique solution u to the problem

$$(P7)\begin{cases} u \in W_0^{1,2}(\Omega,\omega), \\ \displaystyle\sum_{i,j=1}^{N}\int_\Omega a_{ij}(x)D_iuD_j\varphi\,dx + \int_\Omega b(x)\,u\,\varphi\,dx \\ \displaystyle = \sum_{j=1}^{N}\int_\Omega \tilde{\Phi}_j(v)\,D_j\varphi\,dx + \int_\Omega g\,\varphi\,dx + \sum_{j=1}^{N}\int_\Omega f_jD_j\varphi\,dx, \end{cases}$$

for all $\varphi \in W_0^{1,2}(\Omega,\omega)$. In fact, we define
$\mathcal{B} : W_0^{1,2}(\Omega,\omega)\times W_0^{1,2}(\Omega,\omega) \to \mathbb{R}$ and $\mathcal{T} : W_0^{1,2}(\Omega,\omega)\to\mathbb{R}$ by

$$\mathcal{B}(u,\varphi) = \sum_{i,j=1}^{N}\int_\Omega a_{ij}(x)D_iu\,D_j\varphi\,dx + \int_\Omega b(x)\,u\,\varphi\,dx,$$

$$\mathcal{T}(\varphi) = \sum_{j=1}^{N} \int_{\Omega} \tilde{\Phi}_j(v)\, D_j\varphi\, dx + \int_{\Omega} g\,\varphi\, dx + \sum_{j=1}^{N} \int_{\Omega} f_j D_j\varphi\, dx.$$

We have that
(i) $\mathcal{B}$ is continuous (by (4.6) and (H4))

$$\begin{aligned} |\mathcal{B}(u,\varphi)| &\leq \sum_{i,j=1}^{N} \int_{\Omega} |a_{ij}(x)|\, |D_i u|\, |D_j\varphi|\, dx + \int_{\Omega} \frac{b(x)}{\omega}\, |u|\, |\varphi|\, \omega\, dx \\ &\leq (\Lambda + \|b/\omega\|_{L^\infty(\Omega)}) \|u\|_{W_0^{1,2}(\Omega,\omega)}\, \|\varphi\|_{W_0^{1,2}(\Omega,\omega)}. \end{aligned}$$

and $\mathcal{B}$ is coercive,

$$\mathcal{B}(u,u) = \sum_{i,j=1}^{N} \int_{\Omega} a_{ij}(x) D_i u\, D_j u\, dx + \int_{\Omega} b(x)\, u^2\, dx \geq \lambda \int_{\Omega} |\nabla u|^2 \omega\, dx.$$

(ii) Since the $\tilde{\Phi}_j$ are bounded ($|\tilde{\Phi}_j| \leq \tilde{C}$, $(j = 1, ..., N)$, $\mathcal{T}$ is continuous and

$$\begin{aligned} |\mathcal{T}(\varphi)| &\leq \left[\tilde{C} \left(\int_{\Omega} \omega^{-1}\, dx \right)^{1/2} + \|g/\omega\|_{L^2(\Omega,\omega)} \right. \\ &\quad + \left. \sum_{j=1}^{N} \|f_j/\omega\|_{L^2(\Omega,\omega)} \right] \|\varphi\|_{W_0^{1,2}(\Omega,\omega)}. \end{aligned}$$

Hence, by the Lax-Milgram theorem there exists a unique solution $u \in W_0^{1,2}(\Omega,\omega)$ to $\mathcal{B}(u,\varphi) = \mathcal{T}(\varphi)$, for all $\varphi \in W_0^{1,2}(\Omega,\omega)$ (that is, u is the unique solution of problem $(P7)$).
Therefore let us consider the mapping $T : L^2(\Omega,\omega) \to L^2(\Omega,\omega)$ defined for $v \in L^2(\Omega,\omega)$ by $T(v) = u$ (where u is a solution of problem $(P7)$). By taking $\varphi = u$ in (P7), we obtain

$$\sum_{i,j=1}^{N} \int_{\Omega} a_{ij}(x) D_i u\, D_j u\, dx + \int_{\Omega} b(x)\, u^2\, dx$$

$$= \sum_{j=1}^{N} \int_{\Omega} \tilde{\Phi}_j(v)\, D_i u\, dx + \int_{\Omega} g\, u\, dx + \sum_{j=1}^{N} \int_{\Omega} f_j D_j u\, dx.$$

Using (4.6) and (H4) we have

$$\lambda \int_{\Omega} |\nabla u|^2\, \omega\, dx \leq \sum_{i,j=1}^{N} \int_{\Omega} a_{ij}(x) D_i u\, D_j u\, dx + \int_{\Omega} b(x)\, u^2\, dx, \tag{4.23}$$

and using the fact that $\tilde{\Phi}_j$ are bounded (i.e., $|\tilde{\Phi}_j| \leq \tilde{C}$), $\omega \in A_2$ (then $\omega^{-1} \in L^1_{\text{loc}}(\mathbb{R}^N)$), we obtain

$$\begin{aligned} &\left| \sum_{j=1}^{N} \int_{\Omega} \tilde{\Phi}_j(v)\, D_i u\, dx + \int_{\Omega} g\, u\, dx + \sum_{j=1}^{N} \int_{\Omega} f_j\, D_j u\, dx \right| \qquad (4.24) \\ &\leq \left[\tilde{C} \left(\int_{\Omega} \omega^{-1} dx \right)^{1/2} + \left\| \frac{g}{\omega} \right\|_{L^2(\Omega,\omega)} + \sum_{j=1}^{N} \left\| \frac{f_j}{\omega} \right\|_{L^2(\Omega,\omega)} \right] \|u\|_{W_0^{1,2}(\Omega,\omega)}. \end{aligned}$$

Hence, by (4.23) and (4.24) we obtain

$$\begin{aligned} \|u\|_{W_0^{1,2}(\Omega,\omega)} &\leq \frac{1}{\lambda} \left[\tilde{C} \left(\int_{\Omega} \omega^{-1} dx \right)^{1/2} + \left\| \frac{g}{\omega} \right\|_{L^2(\Omega,\omega)} \right. \\ &\left. + \sum_{j=1}^{N} \left\| \frac{f_j}{\omega} \right\|_{L^2(\Omega,\omega)} \right] = C_8, \end{aligned} \tag{4.25}$$

where C_8 is a constant which does not depend on $v \in L^2(\Omega, \omega)$.

Let us denote by $B = B(0, C_8)$ the ball in $L^2(\Omega, \omega)$ of center 0 and radius C_8. From (4.25) we have $T(B) \subset B$. Since $W_0^{1,2}(\Omega, \omega)$ is compactly embedded in $L^2(\Omega, \omega)$ (by Lemma 4.1.2) then $T(B)$ is precompact in B. To prove that T is continuous, let $\{v_m\}$ be a sequence in $L^2(\Omega, \omega)$ such that $v_m \to v$. By (4.25) and Lemma 4.1.2, it is enough to show that $T(v)$ is the only limit point of the sequence $T(v_m)$. Let us assume that a subsequence $T(v_{m_k}) \to u$

when $k \to \infty$. One can extract a subsequence (still denoted by v_{m_k}) such that

$$\begin{aligned} &v_{m_k} \to v \ \mu - a.e. \text{ in } \Omega, \\ &u_{m_k} = T(v_{m_k}) \rightharpoonup u \text{ in } W_0^{1,2}(\Omega, \omega). \end{aligned} \tag{4.26}$$

By (4.26) and the Lebesgue Dominated Convergence Theorem we have
$\tilde{\Phi}_j(v_{m_k}) \to \tilde{\Phi}_j(v)$ in $L^2(\Omega, \omega)$. Now, passing to the limit in

$$\begin{aligned} &\sum_{i,j=1}^{N} \int_\Omega a_{ij}(x) D_i u_{m_k} D_i \varphi \, dx + \int_\Omega b(x)\, u_{m_k}\, \varphi \, dx \\ &= \sum_{j=1}^{N} \int_\Omega \tilde{\Phi}_j(v_{m_k})\, D_j \varphi \, dx + \int_\Omega g\, \varphi \, dx + \sum_{j=1}^{N} \int_\Omega f_j\, D_j \varphi \, dx \end{aligned}$$

we get from (4.26)

$$\begin{aligned} &\sum_{i,j=1}^{N} \int_\Omega a_{ij}(x) D_i u\, D_i \varphi \, dx + \int_\Omega b(x)\, u\, \varphi \, dx \\ &= \sum_{j=1}^{N} \int_\Omega \tilde{\Phi}_j(v)\, D_j \varphi \, dx + \int_\Omega g\, \varphi \, dx + \sum_{j=1}^{N} \int_\Omega f_j\, D_j \varphi \, dx, \end{aligned}$$

for all $\varphi\, W_0^{1,2}(\Omega, \omega)$.

Hence, $u = T(v)$ and T is continuous. Therefore, by the Schauder Fixed Point Theorem (see Theorem 10.1 in [39]), T has a fixed point $u \in B$. Such fixed point is a solution to problem $(P6)$ with $\tilde{\Phi}_j$ instead of Φ_j, but from step 1 we have $\|u\|_{L^\infty(\Omega)} \leq M$ and thus $\tilde{\Phi}_j(u) = \Phi_j(u)$.

Step 3. *Uniqueness of solution*
If u_1 and u_2 are two solutions to problem $(P6)$, then

$$\sum_{i,j=1}^{N}\int_\Omega a_{ij}(x)D_iu_lD_j\varphi\,dx+\int_\Omega b(x)\,u_l\,\varphi\,dx-\int_\Omega\langle\Phi(u_l),\nabla\varphi\rangle\,dx$$
$$=\int_\Omega g\,\varphi\,dx+\sum_{j=1}^{N}\int_\Omega f_jD_j\varphi\,dx$$

$(l=1,2)$ for all $\varphi\in W_0^{1,2}(\Omega,\omega)$. We obtain

$$\sum_{i,j=1}^{N}\int_\Omega a_{ij}(x)\langle\nabla u_1-\nabla u_2,\nabla\varphi\rangle\,dx$$
$$+\int_\Omega b(x)(u_1-u_2)\,\varphi dx-\int_\Omega\langle\Phi(u_1)-\Phi(u_2),\nabla\varphi\rangle\,dx=0,$$

for all $\varphi\in W_0^{1,2}(\Omega,\omega)$. Then, if we define

$$\tilde{L}=-\sum_{i,j=1}^{N}D_j(a_{ij}(x)D_i)+b(x),$$

and integrate by parts we obtain

$$\int_\Omega\Big[(u_1-u_2)\tilde{L}\varphi-\langle\Phi(u_1)-\Phi(u_2),\nabla\varphi\rangle\Big]\,dx=0.\qquad(4.27)$$

We set

$$G_j=\begin{cases}\dfrac{\Phi_j(u_1)-\Phi_j(u_2)}{u_1-u_2},\ \text{if}\ u_1\neq u_2,\\ 0,\ \text{if}\ u_1=u_2.\end{cases}$$

By (H6) we have $G_j/\omega\in L^\infty(\Omega)$ $(j=1,...,N)$. By (4.27) we obtain

$$\int_\Omega(u_1-u_2)\big(\tilde{L}\varphi-G_j\,D_j\varphi\big)\,dx=0,\ \forall\,\varphi\in W_0^{1,2}(\Omega,\omega).\qquad(4.28)$$

But, similar to problem $(P7)$, there exists a unique $\varphi \in W_0^{1,2}(\Omega,\omega)$ satisfying $\tilde{L}\varphi - G_j\, D_j\varphi = (u_1 - u_2)\,\omega$ and for such a φ, (4.28) becomes

$$\int_\Omega (u_1 - u_2)^2 \omega\, dx = 0.$$

Therefore $u_1 = u_2$ a.e. in Ω. □

Example 4.1.3. *Let us have* $\Omega = \{(x,y) \in \mathbb{R}^2 : x^2 + y^2 < 1\}$. *Consider the weight* $\omega(x,y) = (x^2+y^2)^{-1/2}$ $(\omega \in A_{5/4}$, $r = 5/4$, $p = 3$, $p' = 3/2$, $q = 15/11$, $q' = 15/4)$, $0 < a_1 < a_2$ *and the function* $\Phi : \mathbb{R} \to \mathbb{R}^2$, $\Phi(t) = (\sqrt{1+|t|}\,, \sqrt{4+|t|})$, *and the functions* $b, f_1, f_2, g : \Omega \to \mathbb{R}$, $g(x,y) = \dfrac{\arctan(1/(x^2+y^2))}{(x^2+y^2)^{1/2}}$, $f_1(x,y) = \dfrac{\cos(1/(x^2+y^2))}{(x^2+y^2)^{1/3}}$, $f_2(x,y) = \dfrac{\sin(1/(x^2+y^2))}{(x^2+y^2)^{1/3}}$ *and* $b(x,y) = \mathrm{e}^{-(x^2+y^2)}$. *Let us consider the partial differential operator*

$$\begin{aligned} Lu(x,y) &= -\frac{\partial}{\partial x}\left(a_1 (x^2+y^2)^{-1/2}\frac{\partial u}{\partial x}\right) \\ &- \frac{\partial}{\partial y}\left(a_2 (x^2+y^2)^{-1/2}\frac{\partial u}{\partial y}\right) \\ &+ b(x,y)\, u. \end{aligned}$$

By Theorem 4.1.8, the problem

$$\begin{cases} Lu(x,y) + \operatorname{div}(\Phi(u(x,y))) = g(x,y) - \dfrac{\partial f_1}{\partial x} - \dfrac{\partial f_2}{\partial y}, & \text{on } \Omega, \\ u(x) = 0, \quad \text{on } \partial\Omega, \end{cases}$$

has a unique solution $u \in W_0^{1,2}(\Omega,\omega)$.

4.2 Approximation of solutions

Let L be a degenerate elliptic operator in divergence form

$$Lu = -\sum_{i,j=1}^{N} D_j(a_{ij}(x)\, D_i u(x)),\ D = \frac{\partial}{\partial x_j}, \tag{4.29}$$

The main purpose of this section is to establish that a weak solution $u \in W_0^{1,2}(\Omega, \omega)$ for the semilinear Dirichlet problem

$$(P8)\begin{cases} Lu(x) - \gamma\, u(x)\, g_1(x) + h(u(x))\, g_2(x) = f(x) \ \text{ in } \ \Omega, \\ u(x) = 0 \ \text{ on } \ \partial\Omega, \end{cases}$$

where $\gamma \in \mathbb{R}$, can be approximated by a sequence of solutions of non-degenerate semilinear elliptic equations.

Definition 4.2.1. *We say that an element $u \in W_0^{1,2}(\Omega, \omega)$ is weak solution of problem (P8) if*

$$\sum_{i,j=1}^{N} \int_\Omega a_{ij}(x) D_i u(x) D_j \varphi(x)\, dx - \int_\Omega \gamma\, u(x) g_1(x)\, \varphi(x)\, dx$$
$$+ \int_\Omega h(u(x))\, g_2(x)\, \varphi(x)\, dx = \int_\Omega f(x)\, \varphi(x)\, dx,$$

for every $\varphi \in W_0^{1,2}(\Omega, \omega)$.

Theorem 4.2.1. *Suppose that*
(H1) *The function $h : \mathbb{R} \to \mathbb{R}$ is Lipschitz continuous (i.e., there exists a constant $C_h > 0$ such that $|h(t_1) - h(t_2)| \le C_h |t_1 - t_2|$ for all $t_1, t_2 \in \mathbb{R}$) and $h(0) = 0$;*
(H2) *$\omega \in A_2$;*
(H3) *$g_1/\omega \in L^\infty(\Omega)$, $g_2/\omega \in L^\infty(\Omega)$ and $f/\omega \in L^2(\Omega, \omega) \cap L^2(\Omega, \omega^3)$;*
(H4) *$\gamma > 0$ is not an eigenvalue of the linearized problem*

$$(LP)\begin{cases} Lu(x) - \gamma\, u(x)\, \omega(x) = 0 \ \text{ in } \ \Omega, \\ u(x) = 0 \ \text{ on } \ \partial\Omega. \end{cases}$$

(**H5**) *The constant*

$$M = \lambda - \gamma\, C_\Omega^2 \|g_1/\omega\|_{L^\infty(\Omega)} - C_h\, C_\Omega^2 \|g_2/\omega\|_{L^\infty(\Omega)} > 0,$$

(with C_Ω as in Theorem 2.1.5).
Then the problem (P8) has a unique solution $u \in W_0^{1,2}(\Omega, \omega)$ and there exists a constant $C > 0$ such that

$$\|u\|_{W_0^{1,2}(\Omega,\omega)} \le C \left\| \frac{f}{\omega} \right\|_{L^2(\Omega,\omega)}. \tag{4.30}$$

Moreover, u is the weak limit in $W_0^{1,2}(\Omega, \tilde{\omega}_1)$ of a sequence of solutions $u_m \in W_0^{1,2}(\Omega, \omega_m)$ of the problems

$$(P_m) \begin{cases} L_m u_m(x) - \gamma\, u_m(x)\, g_{1m}(x) + h(u_m(x))\, g_{2m}(x) \\ = f_m(x) \ \text{ in } \ \Omega, \\ u_m(x) = 0 \ \text{ on } \ \partial\Omega, \end{cases}$$

with $L_m u_m = -\sum_{i,j=1}^{N} D_j(a_{ij}^{mm}(x) D_i u_m(x))$, $g_{1m} = g_1 \omega_m/\omega$, $g_{2m} = g_2 \omega_m/\omega$, $f_m = f(\omega/\omega_m)^{-1/2}$ and $\omega_m = \omega_{mm}$ (where ω_{mm}, a_{ij}^{mm} and $\tilde{\omega}_1$ are as Lemma 2.2.1 and Lemma 3.1.2).

Proof. **Step 1 Existence.** The existence of solution $u \in W_0^{1,2}(\Omega, \omega)$ for the problem (P8) has been demonstrated in Theorem 4.1.4. In particular, for $\varphi = u$ in Definition 4.2.1, we have

$$\sum_{i,j=1}^{N} \int_\Omega a_{ij}(x) D_i u(x) D_j u(x)\, dx - \int_\Omega \gamma\, u^2(x) g_1(x)\, dx + \int_\Omega h(u(x))\, g_2(x)\, u(x)\, dx = \int_\Omega f(x)\, u(x)\, dx. \tag{4.31}$$

(i) By (4.2) we have

$$\sum_{i,j=1}^{N} \int_\Omega a_{ij}(x) D_i u(x) D_j u(x)\, dx \ge \lambda \int_\Omega |\nabla u|^2 \omega\, dx.$$

(ii) By (H3) and Theorem 2.1.5 (with $p = 2$ and $\theta = 1$) we obtain

$$\begin{aligned}\left|\int_\Omega \gamma\, u^2 g_1\, dx\right| &\le \gamma \int_\Omega u^2 \frac{|g_1|}{\omega}\,\omega\, dx\\ &\le \gamma\, \|g_1/\omega\|_{L^\infty(\Omega)} \int_\Omega u^2\,\omega\, dx\\ &\le \gamma\, C_\Omega^2\, \|g_1/\omega\|_{L^\infty(\Omega)} \int_\Omega |\nabla u|^2\,\omega\, dx,\end{aligned}$$

and

$$\begin{aligned}\left|\int_\Omega f\, u\, dx\right| &\le \int_\Omega \frac{|f|}{\omega}|u|\,\omega\, dx\\ &\le \|f/\omega\|_{L^2(\Omega,\omega)}\|u\|_{L^2(\Omega,\omega)}\\ &\le C_\Omega \|f/\omega\|_{L^2(\Omega,\omega)}\|\, |\nabla u|\,\|_{L^2(\Omega,\omega)}.\end{aligned}$$

(iii) By (H1), since $h(0) = 0$, then $|h(t)| \le C_h |t|$ for all $t \in \mathbb{R}$. By (H3) and Theorem 2.1.5, we obtain

$$\begin{aligned}\left|\int_\Omega h(u)\, u\, g_2\, dx\right| &\le \int_\Omega |h(u)||u| \frac{|g_2|}{\omega}\,\omega\, dx\\ &\le C_h \|g_2/\omega\|_{L^\infty(\Omega)} \int_\Omega |u|^2 \omega\, dx\\ &\le C_h\, C_\Omega^2 \|g_2/\omega\|_{L^\infty(\Omega} \int_\Omega |\nabla u|^2\,\omega\, dx.\end{aligned}$$

Hence, in (4.31), we obtain

$$\begin{aligned}&\lambda \int_\Omega |\nabla u|^2 \omega\, dx - \gamma\, C_\Omega^2 \|g_1/\omega\|_{L^\infty(\Omega)} \int_\Omega |\nabla u|^2\,\omega\, dx\\ &- C_h\, C_\Omega^2\, \|g_2/\omega\|_{L^\infty(\Omega)} \int_\Omega |\nabla u|^2 \omega\, dx\\ &\le C_\Omega\, \|f/\omega\|\, \|\, |\nabla u|\, \|_{L^2(\Omega,\omega)}.\end{aligned}$$

Therefore,

$$\|\, |\nabla u|\, \|_{L^2(\Omega,\omega)} \le \frac{C_\Omega}{M} \|f/\omega\|_{L^2(\Omega,\omega)}$$

where $M = \lambda - \gamma\, C_\Omega^2 \|g_1/\omega\|_{L^\infty(\Omega)} - C_h C_\Omega^2 \|g_2/\omega\|_{l^\infty(\Omega)} > 0$. Consequently, we obtain

$$\begin{aligned}\|u\|^2_{W_0^{1,2}(\Omega,\omega)} &= \int_\Omega u^2\,\omega\,dx + \int_\Omega |\nabla u|^2\,\omega\,dx\\ &\leq (C_\Omega^2+1)\int_\Omega |\nabla u|^2\,\omega\,dx\\ &\leq (C_\Omega^2+1)\frac{C_\Omega^2}{M^2}\,\|f/\omega\|^2_{L^2(\Omega,\omega)}.\end{aligned}$$

Therefore,

$$\begin{aligned}\|u\|_{W_0^{1,2}(\Omega,\omega)} &\leq (C_\Omega^2+1)^{1/2}\frac{C_\Omega}{M}\,\|f/\omega\|_{L^2(\Omega,\omega)}\\ &= C\,\|f/\omega\|_{L^2(\Omega,\omega)}.\end{aligned} \tag{4.32}$$

Step 2 Uniqueness. If $u_1, u_2 \in W_0^{1,2}(\Omega,\omega)$ are solutions of the problem $(P8)$, then

$$\sum_{i,j=1}^N \int_\Omega a_{ij} D_i u_k\, D_j\varphi\,dx - \gamma\int_\Omega u_k\,g_1\,\varphi\,dx + \int_\Omega h(u_k)\,g_2\,\varphi\,dx$$
$$= \int_\Omega f\,\varphi\,dx \ (k=1,2).$$

Hence,

$$\sum_{i,j=1}^N \int_\Omega a_{ij}(D_i u_1 - D_i u_2)\,D_j\varphi\,dx \;-\; \gamma\int_\Omega (u_1-u_2)\,g_1\,\varphi\,dx$$
$$+ \int_\Omega (h(u_1)-h(u_2))\,g_2\,\varphi\,dx = 0, \tag{4.33}$$

for all $\varphi \in W_0^{1,2}(\Omega,\omega)$. In particular, for $\varphi = u_1 - u_2$ in Definition 4.2.1 we obtain:

(i) By (4.2),

$$\sum_{i,j=1}^{N} \int_{\Omega} a_{ij} D_i(u_1 - u_2) D_j(u_1 - u_2)\, dx \geq \lambda \int_{\Omega} |\nabla(u_1 - u_2)|^2\, \omega\, dx.$$

(ii) By (H3) and Theorem 2.1.5 (with $p = 2$ and $\theta = 1$),

$$\begin{aligned} \left| \int_{\Omega} (u_1 - u_2)^2\, g_1\, dx \right| &\leq \int_{\Omega} |u_1 - u_2|^2 \frac{|g_1|}{\omega}\, \omega\, dx \\ &\leq \|g_1/\omega\|_{L^\infty(\Omega)} \int_{\Omega} |u_1 - u_2|^2\, \omega\, dx \\ &\leq C_{\Omega}^2\, \|g_1/\omega\|_{L^\infty(\Omega)} \int_{\Omega} |\nabla(u_1 - u_2)|^2\, \omega\, dx. \end{aligned}$$

(iii) By (H1), (H3) and Theorem 2.1.5,

$$\begin{aligned} &\left| \int_{\Omega} (h(u_1) - h(u_2))\, (u_1 - u_2)\, g_2\, dx \right| \\ &\leq \int_{\Omega} |h(u_1) - h(u_2)|\, |u_1 - u_2|\, \frac{|g_2|}{\omega}\, \omega\, dx \\ &\leq C_h \int_{\Omega} |u_1 - u_2|^2 \frac{|g_2|}{\omega}\, \omega\, dx \\ &\leq C_h\, C_{\Omega}^2\, \|g_2/\omega\|_{L^\infty(\Omega)} \int_{\Omega} |\nabla(u_1 - u_2)|^2\, \omega\, dx. \end{aligned}$$

Hence,

$$\left(\lambda - \gamma\, C_{\Omega}^2 \|g_1/\omega\|_{L^\infty(\Omega)} - C_h\, C_{\Omega}^2 \|g_2/\omega\|_{L^\infty(\Omega)} \right) \int_{\Omega} |\nabla(u_1 - u_2)|^2\, \omega\, dx \leq 0,$$

and since $M = \lambda - \gamma\, C_{\Omega}^2 \|g_1/\omega\|_{L^\infty(\Omega)} - C_h\, C_{\Omega}^2 \|g_2/\omega\|_{L^\infty(\Omega)} > 0$, then

$$\int_{\Omega} |\nabla(u_1 - u_2)|^2\, \omega\, dx = 0,$$

and $\displaystyle\int_\Omega |u_1 - u_2|^2\,\omega\,dx = 0$. Therefore, $u_1 = u_2$ a.e. in Ω.

Step 3. First, if $f_m = f(\omega/\omega_m)^{-1/2}$, $g_{1m} = g_1\omega_m/\omega$ and $g_{2m} = g_2\,\omega_m/\omega$, we note that

$$\left\|\frac{f_m}{\omega_m}\right\|_{L^2(\Omega,\omega_m)} = \left\|\frac{f}{\omega}\right\|_{L^2(\Omega,\omega)}, \quad \left\|\frac{g_{1m}}{\omega_m}\right\|_{L^\infty(\Omega)} = \left\|\frac{g_1}{\omega}\right\|_{L^\infty(\Omega)},$$

$$\text{and}\quad \left\|\frac{g_{2m}}{\omega_m}\right\|_{L^\infty(\Omega)} = \left\|\frac{g_2}{\omega}\right\|_{L^\infty(\Omega)}.$$

Then, we have

$$\begin{aligned} M_m &= \lambda - \gamma\,C_\Omega^2\|g_{1m}/\omega_m\|_{L^\infty(\Omega)} - C_h\,C_\Omega^2\|g_{2m}\omega_m\|_{L^\infty(\Omega)} \\ &= \lambda - \gamma\,C_\Omega^2\|g_1/\omega\|_{L^\infty(\Omega)} - C_h\,C_\Omega^2\|g_2/\omega\|_{L^\infty(\Omega)} = M > 0. \end{aligned}$$

If $u_m \in W_0^{1,2}(\Omega,\omega_m)$ is a solution of problem (P_m) we have (by (4.32))

$$\begin{aligned} \|u_m\|_{W_0^{1,2}(\Omega,\omega_m)} &\le \frac{(C_\Omega^2+1)^{1/2}C_\Omega}{M_m}\,\|f_m/\omega_m\|_{L^2(\Omega,\omega_m)} \\ &= \frac{(C_\Omega^2+1)^{1/2}C_\Omega}{M}\,\|f/\omega\|_{L^2(\Omega,\omega)} = C_1. \end{aligned}$$

Using Lemma 2.2.1, $\tilde{\omega}_1 \le \omega_m$, we obtain

$$\|u_m\|_{W_0^{1,2}(\Omega,\tilde{\omega}_1)} \le \|u_m\|_{W_0^{1,2}(\Omega,\omega_m)} \le C_1. \tag{4.34}$$

Consequently, $\{u_m\}$ is a bounded sequence in $W_0^{1,2}(\Omega,\tilde{\omega}_1)$. Therefore, there is a subsequence, again denoted by $\{u_m\}$, and $\tilde{u} \in W_0^{1,2}(\Omega,\tilde{\omega}_1)$ such that

$$u_m \rightharpoonup \tilde{u} \text{ in } L^2(\Omega,\tilde{\omega}_1), \tag{4.35}$$

$$\nabla u_m \rightharpoonup \nabla\tilde{u} \text{ in } L^2(\Omega,\tilde{\omega}_1), \tag{4.36}$$

$$u_m \to \tilde{u} \text{ a.e. in } \Omega, \tag{4.37}$$

where the symbol "$\rightharpoonup$" denotes weak convergence (see Theorem 1.4.9).

Step 4. We have that $\tilde{u}\in W_0^{1,2}(\Omega,\omega)$. In fact, for F_k fixed, we have by (4.35) and (4.36), for all $\varphi\in W_0^{1,2}(\Omega,\tilde{\omega}_1)$,

$$\int_\Omega u_m\varphi\,\tilde{\omega}_1\,dx \to \int_\Omega \tilde{u}\,\varphi\,\tilde{\omega}_1\,dx,$$

$$\int_\Omega D_iu_mD_i\varphi\,\tilde{\omega}_1\,dx \to \int_\Omega D_i\tilde{u}\,D_i\varphi\,\tilde{\omega}_1\,dx.$$

If $\psi\in W_0^{1,2}(\Omega,\omega)$, then $\varphi=\psi\,\chi_{F_k}\in W_0^{1,2}(\Omega,\tilde{\omega_1})$ (since $\omega\approx\tilde{\omega}_1$ in F_k, i.e., there is a constant $c>0$ such that $\tilde{\omega}_1\le c\,\omega$ in F_k, and χ_E denotes the characteristic function of a measurable set $E\subset\mathbb{R}^n$) and

$$\int_\Omega \varphi^2\,\tilde{\omega}_1\,dx = \int_{F_k}\psi^2\,\tilde{\omega}_1\,dx\le c\int_{F_k}\psi^2\omega\,dx\le c\int_\Omega \psi^2\,\omega\,dx<\infty,$$

$$\begin{aligned}\int_\Omega (D_i\varphi)^2\,\tilde{\omega}_1\,dx &= \int_{F_k}(D_i\psi)^2\tilde{\omega}_1\,dx\\ &\le c\int_{F_k}(D_i\psi)^2\omega\,dx\\ &\le c\int_\Omega (D_i\psi)^2\omega\,dx<\infty.\end{aligned}$$

Consequently,

$$\int_\Omega u_m\psi\chi_{F_k}\,\tilde{\omega}_1\,dx\to\int_\Omega \tilde{u}\,\psi\,\chi_{F_k}\,\tilde{\omega}_1\,dx,$$

$$\int_\Omega D_iu_mD_i\psi\,\chi_{F_k}\,\tilde{\omega}_1\,dx\to\int_\Omega D_i\tilde{u}\,D_i\psi\,\chi_{F_k}\,\tilde{\omega}_1\,dx,$$

for all $\psi\in W_0^{1,2}(\Omega,\omega)$, that is, the sequence $\{u_m\,\chi_{F_k}\}$ is weakly convergent in $W_0^{1,2}(\Omega,\omega)$. Therefore, we have

$$\|\,|\nabla\tilde{u}|\,\|^2_{L^2(F_k,\omega)}=\int_{F_k}|\nabla\tilde{u}|^2\omega\,dx\le\limsup_{m\to\infty}\int_{F_k}|\nabla u_m|^2\,\omega\,dx,$$

and for $m \geq k$ we have $\omega = \omega_m$ in F_k. Hence, by (4.34), we obtain

$$\begin{aligned}\| \, |\nabla \tilde{u}| \, \|^2_{L^2(F_k,\omega)} &\leq \limsup_{m\to\infty} \int_{F_k} |\nabla u_m|^2 \, \omega \, dx \\ &= \limsup_{m\to\infty} \int_{F_k} |\nabla u_m|^2 \, \omega_m \, dx \\ &\leq \limsup_{m\to\infty} \int_{\Omega} |\nabla u_m|^2 \omega_m \, dx \leq C_1^2.\end{aligned}$$

By the Monotone Convergence Theorem we obtain
$\| \, |\nabla \tilde{u}| \, \|_{L^2(\Omega,\omega)} \leq C_1$. Therefore, we have $\tilde{u} \in W_0^{1,2}(\Omega,\omega)$.
Step 5. We need to show that $\tilde{u}$ is a solution of problem $(P8)$, i.e, for every $\varphi \in W_0^{1,2}(\Omega,\omega)$ we have

$$\begin{aligned}&\sum_{i,j=1}^{N} \int_\Omega a_{ij}(x) D_i\tilde{u}(x) D_j\varphi(x)\, dx - \int_\Omega \gamma\, \tilde{u}(x) g_1(x)\, \varphi(x)\, dx \\ &+ \int_\Omega h(\tilde{u}(x))\, g_2(x)\, \varphi(x)\, dx = \int_\Omega f(x)\, \varphi(x)\, dx.\end{aligned}$$

Using the fact that u_m is a solution of (P_m), we have

$$\begin{aligned}&\sum_{i,j=1}^{N} \int_\Omega a_{ij}^{mm}(x) D_i u_m(x) D_j\varphi(x)\, dx - \int_\Omega \gamma\, u_m(x) g_{1m}(x)\, \varphi(x)\, dx \\ &+ \int_\Omega h(u_m(x))\, g_{2m}(x)\, \varphi(x)\, dx \\ &= \int_\Omega f_m(x)\, \varphi(x)\, dx,\end{aligned}$$

for every $\varphi \in W_0^{1,2}(\Omega,\omega_m)$. Moreover, over F_k (for $m \geq k$) we have the following properties:
(i) $\omega = \omega_m$;
(ii) $g_{1m} = g_1$;
(iii) $g_{2m} = g_2$;

(iv) $f_m = f$;
(v) $a_{ij}^{mm}(x) = a_{ij}(x)$.
For $\varphi \in W_0^{1,2}(\Omega, \omega)$ and $k > 0$(fixed), we define
$G_1, G_2 : W_0^{1,2}(\Omega, \tilde{\omega}_1) \to \mathbb{R}$ by

$$\begin{aligned} G_1(u) &= \sum_{i,j=1}^{N} \int_\Omega a_{ij}(x) D_i u(x)\, D_j \varphi(x)\, \chi_{F_k}\, dx \\ &- \gamma \int_\Omega u(x) g_1(x) \varphi(x)\, \chi_{F_k}(x)\, dx, \\ G_2(u) &= \int_\Omega h(u(x))\, g_2(x) \varphi(x)\, \chi_{F_k}(x)\, dx. \end{aligned}$$

(a) We have that G_1 is linear and continuous functional. In fact, since the matrix $\mathcal{A} = (a_{ij})$ is symmetric, we have $|\langle \mathcal{A}\nabla u, \nabla\varphi\rangle| \leq \langle \mathcal{A}\nabla u, \nabla u\rangle^{1/2} \langle \mathcal{A}\nabla\varphi, \nabla\varphi\rangle^{1/2}$, where $\langle .,. \rangle$ denotes here the Euclidean scalar product in $\mathbb{R}^N$. We also have $\omega \approx \tilde{\omega}_1$ in F_k $(\omega \leq c\,\tilde{\omega}_1)$. By (4.2) and (H3) we obtain

$$\begin{aligned} |G_1(u)| &\leq \int_{F_k} |\langle \mathcal{A}\nabla u, \nabla\varphi\rangle|\, dx + \gamma \int_{F_k} |u|\, |g_1|\, |\varphi|\, dx \\ &\leq \int_{F_k} \langle \mathcal{A}\nabla u, \nabla u\rangle^{1/2} \langle \mathcal{A}\nabla\varphi, \nabla\varphi\rangle^{1/2}\, dx + \gamma \int_{F_k} |u| \frac{|g_1|}{\omega}\, |\varphi|\, \omega\, dx \\ &\leq \left(\int_{F_k} \langle \mathcal{A}\nabla u, \nabla u\rangle\, dx \right)^{1/2} \left(\int_{F_k} \langle \mathcal{A}\nabla\varphi, \nabla\varphi\rangle^{1/2}\, dx \right)^{1/2} \\ &+ \gamma \|g_1/\omega\|_{L^\infty(\Omega)} \left(\int_{F_k} |u|^2 \omega\, dx \right)^{1/2} \left(\int_{F_k} |\varphi|^2\, \omega \right)^{1/2} \\ &\leq \Lambda \left(\int_{F_k} |\nabla u|^2\, \omega\, dx \right)^{1/2} \left(\int_\Omega |\nabla\varphi|^2\, \omega\, dx \right)^{1/2} \\ &+ \gamma \left(\int_{F_k} |u|^2\, \omega\, dx \right)^{1/2} \left(\int_\Omega |\varphi|^2 \omega\, dx \right)^{1/2} \end{aligned}$$

$$\le \Lambda\left(\int_{F_k} c\,|\nabla u|^2\tilde{\omega}_1\,dx\right)^{1/2}\left(\int_\Omega |\nabla\varphi|^2\,\omega\,dx\right)^{1/2}$$
$$+\gamma\left(\int_{F_k} c\,|u|^2\tilde{\omega}_1\,dx\right)^{1/2}\left(\int_\Omega |\varphi|^2\omega\,dx\right)^{1/2}$$
$$\le (\Lambda\,c^{1/2}+\gamma\,c^{1/2})\|\varphi\|_{W_0^{1,2}(\Omega,\omega)}\,\|u\|_{W_0^{1,2}(\Omega,\tilde{\omega}_1)}.$$

(b) We have that G_2 is continuous functional. In fact, if $u_1, u_2 \in W_0^{1,2}(\Omega,\tilde{\omega}_1)$, we obtain by (H1) and (H3)

$$|G_2(u_2)-G_2(u_1)|\le \int_{F_k} |h(u_2)-h(u_1)|\,|g_2|\,|\varphi|\,dx$$
$$\le \int_{F_k} C_h\,|u_1-u_2|\frac{|g_1|}{\omega}\,|\varphi|\,\omega\,dx$$
$$\le C_h\,\|g_2/\omega\|_{L^\infty(\Omega)}\left(\int_\Omega |\varphi|^2\omega\,dx\right)^{1/2}\left(\int_{F_k} c\,|u_1-u_2|^2\,\tilde{\omega}_1\,dx\right)^{1/2}$$
$$\le c^{1/2}\,C_h\,\|g_2/\omega\|_{L^\infty(\Omega)}\|\varphi\|_{W_0^{1,2}(\Omega,\omega)}\|u_1-u_2\|_{W_0^{1,2}(\Omega,\tilde{\omega}_1)}.$$

Using (a), (b), properties (i),(ii), (iii), (iv) and (v), and that u_m is solution of (P_m), we obtain

$$\sum_{i,j=1}^N\int_{F_k} a_{ij}\,D_i\tilde{u}\,D_j\varphi\,dx-\gamma\int_{F_k}\tilde{u}\,g_1\,\varphi\,dx+\int_{F_k} h(\tilde{u})\,g_2\,\varphi\,dx$$
$$=\lim_{m\to\infty}\,[G_1(u_m)+G_2(u_m)]$$
$$=\lim_{m\to\infty}\left(\sum_{i,j=1}^N\int_{F_k} a_{ij}^{mm}\,D_iu_m\,D_j\varphi\,dx-\gamma\int_{F_k} u_m\,g_{1m}\,\varphi\,dx\right.$$
$$\left.+\int_{F_k} h(u_m)\,g_{2m}\,\varphi\,dx\right)$$
$$=\lim_{m\to\infty}\left(\sum_{i,j=1}^N\int_\Omega a_{ij}^{mm}\,D_iu_m\,D_j\varphi\,dx-\gamma\int_\Omega u_m\,g_{1m}\,\varphi\,dx\right.$$

$$
+\int_{\Omega} h(u_m)\, g_{2m}\, \varphi\, dx - \sum_{i,j=1}^{N} \int_{\Omega\cap F_k^c} a_{ij}^{mm}\, D_i u_m\, D_j\varphi\, dx
$$

$$
+\gamma \int_{\Omega\cap F_k^c} u_m\, g_{1m}\, \varphi\, dx - \int_{\Omega\cap F_k^c} h(u_m)\, g_{2m}\, \varphi\, dx\Big)
$$

$$
= \lim_{m\to\infty} \Big(\int_{\Omega} f_m\, \varphi\, dx - \sum_{i,j=1}^{N} \int_{\Omega\cap F_k^c} a_{ij}^{mm}\, D_i u_m\, D_j\varphi\, dx
$$

$$
+\gamma \int_{\Omega\cap F_k^c} u_m\, g_{1m}\, \varphi\, dx - \int_{\Omega\cap F_k^c} h(u_m)\, g_{2m}\, \varphi\, dx\Big), \tag{4.38}
$$

where E^c denotes the complement of a set $E\subset\mathbb{R}^n$.
(I) By the Lebesgue Dominated Convergence Theorem we obtain (an analogous argument in Theorem 3.1.3)

$$
\int_{\Omega} f_m\, \varphi\, dx \to \int_{\Omega} f\, \varphi\, dx. \tag{4.39}
$$

(II) Since the matrix $\mathcal{A}^m = (a_{ij}^{mm})$ is symmetric, we have

$$
|\langle \mathcal{A}^m \nabla u_m, \nabla\varphi\rangle| \le \langle \mathcal{A}^m \nabla u_m, \nabla u_m\rangle^{1/2} \langle \mathcal{A}^m \nabla\varphi, \nabla\varphi\rangle^{1/2}.
$$

Then, by (4.2) and (4.34), we obtain

$$
\sum_{i,j=1}^{N} \left| \int_{\Omega\cap F_k^c} a_{ij}^{mm} D_i u_m D_j\varphi\, dx \right| \le \int_{\Omega\cap F_k^c} |\langle \mathcal{A}^m \nabla u_m, \nabla\varphi\rangle|\, dx
$$

$$
\le \Lambda \left(\int_{\Omega\cap F_k^C} |\nabla u_m|^2 \omega_m\, dx \right)^{1/2} \left(\int_{\Omega\cap F_k^c} |\nabla\varphi|^2 \omega_m\, dx \right)^{1/2}
$$

$$
\le \Lambda \|u_m\|_{W_0^{1,2}(\Omega,\omega_m)} \left(\int_{\Omega\cap F_k^c} |\nabla\varphi|^2\, w_m\, dx \right)^{1/2}
$$

$$
\le \Lambda\, C_1 \left(\int_{\Omega\cap F_k^c} |\nabla\varphi|^2\, w_m\, dx \right)^{1/2}. \tag{4.40}
$$

(III) By (H3) and (4.34), we obtain

$$
\begin{aligned}
\left|\int_{\Omega\cap F_k^c} u_m g_{1m}\varphi\,dx\right| &\leq \int_{\Omega\cap F_k^c} |u_m|\frac{|g_{1m}|}{\omega_m}|\varphi|\,\omega_m\,dx\\
&\leq \|g_{1m}/\omega_m\|_{L^\infty(\Omega)}\left(\int_\Omega |u_m|^2\omega_m\,dx\right)^{1/2}\left(\int_{\Omega\cap F_k^c}|\varphi|^2\omega_m\,dx\right)^{1/2}\\
&\leq \|g_1/\omega\|_{L^\infty(\Omega)}\|u_m\|_{W_0^{1,2}(\Omega,\omega_m)}\left(\int_{\Omega\cap F_k^c}|\varphi|^2\,\omega_m\,dx\right)^{1/2}\\
&\leq C_1\,\|g_1/\omega\|_{L^\infty(\Omega)}\left(\int_{\Omega\cap F_k^c}|\varphi|^2\,\omega_m\,dx\right)^{1/2}. \qquad (4.41)
\end{aligned}
$$

(IV) By (H1), (H3) and $|h(t)|\leq C_h|t|$ (for all $t\in\mathbb{R}$) we have

$$
\begin{aligned}
\left|\int_{\Omega\cap F_k^c} h(u_m)g_{2m}\varphi\,dx\right| &\leq \int_{\Omega\cap F_k^c} |h(u_m)|\frac{|g_{2m}|}{\omega_m}\,|\varphi|\,\omega_m\,dx\\
&\leq C_h\,\|g_{2m}/\omega_m\|_{L^\infty(\Omega)}\int_{\Omega\cap F_k^c}|u_m||\varphi|\,\omega_m\,dx\\
&\leq C_h\|g_2/\omega\|_{L^\infty(\Omega)}\left(\int_{\Omega\cap F_k^c}|u_m|^2\omega_m\,dx\right)^{1/2}\left(\int_{\Omega\cap F_k^c}|\varphi|^2\omega_m\,dx\right)^{1/2}\\
&\leq C_h\|g_2/\omega\|_{L^\infty(\Omega)}\,\|u_m\|_{W_0^{1,2}(\Omega,\omega_m)}\left(\int_{\Omega\cap F_k^c}|\varphi|^2\omega_m\,dx\right)^{1/2}\\
&\leq C_h\,C_1\,\|g_2/\omega\|_{L^\infty(\Omega)}\left(\int_{\Omega\cap F_k^c}|\varphi|^2\omega_m\,dx\right)^{1/2}. \qquad (4.42)
\end{aligned}
$$

Using Lemma 2.2.1, we know that $|\Omega\cap F_k^c|\rightarrow 0$ when $k\rightarrow\infty$. Then

$$
\lim_{k\to\infty}\left(\int_{\Omega\cap F_k^c}|\varphi|^2\omega_m\,dx\right)^{1/2}=0
$$

$$
\lim_{k\to\infty}\left(\int_{\Omega\cap F_k^c}|\nabla\varphi|^2\omega_m\,dx\right)^{1/2}=0,
$$

and we obtain in (4.40), (4.41) and (4.42)

$$\lim_{k\to\infty}\int_{\Omega\cap F_k^c} a_{ij}^{mm}(x)D_iu(x)\,D_j\varphi(x)\,dx=0, \tag{4.43}$$

$$\lim_{k\to\infty}\int_{\Omega\cap F_k^c} u_m\,g_{1m}\,\varphi\,dx=0, \tag{4.44}$$

$$\lim_{k\to\infty}\int_{\Omega\cap F_k^c} h(u_m(x))\,g_{2m}(x)\,\varphi(x)\,dx=0. \tag{4.45}$$

Therefore, by (4.38), (4.43), (4.44) and (4.45) we conclude, when $k\to\infty$ (and $m\geq k$),

$$\sum_{i,j=1}^{N}\int_\Omega a_{ij}D_i\tilde{u}\,D_j\varphi(x)\,dx-\gamma\int_\Omega g_1\tilde{u}\varphi\,dx+\int_\Omega h(\tilde{u})\,g_2\,\varphi\,dx$$
$$=\int_\Omega f\,\varphi\,dx,$$

for all $\varphi\in W_0^{1,2}(\Omega,\omega)$, that is, $\tilde{u}$ is a solution of problem (P8). Therefore, $u=\tilde{u}$ (by the uniqueness). □

Example 4.2.1. *Let* $\Omega=\{(x,y)\in\mathbb{R}^2: x^2+y^2<1\}$ *and* $0<a<b$. *By Theorem 4.2.1, with* $h(t)=\sin(t)$, $f(x,y)=x|y|$, $\omega(x,y)=(x^2+y^2)^{-1/2}$, $g_1(x,y)=(x^2+y^2)^{-1/2}\cos(xy)$, $g_2(x,y)=(x^2+y^2)^{-1/4}\sin(xy)$ *and*

$$\mathcal{A}(x,y)=\begin{pmatrix} a(x^2+y^2)^{-1/2} & 0\\ 0 & b(x^2+y^2)^{-1/2}\end{pmatrix},$$

the problem

$$\begin{cases} Lu(x)-\gamma\,u(x)\,g_1(x)+h(u(x))\,g_2(x)=f(x) \ \text{ in } \ \Omega,\\ u(x)=0 \ \text{ on } \ \partial\Omega,\end{cases}$$

where

$$Lu(x)=-\frac{\partial}{\partial x}\left(a(x^2+y^2)^{-1/2}\frac{\partial u}{\partial x}\right)-\frac{\partial y}{\partial x}\left(b(x^2+y^2)^{-1/2}\,\frac{\partial u}{\partial y}\right),$$

has a unique solution $u \in W_0^{1,2}(\Omega, \omega)$ *if* $\gamma > 0$ *is not an eigenvalue of linearized problem (LP), and* u *can be approximated by a sequence of solutions for non-degenerate semilinear elliptic equations.*

Chapter 5

Degenerate Quasilinear Elliptic Equations

In this chapter we prove the existence and uniqueness of (weak) solution for three Dirichlet problems associated to some degenerate quasilinear elliptic equations and the existence of entropy solutions. For this purpose we will use some results from the Theory of Monotone Operators.

5.1 Existence and uniqueness of solutions

In this section we prove the existence of (weak) solutions in the weighted Sobolev spaces $W_0^{1,p}(\Omega, \omega)$ (see Definition 5.1.1) for the Dirichlet problem

$$(P9)\left\{\begin{array}{ll} Lu(x) & = f(x) \ \text{ in } \ \Omega, \\ u(x) & = 0 \ \text{ on } \ \partial\Omega, \end{array}\right.$$

where L is the partial differential operator

$$Lu(x) = -\sum_{j=1}^{N} D_j\big[\omega(x)\mathcal{A}_j(x, u(x), \nabla u(x))\big], \qquad (5.1)$$

where ω is a weight function, Ω is a bounded open set in $\mathbb{R}^N$ and $\mathcal{A}_j : \Omega\times\mathbb{R}\times\mathbb{R}^N \to \mathbb{R}$ are given functions.

Definition 5.1.1. *Let $1 < p < \infty$. We say that $u\in W_0^{1,p}(\Omega,\omega)$ is a (weak) solution of problem (P9) if*

$$\sum_{j=1}^{N}\int_\Omega \mathcal{A}_j(x,u,\nabla u)\, D_j\varphi\, \omega\, dx = \int_\Omega f\,\varphi\, dx$$

for all $\varphi\in W_0^{1,p}(\Omega,\omega)$.

In this chapter we use frequently the following theorem.

Theorem 5.1.1. *Let $A : X \to X^*$ be a monotone, coercive and hemicontinuous operator on the real, separable, reflexive Banach space X. Then the following assertions hold:*
(a) For each $T\in X^*$ the equations $Au = T$ has a solution $u\in X$.
(b) If the operator A is strictly monotone, then equation $Au = T$ is uniquely solvable in X.

Proof. See Theorem 26.A in [61]. □

Then we can show the following result.

Theorem 5.1.2. *Suppose that the following four conditions hold:*
(H1) *$x \mapsto \mathcal{A}_j(x,\eta,\xi)$ is measurable on Ω for all $(\eta,\xi)\in\mathbb{R}\times\mathbb{R}^N$*
$(\eta,\xi) \mapsto \mathcal{A}_j(x,\eta,\xi)$ is continuous on $\mathbb{R}\times\mathbb{R}^N$ for almost all $x\in\Omega$.
(H2) *$\langle\mathcal{A}(x,\eta,\xi) - \mathcal{A}(x,\eta',\xi'),\xi - \xi'\rangle \geq 0$, whenever $\xi,\xi'\in\mathbb{R}^N$, $\xi\neq\xi'$, where*

$$\mathcal{A}(x,\eta,\xi) = (\mathcal{A}_1(x,\eta,\xi),...,\mathcal{A}_N(x,\eta,\xi)).$$

(H3) *$\langle\mathcal{A}(x,\eta,\xi),\xi\rangle \geq \lambda|\xi|^p - g_1(x)|\eta| - g_2(x)$, $1 < p < \infty$, $g_1\in L^{p'}(\Omega,\omega)$ and $g_2\in L^1(\Omega,\omega)$, $\lambda > 0$.*
(H4) *$|\mathcal{A}(x,\eta,\xi)| \leq K(x) + h_1(x)|\eta|^{p/p'} + h_2(x)|\xi|^{p/p'}$, where K, h_1 and h_2 are positive functions, with $h_1\in L^\infty(\Omega)$, $h_2\in L^\infty(\Omega)$ and $K\in L^{p'}(\Omega,\omega)$.*

If $\omega \in A_p$, $1 < p < \infty$, and $f/\omega \in L^{p'}(\Omega, \omega)$ then problem (P9) has a solution $u \in W_0^{1,p}(\Omega, \omega)$. Moreover, if $g_2 = 0$, we have

$$\|u\|_{W_0^{1,p}(\Omega,\omega)} \leq \left[\left(\frac{C_\Omega^p + 1}{\lambda} \right) \left(\|f/\omega\|_{L^{p'}(\Omega,\omega)} + \|g_1/\omega\|_{L^{p'}(\Omega,\omega)} \right) \right]^{1/(p-1)},$$

where C_Ω is the constant in Theorem 2.1.5.

A differential operator L in (5.1) is called a monotone coercive quasi-linear elliptic differential operators iff the conditions (H1) through (H4) are satisfied. A typical example of operator satisfying the assumptions is

$$Lu = -\operatorname{div}(\omega\, |\nabla u|^{p-2}\, \nabla u),$$

where $\omega \in A_p$.

Proof. The basic idea is to reduce the problem (P9) to an operator equation $Au = T$ and apply the Theorem 5.1.1. We define $B : W_0^{1,p}(\Omega, \omega) \times W_0^{1,p}(\Omega, \omega) \to \mathbb{R}$ and $T : W_0^{1,p}(\Omega, \omega) \to \mathbb{R}$ by

$$\begin{aligned} B(u, \varphi) &= \sum_{j=1}^{N} \int_\Omega \mathcal{A}_j(x, u, \nabla u) D_j \varphi\, \omega\, dx \\ &= \int_\Omega \langle \mathcal{A}(x, u, \nabla u), \nabla \varphi \rangle\, \omega\, dx \\ T(\varphi) &= \int_\Omega f \varphi\, dx. \end{aligned}$$

Then $u \in W_0^{1,p}(\Omega, \omega)$ is a (weak) solution to problem (P9) if

$$B(u, \varphi) = T(\varphi),$$

for all $\varphi \in W_0^{1,p}(\Omega, \omega)$.

Step 1. We define the operator (for $j = 1, ..., N$)

$$F_j : W_0^{1,p}(\Omega,\omega) \to L^{p'}(\Omega,\omega)$$
$$(F_j u)(x) = \mathcal{A}_j(x, u(x), \nabla u(x)).$$

We have that the operator F_j is bounded and continuous. Indeed: (i) Using (H4) and the weighted Sobolev inequality (Theorem 2.1.5 with $\theta = 1$) we obtain

$$\begin{aligned}
\|F_j u\|^{p'}_{L^{p'}(\Omega,\omega)} &= \int_\Omega |F_j u(x)|^{p'} \omega\, dx \\
&= \int_\Omega |\mathcal{A}_j(x, u, \nabla u)|^{p'} \omega\, dx \\
&\le \int_\Omega \left(K + h_1 |u|^{p/p'} + h_2 |\nabla u|^{p/p'} \right)^{p'} \omega\, dx \\
&\le C_1 \int_\Omega (K^{p'} + h_1^{p'} |u|^p + h_2^{p'} |\nabla u|^p) \omega\, dx \\
&= C_1 \left[\int_\Omega K^{p'} \omega\, dx + \int_\Omega h_1^{p'} |u|^p \omega\, dx + \int_\Omega h_2^{p'} |\nabla u|^p \omega\, dx \right] \\
&\le C_1 \left[\|K\|^{p'}_{L^{p'}(\Omega,\omega)} + \|h_1\|^{p'}_\infty \int_\Omega |u|^p \omega\, dx \right. \\
&+ \left. \|h_2\|^{p'}_\infty \int_\Omega |\nabla u|^p \omega\, dx \right] \\
&\le C_1 \left[\|K\|^{p'}_{L^{p'}(\Omega,\omega)} + \|h_1\|^{p'}_\infty C_\Omega \int_\Omega |\nabla u|^p \omega\, dx \right. \\
&+ \left. \|h_2\|^{p'}_\infty \int_\Omega |\nabla u|^p \omega\, dx \right] \\
&= C_1 \left[\|K\|^{p'}_{L^{p'}(\Omega,\omega)} + (C_\Omega \|h_1\|^{p'}_\infty + \|h_2\|^{p'}_\infty) \|u\|^p_{W_0^{1,p}(\Omega,\omega)} \right].
\end{aligned}$$

Therefore

$$\|F_j u\|_{L^{p'}(\Omega,\omega)} \le C_2 \left(\|K\|_{L^{p'}(\Omega,\omega)} + C_3 \|u\|^{p/p'}_{W_0^{1,p}(\Omega,\omega)} \right). \tag{5.2}$$

(ii) Let $u_n \to u$ in $W_0^{1,p}(\Omega,\omega)$ as $n \to \infty$. We need to show that $F_j u_n \to F_j u$ in $L^{p'}(\Omega,\omega)$.
If $u_n \to u$ in $W_0^{1,p}(\Omega,\omega)$, then $u_n \to u$ in $L^p(\Omega,\omega)$ (by Theorem 2.1.5 with $\theta = 1$). Using Theorem 1.4.11, there exist a subsequence $\{u_{n_k}\}$ and functions Φ_1 and Φ_2 in $L^p(\Omega,\omega)$ such that

$$\begin{aligned}
&u_{n_k}(x) \to u(x), \quad \mu - \text{a.e. in } \Omega,\\
&|u_{n_k}(x)| \leq \Phi_1(x), \quad \mu - \text{a.e. in } \Omega,\\
&|\nabla u_{n_k}(x)| \to |\nabla u(x)|, \quad \mu - \text{a.e. in } \Omega,\\
&|\nabla u_{n_k}(x)| \leq \Phi_2(x), \quad \mu - \text{a.e. in } \Omega.
\end{aligned}$$

Hence, we obtain

$$\begin{aligned}
& \|F_j u_{n_k} - F_j u\|^{p'}_{L^{p'}(\Omega,\omega)} \\
= & \int_\Omega |F_j u_{n_k}(x) - F_j u(x)|^{p'} \omega \, dx \\
= & \int_\Omega |\mathcal{A}_j(x, u_{n_k}, \nabla u_{n_k}) - \mathcal{A}_j(x, u, \nabla u)|^{p'} \omega \, dx \\
\leq & \ C \int_\Omega \left(|\mathcal{A}_j(x, u_{n_k}, \nabla u_{n_k})|^{p'} + |\mathcal{A}_j(x, u, \nabla u)|^{p'} \right) \omega \, dx \\
\leq & \ C \Bigg[\int_\Omega \left(K + h_1 |u_{n_k}|^{p/p'} + h_2 |\nabla u_{n_k}|^{p/p'} \right)^{p'} \omega \, dx \\
+ & \int_\Omega \left(K + h_1 |u|^{p/p'} + h_2 |\nabla u|^{p/p'} \right)^{p'} \omega \, dx \Bigg] \\
\leq & \ 2C \int_\Omega \left(K + h_1 \Phi_1^{p/p'} + h_2 \Phi_2^{p/p'} \right)^{p'} \omega \, dx \\
\leq & \ C_p \int_\Omega [K^{p'} + h_1^{p'} \Phi_1^p + h_2^{p'} \Phi_2^p] \omega \, dx \\
\leq & \ C_p \left(\|K\|^{p'}_{L^{p'}(\Omega,\omega)} + \|h_1\|^{p'}_\infty \|\Phi_1\|^p_{L^p(\Omega,\omega)} + \|h_2\|^{p'}_\infty \|\Phi_2\|^p_{L^p(\Omega,\omega)} \right).
\end{aligned}$$

By condition (H1) we have

$$\begin{aligned} F_j u_n(x) &= \mathcal{A}_j(x, u_n(x), \nabla u_n(x)) \\ &\to \mathcal{A}_j(x, u(x), \nabla u(x)) \\ &= F_j u(x), \ \text{as } n \to +\infty. \end{aligned}$$

Therefore, by the Lebesgue Dominated Convergence Theorem, we obtain

$$\|F_j u_{n_k} - F_j u_n\|_{L^{p'}(\Omega,\omega)} \to 0,$$

that is, $F_j u_{n_k} \to F_j u$ in $L^{p'}(\Omega, \omega)$. By the Convergence Principle in Banach spaces (see Proposition 10.13 [60]),we have

$$F_j u_n \to F_j u \ \text{ in } \ L^{p'}(\Omega, \omega). \tag{5.3}$$

Step 2. Using the weighted Sobolev inequality (with $\theta = 1$), we obtain

$$\begin{aligned} |T(\varphi)| &\leq \int_\Omega |f||\varphi|\,dx \\ &= \int_\Omega \frac{|f|}{\omega}|\varphi|\omega\,dx \\ &\leq \left[\int_\Omega \left(\frac{|f|}{\omega}\right)^{p'} \omega\,dx\right]^{1/p'} \left[\int_\Omega |\varphi|^p \omega\,dx\right]^{1/p} \\ &\leq C_\Omega \|f/\omega\|_{L^{p'}(\Omega,\omega)} \|\varphi\|_{W_0^{1,p}(\Omega,\omega)}. \end{aligned}$$

Moreover, by (5.2), we also have

$$\begin{aligned} |B(u,\varphi)| &\leq \sum_{j=1}^N \int_\Omega |\mathcal{A}_j(x,u,\nabla u)||D_j\varphi|\omega\,dx \\ &= \sum_{j=1}^N \int_\Omega |F_j u||D_j\varphi|\omega\,dx \end{aligned}$$

$$
\begin{aligned}
&\leq \sum_{j=1}^{N}\left(\int_{\Omega}|F_j u|^{p'}\omega\,dx\right)^{1/p'}\left(\int_{\Omega}|D_j\varphi|^{p}\omega\,dx\right)^{1/p}\\
&\leq C_{\Omega}\left(\sum_{j=1}^{N}\|F_j u\|_{L^{p'}(\Omega,\omega)}\right)\|\varphi\|_{W_0^{1,p}(\Omega,\omega)}\\
&\leq C_{\Omega}C_2\left(\|K\|_{L^{p'}(\Omega,\omega)}+C_3\|u\|^{p/p'}_{W_0^{1,p}(\Omega,\omega)}\right)\|\varphi\|_{W_0^{1,p}(\Omega,\omega)}.
\end{aligned}
$$

Since $B(u,.)$ is linear, for each $u\in W_0^{1,p}(\Omega,\omega)$, there exists a linear and continuous functional on $W_0^{1,p}(\Omega,\omega)$ denoted by Au such that $(Au|\varphi)=B(u,\varphi)$ for all $\varphi\in W_0^{1,p}(\Omega,\omega)$. Moreover

$$
\|Au\|_{*}\leq C_{\Omega}C_2\left(\|K\|_{L^{p'}(\Omega,\omega)}+C_3\|u\|^{p/p'}_{W_0^{1,p}(\Omega,\omega)}\right).
$$

Hence, we obtain the operator

$$
\begin{aligned}
&A: W_0^{1,p}(\Omega,\omega)\to[W_0^{1,p}(\Omega,\omega)]^{*}\\
&u\mapsto Au.
\end{aligned}
$$

Consequently, problem (P9) is equivalent to the operator equation $Au=T$, $u\in W_0^{1,p}(\Omega,\omega)$.

Step 3. Using condition (H2) we have

$$
\begin{aligned}
&(Au_1-Au_2|u_1-u_2)\\
=\;& B(u_1,u_1-u_2)-B(u_2,u_1-u_2)\\
=\;& \int_{\Omega}\langle\mathcal{A}(x,u_1,\nabla u_1),\nabla(u_1-u_2)\rangle\,\omega\,dx\\
-\;& \int_{\Omega}\langle\mathcal{A}(x,u_2,\nabla u_2),\nabla(u_1-u_2)\rangle\,\omega\,dx\\
=\;& \int_{\Omega}\langle\mathcal{A}(x,u_1,\nabla u_1)-\mathcal{A}(x,u_2,\nabla u_2),\nabla(u_1-u_2)\rangle\,\omega\,dx
\end{aligned}
$$

$$\geq \ 0.$$

Therefore the operator A is monotone. Moreover, using (H3) we obtain

$$\begin{aligned}
(Au|u) &= B(u,u) \\
&= \int_\Omega \langle \mathcal{A}(x,u,\nabla u), \nabla u\rangle \,\omega\, dx \\
&\geq \int_\Omega \Big(\lambda|\nabla u|^p - g_1|u| - g_2\Big)\omega\, dx \\
&= \lambda\int_\Omega |\nabla u|^p\omega\, dx - \int_\Omega g_1|u|\omega\, dx - \int_\Omega g_2\omega\, dx \\
&\geq \lambda\|u\|^p_{W^{1,p}_0(\Omega,\omega)} - \|g_1\|_{L^{p'}(\Omega,\omega)}\|u\|_{L^p(\Omega,\omega)} - \|g_2\|_{L^1(\Omega,\omega)} \\
&\geq \lambda\|u\|^p_{W^{1,p}_0(\Omega,\omega)} - C_\Omega\|g_1\|_{L^{p'}(\Omega,\omega)}\|u\|_{W^{1,p}_0(\Omega,\omega)} \\
&- \|g_2\|_{L^1(\Omega,\omega)}.
\end{aligned}$$

Hence, since $1 < p < \infty$, we have

$$\frac{(Au|u)}{\|u\|_{W^{1,p}_0(\Omega,\omega)}} \to +\infty, \quad \text{as} \quad \|u\|_{W^{1,p}_0(\Omega,\omega)} \to +\infty,$$

that is, A is coercive.

Step 4. We need to show that the operator A is continuous. Let $u_n \to u$ in $W^{1,p}_0(\Omega,\omega)$ as $n\to\infty$. We have,

$$\begin{aligned}
&|B(u_n,\varphi) - B(u,\varphi)| \\
&\leq \sum_{j=1}^N \int_\Omega |\mathcal{A}_j(x,u_n,\nabla u_n) - \mathcal{A}_j(x,u,\nabla u)||D_j\varphi|\omega\, dx \\
&= \sum_{j=1}^N \int_\Omega |F_j u_n - F_j u||D_j\varphi|\omega\, dx \\
&\leq \sum_{j=1}^N \|F_j u_n - F_j u\|_{L^{p'}(\Omega,\omega)}\|\varphi\|_{W^{1,p}_0(\Omega,\omega)},
\end{aligned}$$

for all $\varphi \in W_0^{1,p}(\Omega,\omega)$. Hence,

$$\|Au_n - Au\|_* \leq \sum_{j=1}^{N} \|F_j u_n - F_j\|_{L^{p'}(\Omega,\omega)}.$$

Therefore, using (5.3) we have $\|Au_n - Au\|_* \to 0$ as $n \to +\infty$, that is, A is continuous (and this implies that A is hemicontinuous).
Therefore, by Theorem 5.1.1, the operator equation $Au = T$ has a solution $u \in W_0^{1,p}(\Omega,\omega)$ and it is the solution for problem (P9).

Step 5. Estimate for $\|u\|_{W_0^{1,p}(\Omega)}$.
In Definition 5.1.1. with $\varphi = u \in W_0^{1,p}(\Omega,\omega)$ we have

$$\int_\Omega \langle \mathcal{A}(x,u,\nabla u), \nabla u\rangle \, \omega \, dx = \int_\Omega f \, u \, dx.$$

Then we obtain
(i) since $f/\omega \in L^{p'}(\Omega,\omega)$

$$\begin{aligned} \left| \int_\Omega f \, u \, dx \right| &\leq \int_\Omega \frac{|f|}{\omega} \, |u| \, \omega \, dx \\ &\leq \|f/\omega\|_{L^{p'}(\Omega,\omega)} \|u\|_{L^p(\Omega,\omega)} \\ &\leq \|f/\omega\|_{L^{p'}(\Omega,\omega)} \|u\|_{W_0^{1,p}(\Omega,\omega)}. \end{aligned}$$

(ii) By (H3) (with $g_2 = 0$) and Theorem 2.1.5, we have

$$\begin{aligned} &\int_\Omega \langle \mathcal{A}(x,u,\nabla u), \nabla u\rangle \, \omega \, dx \geq \int_\Omega \big(\lambda |\nabla u|^p - g_1 |u|\big) \, \omega \, dx \\ &= \lambda \int_\Omega |\nabla u|^p \, \omega \, dx - \int_\Omega \frac{g_1}{\omega} \, |u| \, \omega \, dx \\ &\geq \frac{\lambda}{C_\Omega^p + 1} \|u\|^p_{W_0^{1,p}(\Omega,\omega)} - \|g_1/\omega\|_{L^{p'}(\Omega,\omega)} \|u\|_{W_0^{1,p}(\Omega,\omega)}. \end{aligned}$$

Hence, by (i) and (ii) we obtain

$$\frac{\lambda}{C_\Omega^p+1}\|u\|^p_{W_0^{1,p}(\Omega,\omega)} - \|g_1/\omega\|_{L^{p'}(\Omega,\omega)}\|u\|_{W_0^{1,p}(\Omega,\omega)} \leq \|f/\omega\|_{L^{p'}(\Omega,\omega)}\|u\|_{W_0^{1,p}(\Omega,\omega)}.$$

Therefore,

$$\|u\|_{W_0^{1,p}(\Omega,\omega)} \leq \left[\left(\frac{C_\Omega^p+1}{\lambda}\right)\left(\|f/\omega\|_{L^{p'}(\Omega,\omega)}+\|g_1/\omega\|_{L^{p'}(\Omega,\omega)}\right)\right]^{1/(p-1)}.$$

□

Example 5.1.1. *Let* $\Omega=\{(x,y)\in\mathbb{R}^2 \ : x^2+y^2<1\}$, $\omega(x,y)=(x^2+y^2)^{-1/2}$ *and consider the operator*

$$\begin{aligned} Lu(x,y) = & -\frac{\partial}{\partial x}\left[\omega(x,y)\left(\frac{\partial u}{\partial x}+\frac{1}{1+|\nabla u|^2}\frac{\partial u}{\partial x}\right)\right] \\ & -\frac{\partial}{\partial y}\left[\omega(x,y)\left(\frac{\partial u}{\partial y}+\frac{1}{1+|\nabla u|^2}\frac{\partial u}{\partial y}\right)\right]. \end{aligned}$$

We have that $\omega\in A_2$ $(p=2)$, *and for* $\xi=(\xi_1,\xi_2)\in\mathbb{R}^2$ *we define*

$$\begin{aligned} \mathcal{A}_1(x,\eta,\xi) &= \xi_1+\frac{\xi_1}{1+|\xi|^2} \\ \mathcal{A}_2(x,\eta,\xi) &= \xi_2+\frac{\xi_2}{1+|\xi|^2}, \end{aligned}$$

that is, $\mathcal{A}(x,\eta,\xi)=(\mathcal{A}_1(x,\eta,\xi),\mathcal{A}_2(x,\eta,\xi)=\xi+\dfrac{\xi}{1+|\xi|^2}$, *satisfies the conditions (H1), (H2), (H3) (with* $\lambda=1$, $p=2$, $g_1(x)=0$ *and* $g_2(x)=0$*) and (H4) (with* $K(x)=0$, $h_1(x)=0$, $h_2(x)=2$, $p=p'=2$*). Therefore, by Theorem 5.1.2, the problem*

$$\begin{cases} Lu &= \cos(x+y) \ \text{ in } \ \Omega, \\ u &= 0, \ \text{ on } \ \partial\Omega, \end{cases}$$

has a solution $u \in W_0^{1,2}(\Omega, \omega)$.

Now we will prove the existence of (weak) solutions in the weighted Sobolev spaces $W_0^{1,2}(\Omega, \omega)$ (see Definition 5.1.2) for the Dirichlet problem

$$(P10) \begin{cases} Lu(x) & = f_0(x) - \sum_{j=1}^{N} D_j f_j(x) \text{ in } \Omega, \\ u(x) & = 0 \text{ on } \partial\Omega, \end{cases}$$

where L is the partial differential operator

$$\begin{aligned} Lu(x) &= -\sum_{j=1}^{N} D_j\big[\omega(x)\mathcal{A}_j(x, u(x), \nabla u(x))\big] + \sum_{j=1}^{N} b_j(x)\, D_j u(x) \\ &+ \alpha\, g(x)\, u(x), \end{aligned}$$

where $D_j = \partial/\partial x_j$, Ω is a bounded open set in $\mathbb{R}^n$, α is a constant and $\mathcal{A}_j : \Omega \times \mathbb{R} \times \mathbb{R}^N \longrightarrow \mathbb{R}$ $(j = 1, ..., N)$, $b_j, g : \Omega \to \mathbb{R}$ are given functions.

Definition 5.1.2. *We say that an element* $u \in W_0^{1,2}(\Omega, , \omega)$ *is a (weak) solution of problem (P10) if*

$$\sum_{j=1}^{N} \int_\Omega \mathcal{A}_j(x, u, \nabla u)\, D_j\varphi\, \omega\, dx + \sum_{j=1}^{N} \int_\Omega b_j\, \varphi\, D_j u\, dx$$

$$+\alpha \int_\Omega g\, \varphi\, u\, dx = \int_\Omega f_0\, \varphi\, dx + \sum_{j=1}^{N} f_j\, D_j\varphi\, dx,$$

for all $\varphi \in W_0^{1,2}(\Omega, \omega)$.

Theorem 5.1.3. *Assume that the following five conditions hold.*
(H1) $x \mapsto \mathcal{A}_j(x, \eta, \xi)$ *is measurable on* Ω *for all* $(\eta, \xi) \in \mathbb{R} \times \mathbb{R}^N$, $(\eta, \xi) \mapsto \mathcal{A}_j(x, \eta, \xi)$ *is continuous on* $\mathbb{R} \times \mathbb{R}^N$ *for almost all* $x \in \Omega$.

(**H2**) *there exist a constant* $\beta > 0$ *such that*

$$\langle \mathcal{A}(x,\eta,\xi) - \mathcal{A}(x,\eta',\xi'), \xi - \xi' \rangle \geq \beta\, |\xi - \xi'|^2,$$

whenever $\xi, \xi' \in \mathbb{R}^N$, $\xi \neq \xi'$, *where*

$$\mathcal{A}(x,\eta,\xi) = (\mathcal{A}_1(x,\eta,\xi), ..., \mathcal{A}_N(x,\eta,\xi))$$

(where $\langle .,. \rangle$ *denote here the Euclidean scalar product in* $\mathbb{R}^N$*).*
(**H3**) $\langle \mathcal{A}(x,\eta,\xi), \xi \rangle \geq \lambda|\xi|^2 + \Lambda|\eta|^2 - g_1(x)|\eta| - g_2(x)$, *with* $g_1 \in L^2(\Omega,\omega)$ *and* $g_2 \in L^1(\Omega,\omega)$, *where* λ *and* Λ *are positive constants.*
(**H4**) $|\mathcal{A}(x,\eta,\xi)| \leq K(x) + h_1(x)|\eta|^{1/2} + h_2(x)|\xi|$, *where* K, h_1 *and* h_2 *are positive functions, with* h_1 *and* $h_2 \in L^\infty(\Omega)$, *and* $K \in L^2(\Omega,\omega)$, *and* ω *is a weight function.*
(**H5**) $g/\omega \in L^\infty(\Omega)$ *and* $b_j/\omega \in L^\infty(\Omega)$ $(j = 1, 2, ..., N)$.
If $\omega \in A_2$, $f_j/\omega \in L^2(\Omega,\omega)$ $(j = 0, 1, ..., N)$,
λ and $\beta > C_\Omega \sum_{j=1}^{N} \|b_j/\omega\|_{L^\infty(\Omega)}$ *(with* C_Ω *the constant in Theorem 2.1.5), then there exists a constant* α_0 *such that for all* $\alpha < \alpha_0$ *the problem (P10) has a unique solution* $u \in W_0^{1,2}(\Omega,\omega)$. *Moreover, with* $g_2 = 0$, *we have*

$$\|u\|_{W_0^{1,2}(\Omega,\omega)} \leq \frac{1}{\tilde{C}} \left(C_\Omega \|f_0/\omega\|_{L^2(\Omega,\omega)} + \sum_{j=1}^{N} \|f_j/\omega\|_{L^2(\Omega,\omega)} + \|g_1/\omega\|_{L^2(\Omega,\omega)} \right).$$

Proof. The basic idea is to reduce the problem (P10) to an operator equation $Au = T$ and apply the Theorem 5.1.1.

We define $B, B_1, B_2 : W_0^{1,p}(\Omega,\omega) \times W_0^{1,p}(\Omega,\omega) \to \mathbb{R}$ and $T : W_0^{1,p}(\Omega,\omega) \to \mathbb{R}$ by

$$B(u,\varphi) = B_1(u,\varphi) + B_2(u,\varphi),$$

$$\begin{aligned} B_1(u,\varphi) &= \sum_{j=1}^{N}\int_\Omega \mathcal{A}_j(x,u,\nabla u)D_j\varphi\,\omega\,dx \\ &= \int_\Omega \langle \mathcal{A}(x,u,\nabla u),\nabla\varphi\rangle\,\omega\,dx \\ B_2(u,\varphi) &= \sum_{j=1}^{N}\int_\Omega b_j\,\varphi\,D_j u\,dx + \alpha\int_\Omega g\,\varphi\,u\,dx, \end{aligned}$$

$$T(\varphi) = \int_\Omega f_0\,\varphi(x)\,dx + \sum_{j=1}^{N}\int_\Omega f_j\,D_j\varphi\,dx.$$

Then $u\in W_0^{1,2}(\Omega,\omega)$ is a (weak) solution to problem (P10) if

$$B(u,\varphi) = B_1(u,\varphi) + B_2(u,\varphi) = T(\varphi), \quad \text{for all } \varphi\in W_0^{1,2}(\Omega,\omega).$$

Step 1. For $j=1,...,N$ we define the operator

$$F_j : W_0^{1,2}(\Omega,\omega)\rightarrow L^2(\Omega,\omega)$$

by $(F_j u)(x) = \mathcal{A}_j(x,u(x),\nabla u(x))$. We now show that the operator F_j is bounded and continuous.

(i) Using (H4) we obtain

$$\begin{aligned} \|F_j u\|^2_{L^2(\Omega,\omega)} &= \int_\Omega |F_j u(x)|^2\omega\,dx = \int_\Omega |\mathcal{A}_j(x,u,\nabla u)|^2\omega\,dx \\ &\le \int_\Omega \Big(K + h_1|u|^{1/2} + h_2|\nabla u|\Big)^2\omega\,dx \\ &\le C_1\int_\Omega \Big[(K^2 + h_1^2|u| + h_2^2|\nabla u|^2)\omega\Big]dx \\ &= C_1\Big[\int_\Omega K^2\omega\,dx + \int_\Omega h_1^2|u|\,\omega\,dx \\ &+ \int_\Omega h_2^2|\nabla u|^2\omega\,dx\Big]. \end{aligned} \tag{5.4}$$

We have, by Theorem 2.1.5 (with $p = 2$ and $\theta = 1$),

$$\begin{aligned}\int_\Omega h_1^2|u|\,\omega\,dx &= \|h_1\|^2_{L^\infty(\Omega)}\int_\Omega |u|\,\omega\,dx\\ &\le \|h_1\|^2_{L^\infty(\Omega)}\left(\int_\Omega |u|^2\,\omega\,dx\right)^{1/2}\left(\int_\Omega \omega\,dx\right)^{1/2}\\ &\le C_\Omega\,\|h_1\|^2_{L^\infty(\Omega)}[\omega(\Omega)]^{1/2}\,\|u\|_{W_0^{1,p}(\Omega,\omega)},\end{aligned}$$

and

$$\begin{aligned}\int_\Omega h_2^2|\nabla u|^2\omega\,dx &\le \|h_2\|^2_{L^\infty(\Omega)}\int_\Omega |\nabla u|^2\,\omega\,dx\\ &= \|h_2\|^2_{L^\infty(\Omega)}\|u\|^2_{W_0^{1,2}(\Omega,\omega)}.\end{aligned}$$

Therefore, in (5.4) we obtain

$$\begin{aligned}&\|F_j u\|_{L^2(\Omega,\omega)}\\ &\le C_1\Bigg(\|K\|_{L^2(\Omega,\omega)} + C_\Omega^{1/2}\,\|h_1\|_{L^\infty(\Omega)}\,[\omega(\Omega)]^{1/4}\,\|u\|^{1/2}_{W_0^{1,2}(\Omega,\omega)}\\ &+\|h_2\|_{L^\infty(\Omega)}\|u\|_{W_0^{1,2}(\Omega,\omega)}\Bigg).\end{aligned}$$

(ii) Let $u_m \to u$ in $W_0^{1,p}(\Omega,\omega)$ as $m \to \infty$. We need to show that $F_j u_m \to F_j u$ in $L^2(\Omega,\omega)$. If $u_m \to u$ in $W_0^{1,2}(\Omega,\omega)$, then $u_m \to u$ in $L^2(\Omega,\omega)$ (by Theorem 2.1.5) and $|\nabla u_m| \to |\nabla u|$ in $L^2(\Omega,\omega)$. Using Theorem 1.4.11, there exist a subsequence $\{u_{m_k}\}$ and functions Φ_1 and Φ_2 with $\Phi_1, \Phi_2 \in L^2(\Omega,\omega)$ such that

$$\begin{aligned}&u_{m_k}(x) \to u(x),\ \mu-\text{a.e. in }\Omega,\\ &|u_{m_k}(x)| \le \Phi_1(x),\ \mu-\text{a.e. in }\Omega,\\ &|\nabla u_{m_k}(x)| \to |\nabla u(x)|,\ \mu-\text{a.e. in }\Omega,\\ &|\nabla u_{m_k}(x)| \le \Phi_2(x),\ \mu-\text{a.e. in }\Omega.\end{aligned}$$

Hence, using (H4), we obtain

$$\|F_j u_{m_k} - F_j u\|^2_{L^2(\Omega,\omega)} = \int_\Omega |F_j u_{m_k}(x) - F_j u(x)|^2 \omega\, dx$$
$$= \int_\Omega |\mathcal{A}_j(x, u_{m_k}, \nabla u_{m_k}) - \mathcal{A}_j(x, u, \nabla u)|^2\, \omega\, dx$$
$$\leq 2 \int_\Omega \Big(|\mathcal{A}_j(x, u_{m_k}, \nabla u_{m_k})|^2 + |\mathcal{A}_j(x, u, \nabla u)|^2 \Big) \omega\, dx$$
$$\leq 2 \bigg[\int_\Omega \Big(K + h_1 |u_{m_k}|^{1/2} + h_2 |\nabla u_{m_k}| \Big)^2 \omega\, dx$$
$$+ \int_\Omega \Big(K + h_1 |u|^{1/2} + h_2 |\nabla u| \Big)^2 \omega\, dx \bigg]$$
$$\leq 4 \int_\Omega \Big(K + h_1 \Phi_1^{1/2} + h_2 \Phi_2 \Big)^2 \omega\, dx$$
$$= C_2 \left[\int_\Omega K^2\, \omega\, dx + \int_\Omega h_1^2 \Phi_1\, \omega\, dx + \int_\Omega h_2^2 \Phi_2^2\, \omega\, dx \right]$$
$$\leq C_2 \bigg[\|K\|^2_{L^2(\Omega,\omega)} + \|h_1\|^2_{L^\infty(\Omega)} \int_\Omega \Phi_1 \omega\, dx$$
$$+ \|h_2\|^2_{L^\infty(\Omega)} \int_\Omega \Phi_2^2\, \omega\, dx \bigg]$$
$$\leq C_2 \bigg[\|K\|^2_{L^2(\Omega,\omega)} + \|h_1\|^2_{L^\infty(\Omega)} \left[\omega(\Omega)\right]^{1/2} \|\Phi_1\|_{L^2(\Omega,\omega)}$$
$$+ \|h_2\|^2_{L^\infty(\Omega)} \|\Phi_2\|^2_{L^2(\Omega,\omega)} \bigg].$$

By condition (H1), we have, as $m \to \infty$

$$F_j u_m(x) = \mathcal{A}_j(x, u_m(x), \nabla u_m(x)) \to \mathcal{A}_j(x, u(x), \nabla u(x)) = F_j u(x).$$

Therefore, by the Lebesgue Dominated Convergence Theorem, we obtain

$$\|F_j u_{m_k} - F_j u\|_{L^2(\Omega,\omega)} \to 0,$$

that is, $F_j u_{m_k} \to F_j u$ in $L^2(\Omega, \omega)$. By the Convergence Principle in Banach spaces (see Proposition 10.13 in [59]), we have

$$F_j u_m \to F_j u \quad \text{in} \quad L^2(\Omega, \omega). \tag{5.5}$$

Step 2. We have, by Theorem 2.1.5 (with $p = 2$ and $\theta = 1$),

$$\begin{aligned} |T(\varphi)| &\leq \int_\Omega |f_0||\varphi|\,dx + \sum_{j=1}^N \int_\Omega |f_j||D_j\varphi|\,dx \\ &= \int_\Omega \frac{|f_0|}{\omega}|\varphi|\omega\,dx + \sum_{j=1}^N \int_\Omega \frac{|f_j|}{\omega}|D_j\varphi|\,\omega\,dx \\ &\leq \left(C_\Omega\,\|f_0/\omega\|_{L^2(\Omega,\omega)} + \sum_{j=1}^N \|f_j/\omega\|_{L^2(\Omega,\omega)}\right)\|\varphi\|_{W_0^{1,2}(\Omega,\omega)}. \end{aligned}$$

Moreover, using (H4) and (H5), we also have

$$\begin{aligned} &|B(u,\varphi)| \leq |B_1(u,\varphi)| + |B_2(u,\varphi)| \\ \leq\;& \sum_{j=1}^N \int_\Omega |\mathcal{A}_j(x,u,\nabla u)||D_j\varphi|\,\omega\,dx + \sum_{j=1}^N \int_\Omega |b_j|\,|\varphi|\,|D_j u|\,dx \\ +\;& |\alpha| \int_\Omega |g|\,|u|\,|\varphi|\,dx \\ \leq\;& \int_\Omega \left(K + h_1|u|^{1/2} + h_2|\nabla u|\right)|\nabla\varphi|\,\omega\,dx \\ +\;& \sum_{j=1}^N \|b_j/\omega\|_{L^\infty(\Omega)} \int_\Omega |\varphi|\,|D_j u|\,\omega\,dx \\ +\;& |\alpha|\,\|g/\omega\|_{L^\infty(\Omega)} \int_\Omega |u||\varphi|\,\omega\,dx \\ \leq\;& \int_\Omega K|\nabla\varphi|\,\omega\,dx + \int_\Omega h_1|u|^{1/2}|\nabla\varphi|\,\omega\,dx + \int_\Omega h_2|\nabla u||\nabla\varphi|\,\omega\,dx \end{aligned}$$

$$
\begin{aligned}
&+ \quad C_\Omega \left(\sum_{j=1}^{N} \|b_j/\omega\|_{L^\infty(\Omega)} \right) \|\varphi\|_{W_0^{1,2}(\Omega,\omega)} \|u\|_{W_0^{1,2}(\Omega,\omega)} \\
&+ \quad C_\Omega^2 \, |\alpha| \, \|g/\omega\|_{L^\infty(\Omega)} \|\varphi\|_{W_0^{1,2}(\Omega,\omega)} \|u\|_{W_0^{1,2}(\Omega,\omega)}
\end{aligned}
\tag{5.6}
$$

In (5.6), we have $\int_\Omega K |\nabla\varphi| \, \omega \, dx \leq \|K\|_{L^2(\Omega,\omega)} \|\varphi\|_{W_0^{1,2}(\Omega,\omega)}$, and (using Theorem 2.1.5 with $p = 2$ and $\theta = 1$)

$$
\begin{aligned}
&\int_\Omega h_1 |u|^{1/2} |\nabla\varphi| \, \omega \, dx \\
&\leq \|h_1\|_{L^\infty(\Omega)} \left(\int_\Omega |u| \, \omega \, dx \right)^{1/2} \left(\int_\Omega |\nabla\varphi|^2 \, \omega \, dx \right)^{1/2} \\
&\leq \|h_1\|_{L^\infty(\Omega)} [\omega(\Omega)]^{1/4} \|u\|_{L^2(\Omega,\omega)}^{1/2} \|\varphi\|_{W_0^{1,2}(\Omega,\omega)} \\
&\leq C_\Omega^{1/2} \, \|h_1\|_{L^\infty(\Omega)} [\omega(\Omega)]^{1/4} \|u\|_{W_0^{1,2}(\Omega,\omega)}^{1/2} \, \|\varphi\|_{W_0^{1,2}(\Omega,\omega)};
\end{aligned}
$$

and

$$
\begin{aligned}
&\int_\Omega h_2 |\nabla u| \, |\nabla\varphi| \, \omega \, dx \\
&\leq \|h_2\|_{L^\infty(\Omega)} \left(\int_\Omega |\nabla u|^2 \, \omega \, dx \right)^{1/2} \left(\int_\Omega |\nabla\varphi|^2 \omega \, dx \right)^{1/2} \\
&= \; \|h_2\|_{L^\infty(\Omega)} \|u\|_{W_0^{1,2}(\Omega,\omega)} \, \|\varphi\|_{W_0^{1,2}(\Omega,\omega)}.
\end{aligned}
$$

Therefore, in (5.6) we obtain

$$
\begin{aligned}
|B(u,\varphi)| \leq &\Bigg[\|K\|_{L^2(\Omega,\omega)} + C_\Omega^{1/2} \, \|h_1\|_{L^\infty(\Omega)} [\omega(\Omega)]^{1/4} \|u\|_{W_0^{1,2}(\Omega,\omega)}^{1/2} \\
&+ \|h_2\|_{L^\infty(\Omega)} \|u\|_{W_0^{1,2}(\Omega,\omega)} \; + C_\Omega^2 \, |\alpha| \, \|g/\omega\|_{L^\infty(\Omega)} \|u\|_{W_0^{1,2}(\Omega,\omega)} \\
&+ C_\Omega \left(\sum_{j=1}^{N} \|b_j/\omega\|_{L^\infty(\Omega)} \right) \|u\|_{W_0^{1,2}(\Omega,\omega)} \Bigg] \|\varphi\|_{W_0^{1,2}(\Omega,\omega)} \\
= &\, M \, \|\varphi\|_{W_0^{1,2}(\Omega,\omega)},
\end{aligned}
$$

where

$$\begin{aligned} M &= \|K\|_{L^2(\Omega,\omega)} + C_\Omega^{1/2}\,\|h_1\|_{L^\infty(\Omega)}[\omega(\Omega)]^{1/4}\|u\|^{1/2}_{W_0^{1,2}(\Omega,\omega)} \\ &+ \|h_2\|_{L^\infty(\Omega)}\|u\|_{W_0^{1,2}(\Omega,\omega)} + C_\Omega^2\,|\alpha|\,\|g/\omega\|_{L^\infty(\Omega)}\|u\|_{W_0^{1,2}(\Omega,\omega)} \\ &+ C_\Omega\left(\sum_{j=1}^N \|b_j/\omega\|_{L^\infty(\Omega)}\right)\|u\|_{W_0^{1,2}(\Omega,\omega)}. \end{aligned}$$

Since $B(u,.)$ is linear, for each $u\in W_0^{1,2}(\Omega,\omega)$, there exists a linear and continuous functional on $W_0^{1,2}(\Omega,\omega)$ denoted by Au such that $(Au|\varphi) = B(u,\varphi)$ for all $\varphi\in W_0^{1,2}(\Omega,\omega)$ (where $(f|x)$ denotes the value of the linear functional f at the point x). Moreover, $\|Au\|_* \leq M$. Hence, we obtain the operator

$$\begin{aligned} A &: W_0^{1,2}(\Omega,\omega) \to [W_0^{1,2}(\Omega,\omega)]^* \\ u &\mapsto Au. \end{aligned}$$

Consequently, problem (P10) is equivalent to the operator equation $Au = T,\ u\in W_0^{1,2}(\Omega,\omega)$.

Step 3. Using condition (H2) and Theorem 2.1.5 (with $p = 2$ and $\theta = 1$), we have

$$\begin{aligned} &(Au_1 - Au_2|u_1-u_2) = B(u_1,u_1-u_2) - B(u_2,u_1-u_2) \\ &= \int_\Omega \langle \mathcal{A}(x,u_1,\nabla u_1) - \mathcal{A}(x,u_2,\nabla u_2), \nabla(u_1-u_2)\rangle\,\omega\,dx \\ &+ \sum_{j=1}^N \int_\Omega b_j(u_1-u_2)D_j(u_1-u_2)\,dx + \alpha\int_\Omega g\,(u_1-u_2)^2\,dx \\ &\geq \beta\int_\Omega |\nabla(u_1-u_2)|^2\,\omega\,dx \\ &-\left(\sum_{j=1}^N \|b_j/\omega\|_{L^\infty(\Omega)}\right)\int_\Omega |u_1-u_2|\,|D_j(u_1-u_2)|\,\omega\,dx \\ &-\alpha\|g/\omega\|_{L^\infty(\Omega)}\int_\Omega |u_1-u_2|^2\,\omega\,dx \end{aligned}$$

$$\geq \beta \, \|\nabla u_1 - u_2\|^2_{L^2(\Omega,\omega)} - \alpha \, \|g/\omega\|_{L^\infty(\Omega)} \|u_1 - u_2\|^2_{L^2(\Omega,\omega)}$$
$$- \left(\sum_{j=1}^{N} \|b_j/\omega\|_{L^\infty(\Omega)} \right) \|u_1 - u_2\|_{L^2(\Omega,\omega)} \|D_j(u_1 - u_2)\|_{L^2(\Omega,\omega)}$$
$$\geq \beta \, \|u_1 - u_2\|^2_{W_0^{1,2}(\Omega,\omega)}$$
$$- \left(\sum_{j=1}^{N} \|b_j/\omega\|_{L^\infty(\Omega)} \right) C_\Omega \, \|u_1 - u_2\|^2_{W_0^{1,2}(\Omega,\omega)}$$
$$- \alpha \, \|g/\omega\|_{L^\infty(\Omega)} \, C_\Omega^2 \, \|u_1 - u_2\|^2_{W_0^{1,2}(\Omega,\omega)}$$
$$= C_1 \, \|u_1 - u_2\|^2_{W_0^{1,2}(\Omega,\omega)},$$

where $C_1 = \beta - C_\Omega \left(\sum_{j=1}^{N} \|b_j/\omega\|_{L^\infty(\Omega)} \right) - \alpha \, \|g/\omega\|_{L^\infty(\Omega)} C_\Omega^2 > 0$, if

$$\alpha \; < \; \frac{\beta - \left(\sum_{j=1}^{N} \|b_j/\omega\|_{L^\infty(\Omega)} \right) C_\Omega}{C_\Omega^2 \, \|g/\omega\|_{L^\infty(\Omega)}} = \gamma_1.$$

Therefore, if $\alpha < \gamma_1$, the operator A is strongly monotone, and this implies that the operator A is strictly monotone. Moreover, using (H3) and the weighted Sobolev inequality (Theorem 2.1.5 with $p = 2$ and $k = 1$), we obtain

$$(Au|u) = B(u,u) = B_1(u,u) + B_2(u,u)$$
$$= \int_\Omega \langle \mathcal{A}(x,u,\nabla u), \nabla u \rangle \, \omega \, dx + \sum_{j=1}^{N} \int_\Omega b_j \, u \, D_j u \, dx + \alpha \int_\Omega g \, u^2 \, dx$$
$$\geq \int_\Omega \left(\Lambda |u|^2 + \lambda |\nabla u|^2 - g_1 |u| - g_2 \right) \omega \, dx$$
$$- \alpha \, \|g/\omega\|_{L^\infty(\Omega)} \int_\Omega |u|^2 \, \omega \, dx$$

$$
\begin{aligned}
&- \left(\sum_{j=1}^{N} \|b_j/\omega\|_{L^\infty(\Omega)} \right) \left(\int_\Omega |u|^2 \,\omega\, dx \right)^{1/2} \left(\int_\Omega |D_j u|^2 \omega\, dx \right)^{1/2} \\
&\geq \Lambda \int_\Omega |u|^2 \omega\, dx + \lambda \int_\Omega |\nabla u|^2 \,\omega\, dx - \int_\Omega g_1 |u| \,\omega\, dx - \int_\Omega g_2 \,\omega\, dx \\
&- \alpha \|g/\omega\|_{L^\infty(\Omega)} \, C_\Omega^2 \, \|u\|^2_{W_0^{1,2}(\Omega,\omega)} \\
&- C_\Omega \left(\sum_{j=1}^{N} \|b_j/\omega\|_{L^\infty(\Omega)} \right) \|u\|^2_{W_0^{1,2}(\Omega,\omega)} \\
&\geq \lambda \int_\Omega |\nabla u|^2 \,\omega\, dx - \left(\int_\Omega g_1^2 \omega\, dx \right)^{1/2} \left(\int_\Omega |u|^2 \,\omega\, dx \right)^{1/2} \\
&- \|g_2\|_{L^1(\Omega,\omega)} \\
&- \alpha \, \|g/\omega\|_{L^\infty(\Omega)} \, C_\Omega^2 \, \|u\|^2_{W_0^{1,2}(\Omega,\omega)} \\
&- C_\Omega \left(\sum_{j=1}^{N} \|b_j/\omega\|_{L^\infty(\Omega)} \right) \|u\|^2_{W_0^{1,2}(\Omega,\omega)} \\
&\geq C_2 \, \|u\|^2_{W_0^{1,2}(\Omega,\omega)} - C_\Omega \, \|g_1\|_{L^2(\Omega,\omega)} \, \|u\|_{W_0^{1,2}(\Omega,\omega)} - \|g_2\|_{L^1(\Omega,\omega)},
\end{aligned}
$$

where $C_2 = \lambda - \alpha \|g/\omega\|_{L^\infty(\Omega)} \, C_\Omega^2 - C_\Omega \left(\sum_{j=1}^{N} \|b_j/\omega\|_{L^\infty(\Omega)} \right) > 0$, if

$$
\alpha < \frac{\lambda - C_\Omega \left(\sum_{j=1}^{N} \|b_j/\omega\|_{L^\infty(\Omega)} \right)}{C_\Omega^2 \, \|g/\omega\|_{L^\infty(\Omega)}} = \gamma_2.
$$

Hence, we have

$$
\frac{(Au|u)}{\|u\|_{W_0^{1,2}(\Omega,\omega)}} \to +\infty, \text{ as } \|u\|_{W_0^{1,2}(\Omega,\omega)} \to +\infty,
$$

that is, A is coercive.

Therefore, if $\alpha < \min\{\gamma_1, \gamma_2\} = \alpha_0$, we have that A is strictly monotone and coercive.

Step 4. We need to show that the operator A is continuous. Let $u_m \to u$ in $W_0^{1,2}(\Omega,\omega)$ as $m \to \infty$. We have,

$$\begin{aligned}
&|B_1(u_m,\varphi) - B_1(u,\varphi)| \\
&\leq \sum_{j=1}^{N} \int_\Omega |\mathcal{A}_j(x,u_m,\nabla u_m) - \mathcal{A}_j(x,u,\nabla u)||D_j\varphi|\,\omega\,dx \\
&= \sum_{j=1}^{N} \int_\Omega |F_j u_m - F_j u||D_j\varphi|\,\omega\,dx \\
&\leq \sum_{j=1}^{N} \|F_j u_m - F_j u\|_{L^2(\Omega,\omega)} \|D_j\varphi\|_{L^2(\Omega,\omega)} \\
&\leq \sum_{j=1}^{N} \|F_j u_m - F_j u\|_{L^2(\Omega,\omega)} \|\varphi\|_{W_0^{1,2}(\Omega,\omega)},
\end{aligned}$$

and

$$\begin{aligned}
&|B_2(u_m,\varphi) - B_2(u,\varphi)| \\
&= \left| \sum_{j=1}^{N} \int_\Omega b_j\,\varphi\,D_j(u_m - u)\,dx + \alpha \int_\Omega g\,\varphi\,(u_m - u)\,dx \right| \\
&\leq \Bigg(C_\Omega \sum_{j=1}^{N} \|b_j/\omega\|_{L^\infty(\Omega)} \\
&+ C_\Omega^2\,|\,\alpha\,|\,\|g/\omega\|_{L^\infty(\Omega)} \Bigg) \|\varphi\|_{W_0^{1,2}(\Omega,\omega)}\,\|u_m - u\|_{W_0^{1,2}(\Omega,\omega)},
\end{aligned}$$

for all $\varphi \in W_0^{1,2}(\Omega,\omega)$. Hence,

$$\begin{aligned}
&|B(u_m,\varphi) - B(u,\varphi)| \\
&\leq |B_1(u_m,\varphi) - B_1(u,\varphi)| + |B_2(u_m,\varphi) - B_2(u,\varphi)| \\
&\leq \Bigg[\sum_{j=1}^{N} \|F_j u_m - F_j u\|_{L^2(\Omega,\omega)} + \Bigg(C_\Omega \sum_{j=1}^{N} \left\| \frac{b_j}{\omega} \right\|_{L^\infty(\Omega)}
\end{aligned}$$

$$+C_\Omega^2\,|\,\alpha\,|\,\|g/\omega\|_{L^\infty(\Omega)}\Big)\,\|u_m-u\|_{W_0^{1,2}(\Omega,\omega)}\Big]\,\|\varphi\|_{W_0^{1,2}(\Omega,\omega)},$$

and

$$\|Au_m-Au\|_*\leq\sum_{j=1}^N\|F_ju_m-F_ju\|_{L^2(\Omega,\omega)}$$
$$+\left(C_\Omega\sum_{j=1}^N\left\|\frac{b_j}{\omega}\right\|_{L^\infty(\Omega)}+C_\Omega^2\,|\,\alpha\,|\,\|g/\omega\|_{L^\infty(\Omega)}\right)\|u_m-u\|_{W_0^{1,2}(\Omega,\omega)},.$$

Therefore, using (5.5) we have $\|Au_m-Au\|_*\to 0$ as $m\to+\infty$, that is, A is continuous (and this implies that A is hemicontinuous).

Therefore, by Theorem 5.1.1, the operator equation $Au=T$ has a unique solution $u\in W_0^{1,2}(\Omega,\omega)$ and it is the unique solution for problem (P10).

Step 5. Estimate for $\|u\|_{W_0^{1,2}(\Omega,\omega)}$.

In Definition 5.1.2 with $\varphi=u\in W_0^{1,2}(\Omega,\omega)$ we obtain

$$\int_\Omega\langle\mathcal{A}(x,u,\nabla u),\nabla u\rangle\,\omega\,dx+\sum_{j=1}^N\int_\Omega b_j\,u\,D_ju\,dx$$
$$+\,\alpha\int_\Omega u^2\,g\,dx=\int_\Omega f_0\,u\,dx+\sum_{j=1}^N\int_\Omega f_j\,D_ju\,dx.$$

Then we obtain

(i) By Theorem 2.1.5 (with $p=2$ and $\theta=1$),

$$\left|\int_\Omega f_0\,u\,dx+\sum_{j=1}^N f_j\,D_ju\,dx\right|$$
$$\leq\left(C_\Omega\|f_0/\omega\|_{L^2(\Omega,\omega)}+\sum_{j=1}^N\|f_j/\omega\|_{L^2(\Omega,\omega)}\right)\|u\|_{W_0^{1,2}(\Omega,\omega)}.$$

(ii) By (H3) (with $g_2 = 0$), $C = \min\{\lambda, \Lambda\}$,

$$\int_\Omega \langle \mathcal{A}(x,u,\nabla u), \nabla u\rangle \,\omega\, dx \geq \int_\Omega \Big(\lambda|\nabla u|^2 + \Lambda|u|^2 - g_1|u|\Big)\,\omega\, dx$$
$$= \lambda \int_\Omega |\nabla u|^2\,\omega\, dx + \Lambda \int_\Omega |u|^2\,\omega\, dx - \int_\Omega g_1|u|\,\omega\, dx$$
$$\geq C\,\|u\|^2_{W_0^{1,2}(\Omega,\omega)} - \|g_1/\omega\|_{L^2(\Omega,\omega)}\|u\|_{W_0^{1,2}(\Omega,\omega)},$$

and

$$\sum_{j=1}^N \int_\Omega b_j\, u\, D_j u\, dx \geq -\left(\max_{1\leq j\leq N}\|b_j/\omega\|_{L^\infty(\Omega)}\right)\|u\|^2_{W_0^{1,2}(\Omega,\omega)},$$

and we also have

$$\alpha\int_\Omega u^2\, g\, dx \geq -|\alpha|\,\|g/\omega\|_{L^\infty(\Omega)}\|u\|^2_{W_0^{1,2}(\Omega,\omega)}.$$

Hence, by (i) and (ii), we obtain

$$C\|u\|^2_{W_0^{1,2}(\Omega,\omega)} - \|g_1/\omega\|_{L^2(\Omega,\omega)}\|u\|_{W_0^{1,2}(\Omega,\omega)}$$
$$-\left(|\alpha|\|g/\omega\|_{L^\infty(\Omega)} + \max_{1\leq j\leq N}\|b_j/\omega\|_{L^\infty(\Omega)}\right)\|u\|^2_{W_0^{1,2}(\Omega,\omega)}$$
$$\leq\left(C_\Omega\|f_0/\omega\|_{L^2(\Omega,\omega)} + \sum_{j=1}^N\|f_j/\omega\|_{L^2(\Omega,\omega)}\right)\|u\|_{W_0^{1,2}(\Omega,\omega)}.$$

Therefore, if $\tilde{C} = C - \max_{1\leq j\leq N}\|b_j/\omega\|_{L^\infty(\Omega)} - |\alpha|\|g/\omega\|_{L^2(\Omega,\omega)} > 0$, then

$$\|u\|_{W_0^{1,2}(\Omega,\omega)}$$
$$\leq\frac{1}{\tilde{C}}\left(C_\Omega\,\|f_0/\omega\|_{L^2(\Omega,\omega)} + \sum_{j=1}^N\|f_j/\omega\|_{L^2(\Omega,\omega)} + \|g_1/\omega\|_{L^2(\Omega,\omega)}\right).$$

□

Example 5.1.2. *Let* $\Omega = \{(x,y)\in\mathbb{R}^2 \ : \ x^2+y^2 < 1\}$, *and consider the weight function* $\omega(x,y) = (x^2+y^2)^{-1/2}$ *(*ω *is a* A_2*-weight), and the function*

$$\mathcal{A} \,:\, \Omega\times\mathbb{R}\times\mathbb{R}^2\to\mathbb{R}^2$$

$$\mathcal{A}((x,y),\eta,\xi) = h_2(x,y)\xi,$$

where $h_2(x,y) = 2\mathrm{e}^{x^2+y^2}$, *and the functions* $b_1, b_2, g : \Omega\to\mathbb{R}$

$$b_1(x,y) = \frac{\sin(1/(x^2+y^2))}{(x^2+y^2)^{1/2}}, \ b_2(x,y) = \frac{\mathrm{e}^{-1/(x^2+y^2)}}{(x^2+y^2)^{1/2}}$$
$$g(x,y) = \frac{\cos(1/(x^2+y^2))}{(x^2+y^2)^{1/2}}.$$

Let us consider the partial differential operator

$$\begin{aligned} Lu(x,y) &= -\operatorname{div}\Big[\omega(x,y)\mathcal{A}((x,y),u,\nabla u)\Big] + b_1(x,y)\frac{\partial u}{\partial x} \\ &+ b_2(x,y)\,\frac{\partial u}{\partial y} + \alpha\, g(x,y)\, u \\ &= -\frac{\partial}{\partial x}\left[\omega(x,y)h_2(x,y)\frac{\partial u}{\partial x}\right] - \frac{\partial}{\partial y}\left[\omega(x,y)h_2(x,y)\frac{\partial u}{\partial y}\right] \\ &+ b_1(x,y)\frac{\partial u}{\partial x} + b_2(x,y)\,\frac{\partial u}{\partial y} + \alpha\, g(x,y)\, u. \end{aligned}$$

Therefore, by Theorem 5.1.3, the problem

$$\begin{cases} Lu(x,y) &= \dfrac{\cos(xy)}{(x^2+y^2)^{1/5}} - \dfrac{\partial}{\partial x}\left(\dfrac{\sin(xy)}{(x^2+y^2)^{1/9}}\right) \\ &\quad -\dfrac{\partial}{\partial y}\left(\dfrac{\sin(xy)}{(x^2+y^2)^{1/9}}\right) \text{ in } \Omega, \\ u(x,y) &= 0 \text{ on } \partial\Omega, \end{cases}$$

has a unique solution $u\in W_0^{1,2}(\Omega,\omega)$, *if* $\alpha < \alpha_0$.

Now we prove the existence of (weak) solutions in the weighted Sobolev spaces $\mathrm{W}_0^{1,p}(\Omega,\omega)$ (see Definition 5.1.3) for the Dirichlet problem

$$(P11)\begin{cases} Lu(x) &= f_0(x) - \sum_{j=1}^{N} D_j f_j(x) \ \text{ in } \ \Omega, \\ u(x) &= 0 \ \text{ on } \ \partial\Omega, \end{cases}$$

where L is the partial differential operator

$$\begin{aligned} Lu(x) &= -\sum_{j=1}^{N} D_j\big[\omega(x)\mathcal{A}_j(x,u(x),\nabla u(x))\big] \\ &+ \omega(x)g(x,u(x)\nabla u(x)) \end{aligned} \tag{5.7}$$

where $D_j = \partial/\partial x_j$, Ω is a bounded open set in $\mathbb{R}^N$, ω is a weight function, and $\mathcal{A}_j : \Omega\times\mathbb{R}\times\mathbb{R}^N\to\mathbb{R}$ $(j = 1,...,N)$ and $g : \Omega\times\mathbb{R}\times\mathbb{R}^N\to\mathbb{R}$ are given functions.

Equations like (5.7) have been studied by many authors in the non-degenerate case (i.e. with $\omega(x)\equiv 1$) (see e.g. [22] and the references therein). The degenerate case with different conditions haven been studied by many authors. In [31] Drabek, Kufner and Mustonen proved that under certain condition, the Dirichlet problem associated with the equation

$$-\mathrm{div}(a(x,u,\nabla u)) = h,$$

$h\in[W_0^{1,p}(\Omega,\omega)]^*$ has at least one solution $u\in W_0^{1,p}(\Omega,\omega)$.

Definition 5.1.3. *Let $1 < p < \infty$. We say that $u\in W_0^{1,p}(\Omega,\omega)$ is a (weak) solution of problem (P11) if*

$$\sum_{j=1}^{N}\int_\Omega \mathcal{A}_j(x,u,\nabla u)D_j\varphi\,\omega\,dx + \int_\Omega g(x,u,\nabla u)\varphi\,\omega\,dx$$

$$= \int_\Omega f_0\varphi\,dx + \sum_{j=1}^{N}\int_\Omega f_jD_j\varphi\,dx,$$

for all $\varphi \in W_0^{1,p}(\Omega,\omega)$.

Theorem 5.1.4. *Assume (H1)-(H8).*
(**H1**) $x \mapsto \mathcal{A}_j(x,\eta,\xi)$ *is measurable in* Ω *for all* $(\eta,\xi) \in \mathbb{R} \times \mathbb{R}^N$, $(\eta,\xi) \mapsto \mathcal{A}_j(x,\eta,\xi)$ *is continuous in* $\mathbb{R} \times \mathbb{R}^N$ *for almost all* $x \in \Omega$.
(**H2**) $\langle \mathcal{A}(x,\eta,\xi) - \mathcal{A}(x,\eta',\xi'), \xi - \xi' \rangle \geq 0$, *whenever* $\xi, \xi' \in \mathbb{R}^N$, $\xi \neq \xi'$, *where*

$$\mathcal{A}(x,\eta,\xi) = (\mathcal{A}_1(x,\eta,\xi), ..., \mathcal{A}_N(x,\eta,\xi)).$$

(**H3**) $\langle \mathcal{A}(x,\eta,\xi), \xi \rangle \geq \lambda|\xi|^p + \Lambda|\eta|^p - h_1(x)|\eta| - h_2(x)$, *with* $1 < p < \infty$, *where* λ *and* Λ *are positive constants,* $h_1 \in L^{p'}(\Omega,\omega)$ *and* $h_2 \in L^1(\Omega,\omega)$ *(we denote by* p' *the real number such that* $1/p + 1/p' = 1$*).*
(**H4**) $|\mathcal{A}(x,\eta,\xi)| \leq K_1(x) + h_3(x)|\eta|^{p/p'} + h_4(x)|\xi|^{p/p'}$, *where* K_1, h_3 *and* h_4 *are positive functions, with* h_3 *and* $h_4 \in L^\infty(\Omega)$, *and* $K_1 \in L^{p'}(\Omega,\omega)$.
(**H5**) $x \mapsto g(x,\eta,\xi)$ *is measurable in* Ω *for all* $(\eta,\xi) \in \mathbb{R} \times \mathbb{R}^N$ $(\eta,\xi) \mapsto g(x,\eta,\xi)$ *is continuous in* $\mathbb{R} \times \mathbb{R}^N$ *for almost all* $x \in \Omega$.
(**H6**) $|g(x,\eta,\xi)| \leq K_2(x) + h_5(x)|\eta|^{p/p'} + h_6(x)|\xi|^{p/p'}$, *where* K_2, h_5 *and* h_6 *are positive functions, with* $h_5, h_6 \in L^\infty(\Omega)$ *and* $K_2 \in L^{p'}(\Omega,\omega)$.
(**H7**) $g(x,\eta,\xi)\,\eta \geq 0$, *for all* $\eta \in \mathbb{R}$.
(**H8**) $(g(x,\eta,\xi) - g(x,\eta',\xi'))(\eta - \eta') > 0$, *whenever* $\eta, \eta' \in \mathbb{R}$, $\eta \neq \eta'$.

If $\omega \in A_p$, *with* $1 < p < \infty$, *and* $f_j/\omega \in L^{p'}(\Omega,\omega)$ $(j = 0, 1, ..., N)$ *then problem (P11) has a solution* $u \in W_0^{1,p}(\Omega,\omega)$. *Moreover, with* $h_2 = 0$ *we have*

$$\|u\|_{W_0^{1,p}(\Omega,\omega)} \leq \frac{1}{C^{1/(p-1)}} \left(C_\Omega \|f_0/\omega\|_{L^{p'}(\Omega,\omega)} + \sum_{j=1}^N \|f_j/\omega\|_{L^{p'}(\Omega,\omega)} \right.$$

$$+ \quad \|h_1\|_{L^{p'}(\Omega,\omega)}\Bigg)^{1/(p-1)},$$

where $C = \min\{\lambda, \Lambda\}$ *and* C_Ω *is the constant in Theorem 2.1.5.*

Proof. We define $B : W_0^{1,p}(\Omega,\omega) \times W_0^{1,p}(\Omega,\omega) \to \mathbb{R}$ and $T : W_0^{1,p}(\Omega,\omega) \to \mathbb{R}$ by

$$\begin{aligned} B(u,\varphi) &= \sum_{j=1}^{N} \int_\Omega \mathcal{A}_j(x,u,\nabla u) D_j\varphi\,\omega\,dx + \int_\Omega g(x,u,\nabla u)\,\varphi\,\omega\,dx \\ &= \int_\Omega \langle \mathcal{A}(x,u,\nabla u), \nabla\varphi\rangle\,\omega\,dx + \int_\Omega g(x,u,\nabla u)\,\varphi\,\omega\,dx; \\ T(\varphi) &= \int_\Omega f_0\varphi\,dx + \sum_{j=1}^{n} \int_\Omega f_j D_j\varphi\,dx. \end{aligned}$$

Then $u \in W_0^{1,p}(\Omega,\omega)$ is a (weak) solution to problem (P11) if

$$B(u,\varphi) = T(\varphi), \quad \text{for all } \varphi \in W_0^{1,p}(\Omega,\omega).$$

Step 1. We define the operators (for $j = 1, ..., N$)

$$\begin{aligned} &F_j : W_0^{1,p}(\Omega,\omega) \to L^{p'}(\Omega,\omega) \\ &(F_j u)(x) = \mathcal{A}_j(x,u(x),\nabla u(x)), \end{aligned}$$

$$\begin{aligned} &G : W_0^{1,p}(\Omega,\omega) \to L^{p'}(\Omega,\omega) \\ &(Gu)(x) = g(x,u(x),\nabla u). \end{aligned}$$

We have that the operators F_j and G are bounded and continuous. In fact,
(i) Using (H4) and Theorem 2.1.5 (with $\theta = 1$), we obtain

$$\|F_j u\|^{p'}_{L^{p'}(\Omega,\omega)} = \int_\Omega |F_j u(x)|^{p'}\,\omega\,dx$$

$$\begin{aligned}
&= \int_\Omega |\mathcal{A}_j(x,u,\nabla u)|^{p'}\omega\,dx \\
&\le \int_\Omega \left(K_1 + h_3|u|^{p/p'} + h_4|\nabla u|^{p/p'}\right)^{p'}\omega\,dx \\
&\le C\int_\Omega \left[(K_1^{p'} + h_3^{p'}|u|^p + h_4^{p'}|\nabla u|^p)\omega\right]dx \\
&= C\left[\int_\Omega K_1^{p'}\omega\,dx + \int_\Omega h_3^{p'}|u|^p\omega\,dx + \int_\Omega h_4^{p'}|\nabla u|^p\omega\,dx\right] \\
&\le C\Big[\|K_1\|^{p'}_{L^{p'}(\Omega,\omega)} + (C_\Omega\|h_3\|^{p'}_{L^\infty(\Omega)} \\
&+ \|h_4\|^{p'}_{L^\infty(\Omega)})\|u\|^p_{W_0^{1,p}(\Omega,\omega)}\Big]. \qquad (5.8)
\end{aligned}$$

Therefore, in (5.8) we obtain

$$\|F_ju\|_{L^{p'}(\Omega,\omega)} \le C\left(\|K_1\|_{L^{p'}(\Omega,\omega)} + \|u\|^{p/p'}_{W_0^{1,p}(\Omega,\omega)}\right). \qquad (5.9)$$

Analogously, by condition (H6), we have

$$\begin{aligned}
&\|Gu\|^{p'}_{L^{p'}(\Omega,\omega)} = \int_\Omega |g(x,u,\nabla u)|^{p'}\omega\,dx \\
&\le \int_\Omega \left(K_2 + h_5|u|^{p/p'} + h_6|\nabla u|^{p/p'}\right)^{p'}\omega\,dx \\
&\le C\left[\|K_2\|^{p'}_{L^{p'}(\Omega,\omega)} + (C_\Omega\|h_5\|^{p'}_{L^{p'}(\Omega,\omega)} + \|h_6\|^{p'}_{L^\infty(\Omega)})\|u\|^p_{W_0^{1,p}(\Omega,\omega)}\right].
\end{aligned}$$

Hence,

$$\|Gu\|_{L^{p'}(\Omega,\omega)} \le C\left(\|K_2\|_{L^{p'}(\Omega,\omega)} + \|u\|^{p/p'}_{W_0^{1,p}(\Omega,\omega)}\right). \qquad (5.10)$$

(ii) Let $u_n \to u$ in $W_0^{1,p}(\Omega,\omega)$ as $n\to\infty$. We need to show that $F_ju_n \to F_ju$ and $Gu_n \to Gu$ in $L^{p'}(\Omega,\omega)$.

If $u_n \to u$ in $W_0^{1,p}(\Omega,\omega)$, then $|\nabla u_n| \to |\nabla u|$ in $L^p(\Omega,\omega)$ and $u_n \to u$ in $L^p(\Omega,\omega)$. Using Theorem 1.4.11, there exist a subsequence $\{u_{n_k}\}$ and functions Φ_1 and Φ_2 in $L^p(\Omega,\omega)$ such that

$$\begin{aligned}
&u_{n_k}(x) \to u(x), \quad \mu - \text{a.e. in } \Omega,\\
&|u_{n_k}(x)| \le \Phi_1(x), \quad \mu - \text{a.e. in } \Omega,\\
&|\nabla u_{n_k}(x)| \to |\nabla u(x)|, \quad \mu - \text{a.e. in } \Omega,\\
&|\nabla u_{n_k}(x)| \le \Phi_2(x), \quad \mu - \text{a.e. in } \Omega.
\end{aligned}$$

Hence, we obtain

$$\begin{aligned}
\|F_j u_{n_k} - F_j u\|^{p'}_{L^{p'}(\Omega,\omega)} &= \int_\Omega |F_j u_{n_k}(x) - F_j u(x)|^{p'} \omega\, dx\\
&= \int_\Omega |\mathcal{A}_j(x, u_{n_k}, \nabla u_{n_k}) - \mathcal{A}_j(x, u, \nabla u)|^{p'} \omega\, dx\\
&\le C \int_\Omega \Big(|\mathcal{A}_j(x, u_{n_k}, \nabla u_{n_k})|^{p'} + |\mathcal{A}_j(x, u, \nabla u)|^{p'} \Big) \omega\, dx\\
&\le C \Big[\int_\Omega \Big(K_1 + h_3 |u_{n_k}|^{p/p'} + h_4 |\nabla u_{n_k}|^{p/p'} \Big)^{p'} \omega\, dx\\
&+ \int_\Omega \Big(K_1 + h_3 |u|^{p/p'} + h_4 |\nabla u|^{p/p'} \Big)^{p'} \omega\, dx \Big]\\
&\le 2C \int_\Omega \Big(K_1 + h_3 \Phi_1^{p/p'} + h_4 \Phi_2^{p/p'} \Big)^{p'} \omega\, dx\\
&= \tilde{C} \Big[\int_\Omega K_1^{p'} \omega\, dx + \int_\Omega h_3^{p'} \Phi_1^p \omega\, dx + \int_\Omega h_4^{p'} \Phi_2^p \omega\, dx \Big]\\
&\le \tilde{C} \Big[\|K_1\|^{p'}_{L^{p'}(\Omega,\omega)} + \|h_3\|^{p'}_{L^\infty(\Omega)} \int_\Omega \Phi_1^p \omega\, dx\\
&+ \|h_4\|^{p'}_{L^\infty(\Omega)} \int_\Omega \Phi_2^p \omega\, dx \Big]\\
&\le \tilde{C} \Big[\|K_1\|^{p'}_{L^{p'}(\Omega,\omega)} + \|h_3\|^{p'}_{L^\infty(\Omega)} \|\Phi_1\|^p_{L^p(\Omega,\omega)}
\end{aligned}$$

$$+ \quad \|h_4\|^{p'}_{L^\infty(\Omega)}\|\Phi_2\|^p_{L^p(\Omega,\omega)}\Big].$$

Analogously, we have

$$\begin{aligned}\|Gu_{n_k} - Gu\|^{p'}_{L^{p'}(\Omega,\omega)} &\leq \tilde{C}\Big[\|K_2\|^{p'}_{L^{p'}(\Omega,\omega)} + \|h_5\|^{p'}_{L^\infty(\Omega)}\|\Phi_1\|^p_{L^p(\Omega,\omega)} \\ &+ \quad \|h_6\|^{p'}_{L^\infty(\Omega)}\|\Phi_2\|^p_{L^p(\Omega,\omega)}\Big].\end{aligned}$$

By condition (H1) and (H5), we have

$$\begin{aligned} F_j u_n(x) &= \mathcal{A}_j(x, u_n(x), \nabla u_n(x)) \\ &\to \mathcal{A}_j(x, u(x), \nabla u(x)) = F_j u(x), \quad \text{as } n \to \infty, \\ G u_n(x) &= g(x, u_n(x), \nabla u_n(x)) \\ &\to g(x, u(x), \nabla u(x)) = Gu(x), \quad \text{as } n \to \infty, \end{aligned}$$

for almost all $x\in\Omega$. Therefore, by the Lebesgue Dominated Convergence Theorem, we obtain

$$\|F_j u_{n_k} - F_j u\|_{L^{p'}(\Omega,\omega)} \to 0,$$

$$\|Gu_{n_k} - Gu\|_{L^{p'}(\Omega,\omega)} \to 0,$$

that is, $F_j u_{n_k} \to F_j u$ and $Gu_{n_k} \to Gu$ in $L^{p'}(\Omega,\omega)$. By the Convergence Principle in Banach spaces (see Proposition 10.13 in [59], we have

$$F_j u_n \to F_j u \quad \text{in} \quad L^{p'}(\Omega,\omega), \tag{5.11}$$

$$Gu_n \to Gu \quad \text{in} \quad L^{p'}(\Omega,\omega). \tag{5.12}$$

Step 2. We have (by Theorem 2.1.5 with $\theta = 1$)

$$|T(\varphi)| \leq \int_\Omega |f_0||\varphi|\,dx + \sum_{j=1}^N \int_\Omega |f_j||D_j\varphi|\,dx$$

$$
\begin{aligned}
&= \int_\Omega \frac{|f_0|}{\omega}|\varphi|\omega\,dx + \sum_{j=1}^N \int_\Omega \frac{|f_j|}{\omega}|D_j\varphi|\omega\,dx \\
&\leq \|f_0/\omega\|_{L^{p'}(\Omega,\omega)}\|\varphi\|_{L^p(\Omega,\omega)} \\
&+ \sum_{j=1}^N \|f_j/\omega\|_{L^{p'}(\Omega,\omega)}\|D_j\varphi\|_{L^p(\Omega,\omega)} \\
&\leq \left(C_\Omega\|f_0/\omega\|_{L^{p'}(\Omega,\omega)} + \sum_{j=1}^N \|f_j/\omega\|_{L^{p'}(\Omega,\omega)}\right)\|\varphi\|_{W_0^{1,p}(\Omega,\omega)}.
\end{aligned}
$$

Moreover, we also have

$$
\begin{aligned}
|B(u,\varphi)| &\leq \sum_{j=1}^N \int_\Omega |\mathcal{A}_j(x,u,\nabla u)||D_j\varphi|\omega\,dx \\
&+ \int_\Omega |g(x,u,\nabla u)|\varphi\omega\,dx \\
&\leq \int_\Omega \left(K_1 + h_3|u|^{p/p'} + h_4|\nabla u|^{p/p'}\right)|\nabla\varphi|\omega\,dx \\
&+ \int_\Omega \left(K_2 + h_5|u|^{p/p'} + h_6|\nabla u|^{p/p'}\right)|\varphi|\omega\,dx \\
&\leq C\Bigg[\|K_1\|_{L^{p'}(\Omega,\omega)} + \|K_2\|_{L^{p'}(\Omega,\omega)} \\
&+ \Big(C_\Omega(\|h_3\|_{L^\infty(\Omega)} + \|h_5\|_{L^\infty(\Omega)}) \qquad (5.13) \\
&+ \|h_4\|_{L^\infty(\Omega)} + \|h_6\|_{L^\infty(\Omega)}\Big)\|u\|^{p/p'}_{W_0^{1,p}(\Omega,\omega)}\Bigg]\|\varphi\|_{W_0^{1,p}(\Omega,\omega)}.
\end{aligned}
$$

Since $B(u,.)$ is linear, for each $u \in W_0^{1,p}(\Omega,\omega)$, there exists a linear and continuous functional on $W_0^{1,p}(\Omega,\omega)$ denoted by Au such that $(Au|\varphi) = B(u,\varphi)$ for all $\varphi \in W_0^{1,p}(\Omega,\omega)$. Moreover

$$\|Au\|_* \leq C\left(\|K_1\|_{L^{p'}(\Omega,\omega)} + \|K_2\|_{L^{p'}(\Omega,\omega)} + \|u\|^{p/p'}_{W_0^{1,p}(\Omega,\omega)}\right). \quad (5.14)$$

Hence, we obtain the operator

$$\begin{aligned} Au \ &: W_0^{1,p}(\Omega,\omega) \to [W_0^{1,p}(\Omega,\omega)]^* \\ u &\mapsto Au \end{aligned}$$

Consequently, problem (P11) is equivalent to the operator equation $Au = T, \quad u \in W_0^{1,p}(\Omega,\omega)$.

Step 3. Using conditions (H2) and (H8), we have

$$\begin{aligned} &(Au_1 - Au_2|u_1 - u_2) = B(u_1, u_1 - u_2) - B(u_2, u_1 - u_2) \\ &= \int_\Omega \langle \mathcal{A}(x,u_1,\nabla u_1), \nabla(u_1 - u_2)\rangle\, \omega\, dx \\ &- \int_\Omega \langle \mathcal{A}(x,u_2,\nabla u_2), \nabla(u_1 - u_2)\rangle\, \omega\, dx \\ &+ \int_\Omega [g(x,u_1,\nabla u_1) - g(x,u_2,\nabla u_2)](u_1 - u_2)\, \omega\, dx \\ &= \int_\Omega \langle \mathcal{A}(x,u_1,\nabla u_1) - \mathcal{A}(x,u_2,\nabla u_2), \nabla(u_1 - u_2)\rangle\, \omega\, dx \\ &+ \int_\Omega [g(x,u_1,\nabla u_1) - g(x,u_2,\nabla u_2)](u_1 - u_2)\, \omega\, dx \geq 0. \end{aligned}$$

Therefore the operator A is monotone.

Moreover, using (H3), (H7), the weighted Sobolev inequality (Theorem 2.1.5 with $\theta = 1$) and $C = \min\{\lambda, \Lambda\}$ we obtain

$$\begin{aligned} &(Au|u) = B(u,u) \\ = &\int_\Omega \langle \mathcal{A}(x,u,\nabla u), \nabla u\rangle\, \omega\, dx + \int_\Omega g(x,u,\nabla u) u\, \omega\, dx \\ \geq &\int_\Omega \Big(\Lambda |u|^p + \lambda |\nabla u|^p - h_1|u| - h_2\Big)\omega\, dx \end{aligned}$$

$$
\begin{aligned}
&= \Lambda \int_\Omega |u|^p \omega\, dx + \lambda \int_\Omega |\nabla u|^p \omega\, dx - \int_\Omega h_1 |u| \omega\, dx - \int_\Omega h_2 \omega\, dx \\
&\geq \Lambda \int_\Omega |u|^p \omega\, dx + \lambda \int_\Omega |\nabla u|^p \omega\, dx \\
&- \left(\int_\Omega h_1^{p'} \omega\, dx \right)^{1/p'} \left(\int_\Omega |u|^p \omega\, dx \right)^{1/p} - \|h_2\|_{L^1(\Omega,\omega)} \\
&\geq C \|u\|^p_{W_0^{1,p}(\Omega,\omega)} - \|h_1\|_{L^{p'}(\Omega,\omega)} \|u\|_{L^p(\Omega,\omega)} - \|h_2\|_{L^1(\Omega,\omega)} \\
&\geq C \|u\|^p_{W_0^{1,p}(\Omega,\omega)} - C_\Omega \|h_1\|_{L^{p'}(\Omega,\omega)} \|\nabla u\|_{L^p(\Omega,\omega)} - \|h_2\|_{L^1(\Omega,\omega)} \\
&\geq C \|u\|^p_{W_0^{1,p}(\Omega,\omega)} - C_\Omega \|h_1\|_{L^{p'}(\Omega,\omega)} \|u\|_{W_0^{1,p}(\Omega,\omega)} - \|h_2\|_{L^1(\Omega,\omega)}.
\end{aligned}
$$

Hence, since $1 < p < \infty$, we have

$$
\frac{(Au|u)}{\|u\|_{W_0^{1,p}(\Omega,\omega)}} = C \|u\|^{p-1}_{W_0^{1,p}(\Omega,\omega)} - C_\Omega \|h_1\|_{L^{p'}(\Omega,\omega)} - \frac{\|h_2\|_{L^1(\Omega,\omega)}}{\|u\|_{W_0^{1,p}(\Omega,\omega)}},
$$

and

$$
\frac{(Au|u)}{\|u\|_{W_0^{1,p}(\Omega,\omega)}} \to +\infty,
$$

as $\|u\|_{W_0^{1,p}(\Omega,\omega)} \to +\infty$, that is, A is coercive.

Step 4. We need to show that the operator A is continuous. Let $u_n \to u$ in $W_0^{1,p}(\Omega,\omega)$ as $n \to \infty$. We have,

$$
\begin{aligned}
& |B(u_n, \varphi) - B(u, \varphi)| \\
& \leq \sum_{j=1}^N \int_\Omega |\mathcal{A}_j(x, u_n, \nabla u_n) - \mathcal{A}_j(x, u, \nabla u)| |D_j \varphi| \omega\, dx \\
& + \int_\Omega |g(x, u_n, \nabla u_n) - g(x, u, \nabla u)| |\varphi| \omega\, dx \\
& = \sum_{j=1}^N \int_\Omega |F_j u_n - F_j u| |D_j \varphi| \omega\, dx + \int_\Omega |G u_n - G u| |\varphi| \omega\, dx
\end{aligned}
$$

$$\leq \sum_{j=1}^{N} \|F_j u_n - F_j u\|_{L^{p'}(\Omega,\omega)} \|D_j \varphi\|_{L^p(\Omega,\omega)}$$
$$+ \|G u_n - G u\|_{L^{p'}(\Omega,\omega)} \|\varphi\|_{L^{p'}(\Omega,\omega)}$$
$$\leq C \bigg(\sum_{j=1}^{N} \|F_j u_n - F_j u\|_{L^{p'}(\Omega,\omega)}$$
$$+ \|G u_n - G u\|_{L^{p'}(\Omega,\omega)} \bigg) \|\varphi\|_{W_0^{1,p}(\Omega,\omega)},$$

for all $\varphi \in W_0^{1,p}(\Omega, \omega)$. Hence,

$$\|Au_n - Au\|_* \quad \leq C \bigg(\sum_{j=1}^{N} \|F_j u_n - F_j u\|_{L^{p'}(\Omega,\omega)} \bigg)$$
$$+ \|G u_n - G u\|_{L^{p'}(\Omega,\omega)}.$$

Therefore, using (5.11) and (5.12), we have $\|Au_n - Au\|_* \to 0$ as $n \to +\infty$. Hence, A is continuous and this implies that A is hemicontinuous.

Therefore, by Theorem 5.1.1, the operator equation $Au = T$ has a solution u in $W_0^{1,p}(\Omega, \omega)$ and u is the solution for problem (P11).

Step 5. In Definition 5.1.3 with $\varphi = u \in W_0^{1,p}(\Omega, \omega)$ we get

$$\int_\Omega \mathcal{A}(x, u, \nabla u), \nabla u \rangle \, \omega \, dx + \int_\Omega g(x, u, \nabla u) \, u \, \omega \, dx$$
$$= \int_\Omega f_0 \, u \, dx + \sum_{j=1}^{N} \int_\Omega f_j \, D_j u \, dx.$$

Then we obtain

(i) By Theorem 2.1.5

$$\left| \int_\Omega f_0 \, u \, dx + \sum_{j=1}^{N} \int_\Omega f_j \, D_j u \, dx \right|$$

$$\leq \left(C_\Omega \| f_0/\omega \|_{L^{p'}(\Omega,\omega)} + \sum_{j=1}^{N} \| f_j/\omega \|_{L^{p'}(\Omega,\omega)} \right) \| u \|_{W_0^{1,p}(\Omega,\omega)}.$$

(ii) By (H3) (with $h_2 = 0$),(H7) and $C = \min\{\lambda, \Lambda\}$, we obtain

$$\begin{aligned} &\int_\Omega \langle \mathcal{A}(x,u,\nabla u), \nabla u \rangle \, \omega \, dx + \int_\Omega g(x,u,\nabla u)\, u \, \omega \, dx \\ &\geq \lambda \int_\Omega |\nabla u|^p \, \omega \, dx + \Lambda \int_\Omega |u|^p \, \omega \, dx - \int_\Omega h_1 |u| \, \omega \, dx \\ &\geq C \, \| u \|^p_{W_0^{1,p}(\Omega,\omega)} - \| h_1 \|_{L^{p'}(\Omega,\omega)} \| u \|_{W_0^{1,p}(\Omega,\omega)}. \end{aligned}$$

Therefore, by (i) and (ii), we get

$$\begin{aligned} \| u \|_{W_0^{1,p}(\Omega,\omega)} \; &\leq \; \frac{1}{C^{1/(p-1)}} \left(C_\Omega \| f_0/\omega \|_{L^{p'}(\Omega,\omega)} + \sum_{j=1}^{N} \| f_j/\omega \|_{L^{p'}(\Omega,\omega)} \right. \\ &+ \; \left. \| h_1 \|_{L^{p'}(\Omega,\omega)} \right). \end{aligned}$$

□

Example 5.1.3. *Let* $\Omega = \{(x,y) \in \mathbb{R}^n \; : x^2 + y^2 < 1\}$, *and consider the weight function* $\omega(x,y) = (x^2+y^2)^{-1/2}$ ($\omega \in A_2$), *the functions*
$\mathcal{A} : \Omega \times \mathbb{R} \times \mathbb{R}^2 \to \mathbb{R}^2$ *and* $g : \Omega \times \mathbb{R} \times \mathbb{R}^2 \to \mathbb{R}$ *defined by*

$$\begin{aligned} \mathcal{A}((x,y),\eta,\xi) &= h_4(x,y)\xi, \\ g((x,y),\eta,\xi) &= \eta \cos^2(xy), \end{aligned}$$

where $h_4(x,y) = 2\mathrm{e}^{x^2+y^2}$. *Let us consider the partial differential operator*

$$\begin{aligned} Lu(x,y) \; &= \; -\,\mathrm{div}\Big[\omega(x,y) \mathcal{A}((x,y),u,\nabla u) \Big] \\ &+ \; \omega(x,y) g((x,y),u,\nabla u) \end{aligned}$$

$$
\begin{aligned}
= & -\frac{\partial}{\partial x}\left[\omega(x,y)h_4(x,y)\frac{\partial u}{\partial x}\right] - \frac{\partial}{\partial y}\left[\omega(x,y)h_4(x,y)\frac{\partial u}{\partial y}\right] \\
+ & \ \omega(x,y)u(x,y)\cos^2(xy).
\end{aligned}
$$

Therefore, by Theorem 5.1.4, the problem

$$
\begin{cases}
Lu(x,y) & = \dfrac{\cos(xy)}{(x^2+y^2)^{1/5}} - \dfrac{\partial}{\partial x}\left(\dfrac{\sin(xy)}{(x^2+y^2)^{1/9}}\right) \\
 & - \dfrac{\partial}{\partial y}\left(\dfrac{\sin(xy)}{(x^2+y^2)^{1/9}}\right) \text{ in } \Omega, \\
u(x,y) & = 0 \text{ on } \partial\Omega,
\end{cases}
$$

has a solution $u \in W_0^{1,2}(\Omega,\omega)$.

5.2 Entropy solutions

The main purpose of this section (see Theorem 5.2.2) is to establish the existence of entropy solutions for the Dirichlet problem

$$
(P12)\begin{cases}
-\text{div}[\omega(x)\mathcal{A}(x,u,\nabla u)] = f(x) \ \text{ in } \ \Omega, \\
u(x) = 0 \ \text{ on } \ \partial\Omega,
\end{cases}
$$

where $\Omega \subset \mathbb{R}^N$ is a bounded open set, $f \in L^1(\Omega)$, ω is a weight function and the function $\mathcal{A} : \Omega\times\mathbb{R}\times\mathbb{R}^N \longrightarrow \mathbb{R}^N$ satisfies certain conditions.

The notion of entropy solution was introduced in [2] where the authors studied the nondegenerate elliptic equation $-\text{div}(a(x,Du)) = f(x)$, with $f \in L^1(\Omega)$. Note that, in the proof of our main result, many ideas have been adapted from [2].

We propose to solve the problem (P12) by approximation with variational solutions: we take $f_n \in C_0^\infty(\Omega)$ such that $f_n \to f$ in $L^1(\Omega)$, we find a solution $u_n \in W_0^{1,p}(\Omega,\omega)$ for the problem with right-hand side f_n and we will try to pass to the limit as $n \to \infty$.

Definition 5.2.1. *Let $f \in L^1(\Omega)$. We say that $u \in \mathcal{T}_0^{1,p}(\Omega,\omega)$ (see Definition 3.2.1) is an entropy solution to problem $(P12)$ if*

$$\int_\Omega \langle \mathcal{A}(x,u,\nabla u), \nabla T_k(u-\varphi)\rangle \, \omega \, dx = \int_\Omega f T_k(u-\varphi)\, dx, \quad (5.15)$$

for all $k>0$ and all $\varphi \in W_0^{1,p}(\Omega,\omega)\cap L^\infty(\Omega)$

We recall that the gradient of u which appears in (5.15) is defined as in Remark 3.2.1, that is to say that $\nabla u = \nabla T_k(u)$ on the set where $|u|<k$.

We need the following results.

Lemma 5.2.1. *Let $\omega \in A_p$, $1<p<\infty$ and a sequence $\{u_n\}$, $u_n \in W_0^{1,p}(\Omega,\omega)$ satisfies*
(i) $u_n \rightharpoonup u$ in $W_0^{1,p}(\Omega,\omega)$ and μ-a.e. in Ω.
(ii) $\displaystyle\int_\Omega \langle \mathcal{A}(x,u_n,\nabla u_n) - \mathcal{A}(x,u_n,\nabla u), \nabla(u_n-u)\rangle \, \omega\, dx \to 0$ with $n\to\infty$.
Then $u_n \to u$ in $W_0^{1,p}(\Omega,\omega)$.

Proof. The proof of this lemma follows the lines of Lemma 5 in [6]. □

For the proof of the next theorem we will use Definition 3.2.4, Lemma 3.2.5 and Lemma 3.2.6.

Theorem 5.2.2. *Let $\omega \in A_p$, $1<p<\infty$, and $\mathcal{A}(x,s,\xi)$ satisfies the following conditions*
(H1) *$x \mapsto \mathcal{A}(x,s,\xi)$ is measurable on Ω for all $(s,\xi)\in \mathbb{R}\times\mathbb{R}^N$, and $(s,\xi)\mapsto \mathcal{A}(x,s,\xi)$ is continuous on $\mathbb{R}\times\mathbb{R}^N$ for almost all $x\in\Omega$.*
(H2) *$\langle \mathcal{A}(x,s,\xi_1)-\mathcal{A}(x,s,\xi_2), \xi_1-\xi_2\rangle > 0$, whenever $\xi_1,\xi_2\in\mathbb{R}^N$, $\xi_1\neq\xi_2$ (where $\langle .,.\rangle$ denotes the usual inner product in $\mathbb{R}^N$).*
(H3) *$\langle \mathcal{A}(x,s,\xi),\xi\rangle \geq \lambda\,|\xi|^p$, with $1<p<\infty$ and $\lambda>0$.*

(H4) $|\mathcal{A}(x,s,\xi)| \leq K(x) + h_1(x)\,|s|^{p/p'} + h_2(x)\,|\xi|^{p/p'}$, *where* K, h_1 *and* h_2 *are positive functions,* $h_1 \in L^\infty(\Omega)$, $h_2 \in L^\infty(\Omega)$ *and* $K \in L^{p'}(\Omega,\omega)$ *(where* $1/p + 1/p' = 1$*).*

Then, there exists an entropy solutions u *of problem (P12). Moreover, we also have* $u \in \mathcal{M}^{p_1}(\Omega,\omega)$ *and* $|\nabla u| \in \mathcal{M}^{p_2}(\Omega,\omega)$, *with* $p_1 = \theta\,(p-1)$ *and* $p_2 = p_1\,p/(p_1+1)$ *(where* θ *is the constant in Theorem 2.1.5).*

Proof. **Step 1.** Considering a sequence $\{f_n\}$, $f_n \in C_0^\infty(\Omega)$, which

$$f_n \to f \text{ in } L^1(\Omega) \quad \text{and} \quad \|f_n\|_{L^1(\Omega)} \leq \|f\|_{L^1(\Omega)}.$$

For each n, there exist a solution $u_n \in W_0^{1,p}(\Omega,\omega)$ of the Dirichlet problem

$$(P_n)\begin{cases} -\text{div}[\omega(x)\mathcal{A}(x,u_n,\nabla u_n)] = f_n(x), & \text{in } \Omega \\ u_n(x) = 0, \quad \text{in } \partial\Omega \end{cases}$$

(see Theorem 5.1.2) that is,

$$\int_\Omega \langle \mathcal{A}(x,u_n,\nabla u_n), \nabla\varphi\rangle\,\omega\,dx = \int_\Omega f_n\varphi\,dx, \tag{5.16}$$

for all $\varphi \in W_0^{1,p}(\Omega,\omega)$. For $\varphi = T_k(u_n)$ we obtain in (5.16) that

$$\int_\Omega \langle \mathcal{A}(x,u_n,\nabla u_n), \nabla T_k(u_n)\rangle\,\omega\,dx = \int_\Omega f_n T_k(u_n)\,dx.$$

We have,

$$\begin{aligned} \left|\int_\Omega f_n T_k(u_n)\,dx\right| &\leq \int_\Omega |f_n||T_k(u_n)|\,dx \\ &\leq k\|f_n\|_{L^1(\Omega)} \leq k\|f\|_{L^1(\Omega)} = C_1. \end{aligned}$$

Using (H3) we have,

$$\int_\Omega \langle \mathcal{A}(x,u_n,\nabla u_n), \nabla T_k(u_n)\rangle\,\omega\,dx$$

$$= \int_\Omega \langle \mathcal{A}(x, u_n, \nabla T_k(u_n)), \nabla T_k(u_n)\rangle \,\omega\, dx$$
$$\geq \lambda \int_\Omega |\nabla T_k(u_n)|^p \omega\, dx.$$

Hence,

$$\int_\Omega |\nabla T_k(u_n)|^p \omega\, dx \leq \frac{k}{\lambda} \|f\|_{L^1(\Omega)}. \tag{5.17}$$

By Lemma 3.2.5 and Lemma 3.2.6, we have that the sequence $\{u_n\}$ is bounded in $\mathcal{M}^{p_1}(\Omega, \omega)$ (with $p_1 = \theta\,(p-1)$ and $\{|\nabla u_n|\}$ is bounded in $\mathcal{M}^{p_2}(\Omega, \omega)$ (with $p_2 = p_1\,p/(p_1 + 1)$). Moreover, $\{u_n\}$ is a Cauchy sequence in μ-measure. Consequently, there exists a function u and a subsequence, that we will still denote by $\{u_n\}$, such that

$$u_n \to u \quad \mu - \text{a.e. in } \Omega, \tag{5.18}$$

and $u_n \to u$ a.e. in Ω (by Lemma 2.1.2). Using (5.17) and (5.18), we have

$$\begin{aligned} T_k(u_n) &\rightharpoonup T_k(u) \text{ weakly in } W_0^{1,p}(\Omega, \omega), \\ T_k(u_n) &\to T_k(u) \text{ strongly in } L^p(\Omega, \omega) \text{ and } \mu - \text{a.e. in } \Omega, \end{aligned} \tag{5.19}$$

for all $k > 0$. Hence $T_k(u) \in W_0^{1,p}(\Omega, \omega)$.
Furthermore, by the weak lower semicontinuity of the norm $W_0^{1,p}(\Omega, \omega)$, we have that (5.17) still holds for u, that is,

$$\int_\Omega |\nabla T_k(u)|^p \omega\, dx \leq \frac{k}{\lambda} \|f\|_{L^1(\Omega)}.$$

Applying Lemma 3.2.5 and Lemma 3.2.6, we have that $u \in \mathcal{M}^{p_1}(\Omega, \omega)$ (with $p_1 = \theta\,(p-1)$) and $|\nabla u| \in \mathcal{M}^{p_2}(\Omega, \omega)$ (with $p_2 = p_1\,p/(p_1 + 1)$).
Step 2. We need to show that $T_k(u_n) \to T_k(u)$ strongly in $W_0^{1,p}(\Omega, \omega)$, for all $k > 0$.

Let $h > k$ and applying (5.16) with function

$$\varphi_n = T_{2k}(u_n - T_h(u_n) + T_k(u_n) - T_k(u)),$$

we get

$$\int_\Omega \langle \mathcal{A}(x, u_n, \nabla u_n), \nabla \varphi_n \rangle \, \omega \, dx = \int_\Omega f_n \varphi_n \, dx. \tag{5.20}$$

If we set $M = 4k + h$, we have $\nabla \varphi_n = 0$ for $|u_n| > M$. Hence, since condition (H3) implies that $\mathcal{A}(x, s, 0) = 0$, we can write

$$\int_\Omega \langle \mathcal{A}(x, T_M(u_n), \nabla T_M(u_n)), \nabla \varphi_n \rangle \, \omega \, dx = \int_\Omega f_n \varphi_n \, dx. \tag{5.21}$$

In the left-hand side of (5.21), we have

$$\begin{aligned}
&\int_\Omega \langle \mathcal{A}(x, T_M(u_n), \nabla T_M(u_n)), \nabla T_{2k}(u_n - T_h(u_n) \\
+ &\quad T_k(u_n) - T_k(u)) \rangle \, \omega \, dx \\
= &\int_{\{|u_n| \leq k\}} \langle \mathcal{A}(x, T_M(u_n), \nabla T_M(u_n)), \nabla T_{2k}(u_n - T_h(u_n) \\
+ &\quad T_k(u_n) - T_k(u)) \rangle \, \omega \, dx \\
+ &\int_{\{|u_n| > k\}} \langle \mathcal{A}(x, T_M(u_n), \nabla T_M(u_n)), \nabla T_{2k}(u_n - T_h(u_n) \\
+ &\quad T_k(u_n) - T_k(u)) \rangle \, \omega \, dx
\end{aligned} \tag{5.22}$$

(a) If $|u_n| \leq k$. Since $h > k$, if $|u_n| \leq k < h$, then $T_h(u_n) = T_k(u_n) = u_n$. Hence, $u_n - T_h(u_n) + T_k(u_n) - T_k(u) = u_n - T_k(u)$. We also have that $|u_n - u| \leq 2k$. Then, since $\nabla T_M(u_n) = \nabla T_k(u_n)$ (because $|u_n| \leq k < M$),

$$\begin{aligned}
&\int_{\{|u_n| \leq k\}} \langle \mathcal{A}(x, T_M(u_n), \nabla T_M(u_n)), \nabla T_{2k}(u_n - T_h(u_n) \\
+ &\quad T_k(u_n) - T_k(u)) \rangle \, \omega \, dx
\end{aligned}$$

$$= \int_{\{|u_n| \leq k\}} \langle \mathcal{A}(x, T_k, \nabla T_k(u_n)), \nabla(T_k(u_n) - T_k(u)) \, \omega \, dx$$
$$= \int_\Omega \langle \mathcal{A}(x, T_k(u_n), \nabla T_k(u_n)), \nabla(T_k(u_n) - T_k(u)) \rangle \, \omega \, dx.$$

(b) If $|u_n| > k$. Since $u_n, T_k(u_n)$ and $T_k(u)$ are in $W_0^{1,p}(\Omega, \omega)$, if

$$|u_n - T_h(u_n) + T_k(u_n) - T_k(u)| \leq 2k,$$

we obtain

$$\begin{aligned} &\nabla T_{2k}(u_n - T_h(u_n) + T_k(u_n) - T_k(u)) \\ &= \nabla(u_n - T_h(u_n) + T_k(u_n) - T_k(u)) \\ &= \nabla u_n - \nabla T_h(u_n) + \nabla T_k(u_n) - \nabla T_k(u) \\ &= \nabla u_n - \nabla T_h(u_n) - \nabla T_k(u) \end{aligned}$$

(because $\nabla T_k(u_n) = 0$ if $|u_n| > k$).

There are two possible cases as follows:

(i) If $k < |u_n| < h$, we have $\nabla T_h(u_n) = \nabla u_n$. Then

$$\nabla T_{2k}(u_n - T_h(u_n) + T_k(u_n) - T_k(u)) = -\nabla T_k(u);$$

(ii) If $h < |u_n| \leq M$, we have that $\nabla T_h(u_n) = 0$. Then

$$\begin{aligned} \nabla T_{2k}(u_n - T_h(u_n) + T_k(u_n) - T_k(u)) &= \nabla u_n - \nabla T_k(u) \\ &= \nabla T_M(u_n) - \nabla T_k(u). \end{aligned}$$

Since $\langle \mathcal{A}(x, s, \xi), \xi \rangle \geq \lambda |\xi|^p \geq 0$, in both cases we obtain

$$\begin{aligned} &\langle \mathcal{A}(x, T_M(u_n), \nabla T_m(u_n)), \nabla T_{2k}(u_n - T_h(u_n) + T_k(u_n) - T_k(u)) \rangle \\ &\geq - \langle \mathcal{A}(x, T_M(u_n), \nabla T_M(u_n), \nabla T_k(u) \rangle \\ &\geq - |\mathcal{A}(x, T_M(x), \nabla T_M(x))||\nabla T_k(u)|. \end{aligned}$$

Therefore we obtain in (5.22)

$$\int_\Omega \langle \mathcal{A}(x, T_M(u_n), \nabla T_M(u_n)), \nabla T_{2k}(u_n - T_h(u_n)$$

$$
\begin{aligned}
+ & \quad T_k(u_n) - T_k(u))\rangle \,\omega\, dx \\
= & \int_{\{|u_n|\leq k\}} \langle \mathcal{A}(x, T_M(u_n), \nabla T_M(u_n)), \nabla T_{2k}(u_n - T_h(u_n) \\
+ & \quad T_k(u_n) - T_k(u))\rangle \,\omega\, dx \\
+ & \int_{\{|u_n|>k\}} \langle \mathcal{A}(x, T_M(u_n), \nabla T_M(u_n)), \nabla T_{2k}(u_n - T_h(u_n) \\
+ & \quad T_k(u_n) - T_k(u))\rangle \,\omega\, dx \\
\geq & \int_\Omega \langle \mathcal{A}(x, T_k(u_n), \nabla T_k(u_n)), \nabla (T_k(u_n) - T_k(u)) \,\omega\, dx \\
- & \int_{\{|u_n|>k\}} |\mathcal{A}(x, T_M(u_n), \nabla T_M(u_n))||\nabla T_k(u)| \,\omega\, dx.
\end{aligned}
$$

Hence, in (5.21) we obtain

$$
\begin{aligned}
& \int_\Omega \langle \mathcal{A}(x, T_k(u_n), \nabla T_k(u_n)) - \mathcal{A}(x, T_k(u_n), \nabla T_k(u)), \\
& \quad \nabla (T_k(u_n) - T_k(u)) \,\rangle \,\omega\, dx \\
\leq & \int_{\{|u_n|>k\}} |\mathcal{A}(x, T_M(u_n), \nabla T_M(u_n))||\nabla T_k(u)| \,\omega\, dx \\
+ & \int_\Omega f_n T_{2k}(u_n - T_h(u_n) + T_k(u_n) - T_k(u))\, dx \\
- & \int_\Omega \langle \mathcal{A}(x, T_k(u_n), \nabla T_k(u)), \nabla (T_k(u_n) - T_k(u))\rangle \,\omega\, dx.
\end{aligned}
\tag{5.23}
$$

Considering the test function $\psi_n = T_{2k}(u_n - T_h(u_n))$ in (5.16), we have

$$
\int_\Omega \langle \mathcal{A}(x, u_n, \nabla u_n), \nabla \psi_n \rangle \,\omega\, dx = \int_\Omega f_n \psi_n \, dx,
$$

and using that

$$
\left| \int_\Omega f_n \psi_n \, dx \right| \;\leq\; \int_\Omega |f_n||\psi_n| \, dx
$$

$$\begin{aligned} &\leq (2k+1)\|f_n\|_{L^1(\Omega)} \\ &\leq (2k+1)\|f\|_{L^1(\Omega)}, \end{aligned}$$

and

$$\begin{aligned} &\int_\Omega \langle \mathcal{A}(x, u_n, \nabla u_n), \nabla \psi_n \rangle \, \omega \, dx = \int_\Omega \langle \mathcal{A}(x, u_n, \nabla \psi_n), \nabla \psi_n \rangle \, \omega \, dx \\ &\geq \lambda \int_\Omega |\nabla \psi_n|^p \omega \, dx \\ &= \lambda \int_\Omega |\nabla T_{2k}(u_n - T_h(u_n))|^p \omega \, dx, \end{aligned}$$

we obtain

$$\int_\Omega |\nabla T_{2k}(u_n - T_h(u_n))|^p \omega \, dx \leq \frac{(2k+1)}{\lambda} \|f\|_{L^1(\Omega)}.$$

Now using that $T_{2k}(u_n - T_h(u_n)) \rightharpoonup T_{2k}(u - T_h(u))$ weakly in $W_0^{1,p}(\Omega, \omega)$ (by (5.19) and Remark 3.2.1(c)), we have

$$\int_\Omega |\nabla T_{2k}(u - T_h(u))|^p \omega \, dx \leq \frac{(2k+1)}{\lambda} \|f\|_{L^1(\Omega)},$$

and then (by Theorem 2.1.5, with $\theta = 1$)

$$\begin{aligned} \int_\Omega |T_{2k}(u - T_h(u))|^p \omega \, dx &\leq C_\Omega \int_\Omega |\nabla T_{2k}(u - T_h(u))|^p \omega \, dx \\ &\leq C_\Omega \frac{(2k+1)}{\lambda} \|f\|_{L^1(\Omega)}. \end{aligned}$$

Moreover, by the Lebesgue Dominated Convergence Theorem, we obtain

$$\lim_{h \to \infty} \int_\Omega f \, T_{2k}(u - T_h(u)) \, dx = 0.$$

We can fix a positive real number h_ε sufficiently large to have

$$\int_\Omega f T_{2k}(u - T_{h_\varepsilon}) \, dx \leq \varepsilon. \tag{5.24}$$

Considering $h = h_\varepsilon$ in (5.23) (and $M = M_\varepsilon = 4k + h_\varepsilon$), by (H4) and (5.17), we have

$$\begin{aligned}
&\int_\Omega |\mathcal{A}(x, T_M(u_n), \nabla T_M(u_n)|^{p'} \omega\, dx \\
&\leq \int_\Omega \left(K(x) + h_1(x)|T_M(u_n)|^{p/p'} + h_2(x)|\nabla T_M(u_n)|^{p/p'} \right)^{p'} \omega\, dx \\
&\leq C\bigg[\int_\Omega K^{p'}(x)\, \omega\, dx + \int_\Omega h_1^{p'}(x)|T_M(u_n)|^p \omega\, dx \\
&+ \int_\Omega h_2^{p'}(x)|\nabla T_M(u_n)|^p \omega\, dx \\
&\leq C\bigg(\|K\|^{p'}_{L^{p'}(\Omega,\omega)} + \|h_1\|^{p'}_{L^\infty(\Omega)} \int_\Omega |T_M(u_n)|^p \omega\, dx \\
&+ \|h_2\|^{p'}_{L^\infty(\Omega)} \int_\Omega |\nabla T_M(u_n)|^p \omega\, dx \bigg) \\
&\leq C\bigg(\|K\|^{p'}_{L^{p'}(\Omega,\omega)} + \|h_1\|^{p'}_{L^\infty(\Omega)} M^p \omega(\Omega) + \|h_2\|^{p'}_{L^\infty(\Omega)} \frac{M}{\lambda} \|f\|_{L^1(\Omega)} \bigg),
\end{aligned}$$

that is, $|\mathcal{A}(x, T_M(u_n), \nabla T_M(u_n))|$ is bounded in $L^{p'}(\Omega, \omega)$. Moreover, we obtain $\chi_{\{|u_n|>k\}}|\nabla T_k(u)| \to 0$ in $L^p(\Omega, \omega)$ as $n \to \infty$. Therefore,

$$\lim_{n\to\infty} \int_{\{|u_n|>k\}} |\mathcal{A}(x, T_M(u_n), \nabla T_M(u_n))||\nabla T_k(u)|\, \omega\, dx = 0. \tag{5.25}$$

Furthermore, we have that

$$T_{2k}(u_n - T_h(u_n) + T_k(u_n) - T_k(u)) \rightharpoonup T_{2k}(u - T_h(u)),$$

weakly in $W_0^{1,p}(\Omega, \omega)$, as $n \to \infty$. Hence, by (5.19), (5.24) and (5.25), passing to the limit in (5.23), we have

$$\lim_{n\to\infty} \int_\Omega \langle \mathcal{A}(x, T_k(u_n), \nabla T_k(u_n)) - \mathcal{A}(x, T_k(u_n), \nabla T_k(u)),$$

$$
\begin{aligned}
&\nabla(T_k(u_n) - T_k(u))\rangle\,\omega\,dx \\
&\leq \int_\Omega fT_{2k}(u - T_{h_\varepsilon})\,dx \\
&\leq \varepsilon,
\end{aligned}
$$

for all $\varepsilon > 0$, that is,

$$
\lim_{n\to\infty}\int_\Omega \langle \mathcal{A}(x, T_k(u_n), \nabla T_k(u_n)) - \mathcal{A}(x, T_k(u_n), \nabla T_k(u)), \\
\nabla(T_k(u_n) - T_k(u))\rangle\,\omega\,dx = 0.
$$

Applying Lemma 5.2.1 we get

$$
T_k(u_n) \to T_k(u) \tag{5.26}
$$

strongly in $W_0^{1,p}(\Omega,\omega)$ for every $k > 0$. This convergence implies that, for every fixed $k > 0$

$$
\mathcal{A}(x, T_k(u_n), \nabla T_k(u_n)) \to \mathcal{A}(x, T_k(u), \nabla T(u)), \tag{5.27}
$$

in $[L^{p'}(\Omega,\omega)]^N = L^{p'}(\Omega,\omega) \times ... \times L^{p'}(\Omega,\omega)$.

Step 3. Finally, we need to show that u is an entropy solution to Dirichlet problem (P12). Let us take $\psi_n = T_k(u_n - \varphi)$ as test function in (5.16), with $\varphi \in W_0^{1,p}(\Omega,\omega) \cap L^\infty(\Omega)$. We obtain,

$$
\int_\Omega \langle \mathcal{A}(x, u_n, \nabla u_n), \nabla\psi_n\rangle\,\omega\,dx = \int_\Omega f_n\psi_n\,dx. \tag{5.28}
$$

If $M = k + \|\varphi\|_{L^\infty(\Omega)}$ and $n > M$, we have

$$
\begin{aligned}
&\int_\Omega \langle \mathcal{A}(x, u_n \nabla u_n), \nabla T_k(u_n - \varphi)\rangle\,\omega\,dx \\
&= \int_\Omega \langle \mathcal{A}(x, T_M(u_n), \nabla T_M(u_n)), \nabla T_k(u_n - \varphi)\rangle\,\omega\,dx.
\end{aligned}
$$

Hence, in (5.28) we obtain

$$
\int_\Omega \langle \mathcal{A}(x, T_M(u_n), \nabla T_M(u_n)), \nabla T_k(u_n - \varphi)\rangle\,\omega\,dx
$$

$$= \int_\Omega f_n T_k(u_n - \varphi)\,dx. \tag{5.29}$$

Therefore, by (5.19) and (5.27), passing to the limit $n \to \infty$ in (5.29), we obtain

$$\int_\Omega \langle \mathcal{A}(x, u, \nabla u), \nabla T_k(u - \varphi)\rangle\, \omega\, dx = \int_\Omega f\, T_k(u - \varphi)\,dx$$

for all $\varphi \in W_0^{1,p}(\Omega, \omega) \cap L^\infty(\Omega)$ and for each $k > 0$.

Therefore u is an entropy solutions of problem $(P12)$. □

Example 5.2.1. *Let* $\Omega = \{(x, y) \in \mathbb{R}^2 : x^2 + y^2 < 1\}$, $\omega(x, y) = (x^2 + y^2)^{\alpha/2}$ *(*$\omega \in A_p$*, with* $-2 < \alpha < 2(p-1)$*),* $f(x, y) = \dfrac{\cos(xy)}{(x^2 + y^2)^{1/3}}$ *(*$f \in L^1(\Omega)$*) and* $\mathcal{A} : \Omega \times \mathbb{R} \times \mathbb{R}^2 \to \mathbb{R}^2$, $\mathcal{A}((x, y), s, \xi) = |\xi|^{p-2}\xi$. *By Theorem 5.2.2, the problem*

$$\begin{cases} -\mathrm{div}[(x^2 + y^2)^{-1/2}|\nabla u|^{p-2}\nabla u] = \dfrac{\cos(xy)}{(x^2 + y^2)^{1/3}} & \text{in } \Omega, \\ u(x, y) = 0 \quad \text{on } \partial\Omega, \end{cases}$$

has an entropy solution.

Chapter 6

Degenerate Nonlinear Elliptic Equations

The goal of this chapter is to introduce degenerate nonlinear problems, i.e., problems whose solution does not depend linearly on the data. Complete expositions about the methods of nonlinear analysis can be found in [30], [60] and [61].

6.1 Strongly nonlinear elliptic equations

In this section we prove the existence of solution in $\mathcal{D}(A) \subseteq W_0^{1,2}(\Omega,\omega)$ for the following nonlinear Dirichlet problem

$$(P13)\begin{cases} -Lu(x) + g(u(x))\,\omega(x) = f_0(x) - \displaystyle\sum_{j=1}^{N} D_j f_j \text{ in } \Omega, \\ u(x) = 0 \text{ on } \partial\Omega, \end{cases}$$

where L is an elliptic operator in divergence form

$$Lu(x) = \sum_{i,j=1}^{N} D_j(a_{ij}(x)D_i u(x)) \tag{6.1}$$

and the coefficients a_{ij} are measurable, real-valued functions whose coefficient matrix $(a_{ij})_{1\le i,j\le N}$ is symmetric and satisfies the degenerate ellipticity condition

$$\lambda\,|\xi|^2\,\omega(x)\le \sum_{i,j=1}^{N} a_{ij}(x)\,\xi_i\xi_j\le \Lambda|\xi|^2\,\omega(x), \tag{6.2}$$

for all $\xi\in\mathbb{R}^N$ and almost every $x\in\Omega\,\mathbb{R}^N$ a bounded open set with piecewise smooth boundary (i.e., $\partial\Omega\in C^{0,1}$), ω is a weight function, λ and Λ are positive constants.

Definition 6.1.1. *Let X and X^+ be Banach spaces over $\mathbb{R}$. Then $\{X,X^+\}$ is called a dual pair iff the following hold:*
(i) There exists a bilinear, bounded map

$$(.|.)_X : X^+\times X\to\mathbb{R}$$
$$(v,u)\mapsto(v\,|u)_X.$$

(ii) If $(v|u)_X=0$ for all $u\in X$ then $v=0$.
(iii) If $(v|u)_X=0$ for all $v\in X^+$ then $u=0$.

Example 6.1.1. *(a) Let X be a Banach space and let X^* be its dual space. Then $\{X,X^*\}$ is a dual pair if we set, in the usual way, $(v|u)_X=v(u)$ for all $v\in X^*$ and $u\in X$.*
(b) Let $\omega\in A_2$. If $X=W^{1,2}(\Omega,\omega)$ and $X^+=L^2(\Omega,\omega)$, then $\{X,X^+\}$ forms a dual pair with respect to $(v|u)_X=\int_\Omega v\,u\,\omega\,dx$

The basic idea is to reduce $(P13)$ to an operator equations

$$Au=T,\ \ u\in\mathcal{D}(A) \tag{6.3}$$

where $\mathcal{D}(A)=\{u\in W_0^{1,2}(\Omega,\omega)\,:\,u\,g(u)\in L^1(\Omega,\omega)\}$ and apply the theorem below.

Theorem 6.1.1. *Suppose that assumptions (H1) through (H7) below are satisfied.*

(**H1**) *Dual pair. Let the dual pair* $\{X, X^+\}$ *and* $\{Y, Y^+\}$ *be given, where* X, X^+, Y *and* Y^+ *are real Banach spaces with corresponding bilinear forms* $(.|.)_X$ *and* $(.|.)_Y$ *and the continuous embeddings* $Y \subseteq X$ *and* $X^+ \subseteq Y^+$. *The dual pair are compatible, i.e.,* $(T|u)_X = (T|u)_Y$ *for all* $T \in, X^+$, $u \in Y$. *Moreover, the Banach spaces* X *and* Y *are separable and* X *is reflexive.*
(**H2**) *Operator* A. *Let the operator* $A : \mathcal{D}(A) \subseteq X \to Y^+$ *be given, and let* K *be a bounded closed convex set in* X *containing the zero point as an interior point and* $K \cap Y \subseteq \mathcal{D}(A)$.
(**H3**) *Local coerciveness. There exists a number* $\alpha \geq 0$ *such that* $(Av|v)_Y \geq \alpha$ *for all* $v \in Y \cap \partial K$, *where* ∂K *denotes the boundary of* K *in the Banach space* X.
(**H4**) *Continuity. For each finite-dimensional subspace* Y_0 *of the Banach space* Y, *the mapping* $u \mapsto (Au|v)_Y$ *is continuous on* $K \cap Y_0$ *for all* $v \in Y_0$.
(**H5**) *Generalized condition (M). Let* $\{u_n\}$ *be a sequence in* $Y \cap K$ *and let* $T \in X^+$. *Then, from*

$$
\begin{aligned}
&u_n \rightharpoonup u \text{ in } X, \text{ as } n \to \infty, \\
&(Au_n|v)_Y \to (T|v)_X, \text{ as } n \to \infty, \text{ for all } v \in Y, \\
&\text{and } \overline{\lim_{n\to\infty}} (Au_n|u_n)_Y \leq (T|u)_X,
\end{aligned}
\tag{6.4}
$$

it follows that $Au = T$.
(**H6**) *Quasi-boundedness. Let* $\{u_n\}$ *be a sequence in* $Y \cap K$. *Then, from (6.4) and* $(Au_n|u_n)_Y \leq C\,\|u\|_X$ *for all* $n \in \mathbb{N}$, *it follows that the sequence* $\{Au_n\}$ *is bounded in* Y^+.
(**H7**) *Global coerciveness.* $\dfrac{(Av|v)_Y}{\|v\|_X} \to \infty$ *as* $\|v\|_X \to \infty$.

Then $X^+ \subseteq \mathcal{R}(A)$, *i.e., the equation* $Au = T$ *has a solution* u *for each* $T \in X^+$.

Proof. See Theorem 27.B and Corollary 27.19 in [61]. □

Let us introduce the following definition of (weak) solution for the problem (P13).

Definition 6.1.2. *We say that $u \in \mathcal{D}(A) \subseteq W_0^{1,2}(\Omega,\omega)$ is a (weak) solution to the problem (P13) if*

$$\sum_{i,j=1}^{N} \int_\Omega a_{ij} D_i u \, D_j\varphi \, dx + \int_\Omega g(u)\, \varphi \, \omega \, dx$$
$$= \int_\Omega f_0 \, \varphi \, dx + \sum_{j=1}^{N} \int_\Omega f_j \, D_j \varphi \, dx,$$

for all $\varphi \in Y = W_0^{1,2}(\Omega,\omega) \cap W^{k,p}(\Omega,\omega)$, where $p = 2q$, $1 < q \leq 2$, $k > N/2$ and $\|\varphi\|_Y = \|\varphi\|_{W^{k,p}(\Omega,\omega)}$, with $\mathcal{D}(A) = \{u \in W_0^{1,2}(\Omega,\omega) : g(u)\, u \in L^1(\Omega,\omega)\}$.

Remark 6.1.1. *Using that $p = 2q > 2$ and $\omega^{1/(1-q)}(\Omega) < \infty$ (because $\omega \in A_q$, $1 < q \leq 2$) we obtain*

$$\|.\|_{L^2(\Omega)} \leq [\omega^{1/(1-q)}(\Omega)]^{(q-1)/2q} \, \|.\|_{L^p(\Omega,\omega)},$$

where $\omega^{1/(1-q)}(\Omega) = \int_\Omega \omega^{1/(1-q)}(x)\, dx$. Thus,

$$W^{k,p}(\Omega) \subset W^{k,2}(\Omega) \subseteq C(\overline{\Omega}).$$

Therefore, the embedding $Y \subset C(\overline{\Omega})$ is continuous.

We will apply Theorem 6.1.1 to a sufficiently large ball K in the Banach spaces $X = W_0^{1,2}(\Omega,\omega)$, $X^+ = [W_0^{1,2}(\Omega,\omega)]^*$ and $Y^+ = Y^*$

Theorem 6.1.2. *Let L be the operator (6.1) and the condition (6.2) hold. Suppose that the following assumptions are satisfied:*
(H1) *$\omega \in A_q$, with $1 < q \leq 2$;*
(H2) *The function $g : \mathbb{R} \to \mathbb{R}$ is continuous with $x\, g(x) \geq 0$ for all $x \in \mathbb{R}$;*
(H3) *$f_j/\omega \in L^2(\Omega,\omega)$, $j = 0, 1, ..., N$.*

Then the problem (P13) has a solution $u \in \mathcal{D}(A) \subseteq W_0^{1,2}(\Omega,\omega)$. *Moreover, we have*

$$\|u\|_{W_0^{1,2}(\Omega,\omega)} \leq \frac{(C_\Omega^2+1)^2}{\lambda}\left(\sum_{j=0}^{N}\|f_j/\omega\|_{L^2(\Omega,\omega)}\right),$$

where C_Ω *is the constant in Theorem 2.1.5.*

Proof. We define for $u \in \mathcal{D}(A)$ and $\varphi \in Y$

$$B_1(u,\varphi) = \sum_{i,j=1}^{N}\int_\Omega a_{ij}\, D_i u\, D_j\varphi\, dx,$$

$$B_2(u,\varphi) = \int_\Omega g(u)\,\varphi\,\omega\, dx,$$

$$T(\varphi) = \int_\Omega f_0\,\varphi\, dx + \sum_{j=1}^{N}\int_\Omega f_j\, D_j\varphi\, dx.$$

Then $u \in \mathcal{D}(A) \subseteq W_0^{1,2}(\Omega,\omega)$ is solution to the problem (P13) if

$$B_1(u,\varphi) + B_2(y,\varphi) = T(\varphi), \text{ for all } \varphi \in Y.$$

Step 1. $T \in [W_0^{1,2}(\Omega,\omega)]^*$. In fact, using the hypothesis (H3), Theorem 2.1.5 and the Hölder inequality, we have

$$\begin{aligned} |T(\varphi)| &\leq \int_\Omega |f_0|\,|\varphi|\, dx + \sum_{j=1}^{N}\int_\Omega |f_j|\,|D_j\varphi|\, dx \\ &= \int_\Omega \frac{|f_0|}{\omega}\,|\varphi|\,\omega\, dx + \sum_{j=1}^{N}\int_\Omega \frac{|f_j|}{\omega}\,|D_j\varphi|\,\omega\, dx \\ &\leq \left\|\frac{f_0}{\omega}\right\|_{L^2(\Omega,\omega)}\|\varphi\|_{L^2(\Omega,\omega)} + \sum_{j=1}^{N}\left\|\frac{f_j}{\omega}\right\|_{L^2(\Omega,\omega)}\|D_j\varphi\|_{L^2(\Omega,\omega)} \end{aligned}$$

$$\leq C_\Omega\bigg(\sum_{j=0}^{N}\bigg\|\frac{f_j}{\omega}\bigg\|_{L^2(\Omega,\omega)}\bigg)\|\varphi\|_{W_0^{1,2}(\Omega,\omega)},$$

for all $\varphi\in W_0^{1,2}(\Omega,\omega)$.

Step 2. By condition (6.2) we obtain $|a_{ij}(x)|\leq C\,\omega(x)$ for almost every $x\in\Omega$. Then, using Remark 6.1.1 (since $p=2q>2$) and Lemma 4.1.2, we obtain

$$\begin{aligned}|B_1(u,\varphi)| &\leq \sum_{i,j=1}^{N}\int_\Omega |a_{ij}|\,|D_iu|\,|D_j\varphi|\,dx\\ &\leq C\sum_{i,j=1}^{N}\int_\Omega |D_iu|\,|D_j\varphi|\,\omega\,dx\\ &\leq C\,[\omega(\Omega)]^{(p-2)/2p}\|u\|_{W_0^{1,2}(\Omega,\omega)}\,\|\varphi\|_Y,\end{aligned}$$

because we have that $\|.\|_{L^2(\Omega,\omega)}\leq[\omega(\Omega)]^{(p-2)/2p}\|.\|_{L^p(\Omega,\omega)}$ if $p>2$. Hence there exists exactly one linear continuous operator

$$A_1:W_0^{1,2}(\Omega,\omega)\to Y^*$$

which

$$(A_1u|\varphi)_Y=B_1(u,\varphi)\text{ for all }u\in W_0^{1,2}(\Omega,\omega),\ \ \varphi\in Y.$$

Step 3. Note that, there exits a constant $C>0$, such that

$$|g(x)|\leq xg(x)+C$$

for all $x\in\mathbb{R}$. Therefore, if $u\in\mathcal{D}(A)$ we have that $g(u)\in L^1(\Omega,\omega)$. Using that $\omega\in A_q\subset A_2$, the hypothesis (H2), Remark 6.1.1 and Lemma 4.1.2, we obtain for $u\in\mathcal{D}(A)$ fixed

$$|B_2(u,\varphi)| \leq \int_\Omega |g(u)|\,|\varphi|\,\omega\,dx$$

$$\begin{aligned} &\leq \|\varphi\|_{C(\bar{\Omega})} \int_{\Omega} |g(u)|\, \omega\, dx \\ &\leq C\, \|\varphi\|_{Y}. \end{aligned}$$

Thus, there exists a uniquely operator

$$A_2 : \mathcal{D}(A) \subseteq W_0^{1,2}(\Omega,\omega) \to Y^*$$

which

$$(A_2u|\varphi)_Y = B_2(u,\varphi), \text{ for all } u \in \mathcal{D}(A),\ \varphi \in Y.$$

Step 4. We define the operator

$$\begin{aligned} A : \mathcal{D}(A) &\subseteq W_0^{1,2}(\Omega,\omega) \to Y^* \\ A &= A_1 + A_2. \end{aligned}$$

We have

$$\begin{aligned} (Au|\varphi)_Y &= (A_1u|\varphi)_Y + (A_2u|\varphi)_Y \\ &= B_1(u,\varphi) + B_2(u,\varphi). \end{aligned}$$

Thus, $u \in \mathcal{D}(A)$ is solution to the problem (P13) is $(Au|\varphi)_Y = T(\varphi)$, for all $\varphi \in Y$. Then the problem (P13) corresponds to the operator equation

$$Au = T,\ u \in \mathcal{D}(A).$$

Step 5. Global coerciveness of operator A. Using the condition (6.2) and the hypothesis (H2), we obtain

$$\begin{aligned} (A\varphi|\varphi)_Y &= B_1(\varphi,\varphi) + B_2(\varphi,\varphi) \\ &= \sum_{i,j=1}^{N} \int_{\Omega} a_{ij}\, D_i\varphi\, D_j\varphi\, dx + \int_{\Omega} g(\varphi)\, \varphi\, \omega\, dx \\ &\geq \lambda \int_{\Omega} |\nabla\varphi|^2\, \omega\, dx \end{aligned}$$

$$\geq \frac{\lambda}{C_\Omega^2 + 1} \|\varphi\|^2_{W_0^{1,2}(\Omega,\omega)}.$$

Thus,

$$\lim_{\|\varphi\|_{W_0^{1,2}(\Omega,\omega)} \to \infty} \frac{(A\varphi|\varphi)_Y}{\|\varphi\|_{W_0^{1,2}(\Omega,\omega)}} = \infty.$$

Step 6. Generalized condition (M). Let $T \in [W_0^{1,2}(\Omega,\omega)]^*$ and $\{u_n\}$ be a sequence in Y with

$$u_n \rightharpoonup u \text{ in } W_0^{1,2}(\Omega,\omega), \tag{6.5}$$

$$(Au_n\varphi)_Y \to T(\varphi), \text{ as } n \to \infty, \text{ for all } \varphi \in Y, \tag{6.6}$$

$$\overline{\lim_{n\to\infty}}\,(Au_n|u_n)_Y \leq T(u). \tag{6.7}$$

We want to show that this implies $Au = T$.
Using that the operator A_1 is linear and continuous, we obtain

$$(A_1u_n|\varphi)_Y \to (A_1u|\varphi)_Y, \text{ for all } \varphi \in Y.$$

Because of (6.6), it is sufficient to prove that $u \in \mathcal{D}(A)$ and

$$(A_2u_n|\varphi)_Y \to (A_2u|\varphi)_Y, \text{ for all } \varphi \in Y.$$

For this, is sufficient to show that

$$\int_\Omega [g(u_n) - g(u)]\,\varphi\,\omega\,dx \to 0, \text{as } n \to \infty.$$

Using the same argument in the Step 3, we obtain

$$\begin{aligned}
\left|\int_\Omega (g(u_n) - g(u))\,\varphi\,\omega\,dx\right| &\leq \int_\Omega |g(u_n) - g(u)|\,|\varphi|\,\omega\,dx \\
&\leq \|\varphi\|_{C(\bar{\Omega})} \int_\Omega |g(u_n) - g(u)|\,\omega\,dx \\
&\leq C\,\|\varphi\|_Y \int_\Omega |g(u_n) - g(u)|\,\omega\,dx.
\end{aligned}$$

Therefore is sufficient to show that

$$g(u_n) \to g(u) \text{ in } L^1(\Omega, \omega). \tag{6.8}$$

Note that is sufficient to prove (6.8) for a subsequence of $\{u_n\}$. Because $\omega \in A_q \subset A_2$ we obtain by Lemma 4.1.2 that

$$W_0^{1,2}(\Omega, \omega) \hookrightarrow L^2(\Omega, \omega)$$

is compact and $\|u\|_{L^2(\Omega,\omega)} \leq C_\Omega \|u\|_{W_0^{1,2}(\Omega,\omega)}$. This implies $u_n \to u$ in $L^2(\Omega, \omega)$. Using again that $\omega \in A_2$, we have $u_n \to u$ in $L^1(\Omega)$. Thus, there exists a subsequence, again denoted by $\{u_n\}$ such that $u_n(x) \to u(x)$ for almost all $x \in \Omega$. The continuity of g implies $g(u_n(x)) \to g(u(x))$ for almost all $x \in \Omega$. Moreover from $u_n \rightharpoonup u$ in $W_0^{1,2}(\Omega, \omega)$, it follow that

$$\sup \|u_n\|_{W_0^{1,2}(\Omega,\omega)} \leq C, \text{ independent of } n.$$

Hence, using (6.2) we obtain

$$(A_1 u_n | u_n)_Y = B_1(u_n, u_n) \leq \Lambda \, \|u_n\|^2_{W_0^{1,2}(\Omega,\omega)} \leq \Lambda \, C^2$$

with C independent of n. Therefore, using (6.6) we obtain

$$\overline{\lim_{n\to\infty}} \, (A_2 u_n | u_n)_Y = \overline{\lim_{n\to\infty}} \int_\Omega g(u_n) \, u_n \, \omega \, dx \leq C, \tag{6.9}$$

with C independent of n. The continuity of g implies that

$$g(u_n(x)) \, u_n(x) \, \omega(x) \to g(u(x)) \, u(x) \, \omega(x)$$

for almost all $x \in \Omega$. Therefore, by Fatou's Lemma, we have

$$\int_\Omega g(u) \, u \, \omega \, dx < \infty,$$

i.e., $u \in \mathcal{D}(A)$.

Now we want to show that $g(u_n) \to g(u)$ in $L^1(\Omega, \omega)$.

Let $a > 0$ be fixed. For each $x \in \Omega$ we have either

$$|u_n(x)| \leq a \text{ or } |g(u_n(x))| \leq a^{-1}\, g(u_n(x))\, u_n(x)$$

(if $x \neq 0$ we can write $g(x) = x^{-1}\,[g(x)\,x]$). We get $|g(x)| \leq c(a)$, if $|x| \leq a$ (because of g is continuous).

Let X be a measurable subset of Ω. Then, by (6.9),

$$\begin{aligned}
\int_X |g(u_n)|\,\omega\,dx &= \int_{X\cap\{x:|u_n(x)|\leq a\}} |g(u_n)|\,\omega\,dx \\
&+ \int_{X\cap\{x:|u_n(x)|> a\}} |g(u_n)|\,\omega\,dx \\
&\leq c(a)\,\omega(X) + a^{-1}\int_X g(u_n)\,u_n\,\omega\,dx \\
&\leq c(a)\,\omega(X) + a^{-1}C.
\end{aligned}$$

Hence, for all $\varepsilon > 0$ we have

$$\int_X |g(u_n)|\,\omega\,dx \leq \frac{\varepsilon}{2}$$

if a is sufficiently large and $\omega(X) = \displaystyle\int_X \omega\,dx$ is sufficiently small. Therefore, for all $\varepsilon > 0$, there exists $\delta = \delta(\varepsilon)$ such that

$$\int_X |g(u_n) - g(u)|\,\omega\,dx \leq \int_X |g(u_n)|\,\omega\,dx + \int_X |g(u)|\,\omega\,dx \leq \varepsilon,$$

with $\omega(X) < \delta$. Thus, the Vitali Convergence Theorem tell us that

$$g(u_n) \to g(u) \text{ in } L^1(\Omega,\omega).$$

Step 7. Quasi-boundedness of the operator A. Let $\{u_n\}$ be a sequence in Y with $u_n \rightharpoonup u$ in $W_0^{1,2}(\Omega,\omega)$ and suppose that

$$(Au_n|u_n)_Y \leq C\|u_n\|_{W_0^{1,2}(\Omega,\omega)}, \text{ for all } n.$$

We want to show that the sequence $\{Au_n\}$ is bounded in Y^*. In fact, the boundedness of $\{u_n\}$ in $W_0^{1,2}(\Omega,\omega)$ implies that

$$\overline{\lim_{n\to\infty}}\,(Au_n|u_n)_Y \leq C.$$

Suppose by contradiction that the sequence $\{Au_n\}$ is unbounded in Y^*. Then there exists a subsequence, again denoted by $\{u_n\}$, such that

$$\|Au_n\|_{Y^*}\to\infty, \text{ as } n\to\infty. \tag{6.10}$$

By the same argument as in Step 6, we obtain that

$$(Au_n|\varphi)_Y\to(Au|\varphi)_Y, \text{ as } n\to\infty, \text{ for all } \varphi\in Y.$$

The Uniform Boundedness Principle tell us that the sequence $\{Au_n\}$ is bounded (contradiction with (6.10)). Therefore, by Theorem 6.1.1, the operator equation

$$Au = T,$$

with $T\in[W_0^{1,2}(\Omega,\omega)]^*$ has a solution $u\in\mathcal{D}(A)\subseteq W_0^{1,2}(\Omega,\omega)$ and it is the solution for the problem (P13).

Step 8. In Definition 6.1.2, with $\varphi = u$ we get

$$\sum_{i,j=1}^{N}\int_\Omega a_{ij}D_iu\,D_ju\,dx + \int_\Omega g(u)\,u\,\omega\,dx = \int_\Omega f_0\,u\,dx + \sum_{j=1}^{N}\int_\Omega f_j\,D_ju\,dx.$$

Then we obtain, by (6.2) and (H2)

$$\sum_{i,j=1}^{N}\int_\Omega a_{ij}D_iu\,D_ju\,dx + \int_\Omega g(u)\,u\,\omega\,dx$$

$$\geq \lambda \int_\Omega |\nabla u|^2 \, \omega \, dx$$
$$\geq \frac{\lambda}{(C_\Omega^2+1)^2} \|u\|^2_{W_0^{1,2}(\Omega,\omega)},$$

and

$$\left| \int_\Omega f_0 \, u \, dx + \sum_{j=1}^N \int_\Omega f_j \, D_j u \, dx \right|$$
$$\leq \|f_0/\omega\|_{L^2(\Omega,\omega)} \|u\|_{L^2(\Omega,\omega)} + \sum_{j=1}^N \|f_j/\omega\|_{L^2(\Omega,\omega)} \| \, |\nabla u| \, \|_{L^2(\Omega,\omega)}$$
$$\leq \left(\sum_{j=0}^N \|f_j/\omega\|_{L^2(\Omega,\omega)} \right) \|u\|_{W_0^{1,2}(\Omega,\omega)}.$$

Therefore, we obtain

$$\|u\|_{W_0^{1,2}(\Omega,\omega)} \leq \frac{(C_\Omega^2+1)^2}{\lambda} \left(\sum_{j=0}^N \|f_j/\omega\|_{L^2(\Omega,\omega)} \right).$$

□

Example 6.1.2. *Consider* $\Omega = \{(x,y) \in \mathbb{R}^2 : x^2+y^2<1\}$, $0 < a < b$, $g(t) = t\,\mathrm{e}^t$, $\omega(x,y) = |x|^{1/2}$ *and* $f_0(x,y) = |x\,y| \cos(x+y)$. *By Theorem 6.1.2, the problem*

$$\begin{cases} -Lu(x,y) + g(u(x,y))\,\omega(x) = |x\,y|\cos(x+y) \text{ in } \Omega, \\ u(x,y) = 0 \text{ on } \partial\Omega, \end{cases}$$

where

$$Lu(x,y) = -\left[\frac{\partial}{\partial x}(a|x|^{1/2}\frac{\partial u}{\partial x}) + \frac{\partial}{\partial y}(b(|x|^{1/2}\frac{\partial u}{\partial y}) \right] + u(x,y)\,\mathrm{e}^{u(x,y)}\,|x|^{1/2},$$

has a solution $u \in \mathcal{D}(A) = \{u \in W_0^{1,2}(\Omega,\omega) : g(u)\,u \in L^1(\Omega,\omega)\}$.

6.2 The Degenerated p-Laplacian and entropy solutions

The main purpose of this section (see Theorem 6.2.3) is to establish the existence of entropy solutions for the Dirichlet problem

$$(P14)\begin{cases} -\operatorname{div}[\omega(x)|\nabla u|^{p-2}\nabla u] = f(x) - \operatorname{div}(G(x)) \text{ in } \Omega, \\ u(x) = 0 \text{ on } \partial\Omega, \end{cases}$$

where $\Omega \subset \mathbb{R}^N$ is a bounded open set, $f \in L^1(\Omega)$, $G/\omega \in [L^{p'}(\Omega,\omega)]^N$, ω is a weight function (i.e., a locally integrable function on $\mathbb{R}^N$ such that $0 < \omega(x) < \infty$ a.e. $x \in \mathbb{R}^N$) and $1 < p < \infty$, $p \neq 2$.

The notion of entropy solution was introduced in [2] where the authors studied the nondegenerate elliptic equation $-\operatorname{div}(a(x,Du)) = f(x)$, with $f \in L^1(\Omega)$. Note that, in the proof of our main result, many ideas have been adapted from [2] and Theorem 3.2.8.

We propose to solve the problem (P14) by approximation with variational solutions: we take $f_n \in C_0^\infty(\Omega)$ such that $f_n \to f$ in $L^1(\Omega)$,
$G_n/\omega \in [L^{p'}(\Omega,\omega)]^N$ such that $G_n/\omega \to G/\omega$ in $[L^{p'}(\Omega,\omega)]^N$, we find a solution $u_n \in W_0^{1,p}(\Omega,\omega)$ for the problem with right-hand side f_n and G_n and we will try to pass to the limit as $n \to \infty$.

Definition 6.2.1. *Let $f \in L^1(\Omega)$, $G/\omega \in [L^{p'}(\Omega,\omega)]^N$ and $u \in \mathcal{T}_0^{1,p}(\Omega,\omega)$ (see Definition 3.2.1). We say that u is an entropy solution to problem $(P14)$ if*

$$\int_\Omega |\nabla u|^{p-2}\langle \nabla u, \nabla T_k(u-\varphi)\rangle\,\omega\,dx = \int_\Omega f\,T_k(u-\varphi)\,dx + \int_\Omega \langle G, \nabla T_k(u-\varphi)\rangle\,dx, \tag{6.11}$$

for all $k > 0$ and all $\varphi \in W_0^{1,p}(\Omega,\omega) \cap L^\infty(\Omega)$, where $\langle .,. \rangle$ denotes the usual inner product in $\mathbb{R}^N$.

We recall that the gradient of u which appears in (6.11) is defined as in Remark 3.2.1, that is to say that $\nabla u = \nabla T_k(u)$ on the set where $|u| < k$.

Firstly, we prove the existence and uniqueness of weak solutions $u \in W_0^{1,p}(\Omega,\omega)$ to the Dirichlet problem

$$(P15)\begin{cases} -\text{div}[\omega(x)|\nabla u|^{p-2}\nabla u] = f(x) - \text{div}(G(x)), & \text{in } \Omega \\ u(x) = 0, \quad \text{in } \partial\Omega \end{cases}$$

where Ω is a bounded open set of $\mathbb{R}^N$ $(N \geq 2)$, $f/\omega \in L^{p'}(\Omega,\omega)$ and $G/\omega\, [L^{p'}(\Omega,\omega)]^N$.

Definition 6.2.2. *We say that $u \in W_0^{1,p}(\Omega,\omega)$ is a weak solution for problem $(P15)$ if*

$$\int_\Omega |\nabla u|^{p-2}\langle \nabla u, \nabla\varphi\rangle\, \omega\, dx = \int_\Omega f\, \varphi\, dx + \int_\Omega \langle G, \nabla\varphi\rangle\, dx, \quad (6.12)$$

for all $\varphi \in W_0^{1,p}(\Omega,\omega)$, with $f/\omega \in L^{p'}(\Omega,\omega)$ and $G/\omega \in [L^{p'}(\Omega,\omega)]^N$.

Lemma 6.2.1. *Let $1 < p < \infty$.*
(a) There exists a constant α_p such that

$$\Big| |x|^{p-2}x - |y|^{p-2}y \Big| \leq \alpha_p\, |x-y|(\,|x| + |y|)^{(p-2)},$$

for all $x, y \in \mathbb{R}^N$.
(b) There exist two positive constants β_p and α_p such that for every $x, y \in \mathbb{R}^N$

$$\begin{aligned} \beta_p\,(\,|x| + |y|)^{p-2}\,|x-y|^2 &\leq \langle |x|^{p-2}x - |y|^{p-2}y, x-y\rangle \\ &\leq \alpha_p\,(\,|x| + |y|)^{p-2}\,|x-y|^2. \end{aligned}$$

Proof. See Proposition 17.2 and Proposition 17.3 in [26]. □

Theorem 6.2.2. *Let* $\omega \in A_p$, $1 < p < \infty$, $\dfrac{f}{\omega} \in L^{p'}(\Omega, \omega)$ *and* $\dfrac{G}{\omega} \in [L^{p'}(\Omega, \omega)]^N$. *Then the problem* $(P15)$ *has a unique (weak) solution* $u \in W_0^{1,p}(\Omega, \omega)$.

Proof. **Step 1.** *Existence.* By Theorem 2.1.5 (with $\theta = 1$), we have that

$$\begin{aligned} \left| \int_\Omega f\,\varphi\,dx \right| &\le \left(\int_\Omega \left| \frac{f}{\omega} \right|^{p'} \omega\,dx \right)^{1/p'} \left(\int_\Omega |\varphi|^p\,\omega\,dx \right)^{1/p} \\ &\le C_\Omega \, \|f/\omega\|_{L^{p'}(\Omega,\omega)} \| \, |\nabla\varphi| \, \|_{L^p(\Omega,\omega)}. \end{aligned} \tag{6.13}$$

Define the functional $J_p : W_0^{1,p}(\Omega, \omega) \to \mathbb{R}$ by

$$J_p(\varphi) = \frac{1}{p} \int_\Omega |\nabla\varphi|^p\,\omega\,dx - \int_\Omega f\,\varphi\,dx - \int_\Omega \langle G, \nabla\varphi \rangle\,dx.$$

Using (6.13) and Young's inequality, we have that

$$\begin{aligned} J_p(\varphi) &\ge \frac{1}{p} \int_\Omega |\nabla\varphi|^p\,\omega\,dx \\ &- \left(C_\Omega \, \|f/\omega\|_{L^{p'}(\Omega,\omega)} + \| |G|/\omega \|_{L^{p'}(\Omega,\omega)} \right) \| \, |\nabla\varphi| \, \|_{L^p(\Omega,\omega)} \\ &\ge \frac{1}{p} \int_\Omega |\nabla\varphi|^p\,\omega\,dx - \frac{1}{p} \| \, |\nabla\varphi| \, \|^p_{L^p(\Omega,\omega)} \\ &- \frac{1}{p'} [C_\Omega \, \|f/\omega\|_{L^{p'}(\Omega,\omega)} + \| |G|/\omega \|_{L^{p'}(\Omega,\omega)}]^{p'} \\ &= -\frac{1}{p'} [C_\Omega \, \|f/\omega\|_{L^{p'}(\Omega,\omega)} + \| |G|/\omega \|_{L^{p'}(\Omega,\omega)}]^{p'}, \end{aligned}$$

that is, J_p is bounded from below.
Let $\{u_n\}$ be a minimizing sequence, that is, a sequence such that

$$J_p(u_n) \to \inf_{\varphi \in W_0^{1,p}(\Omega,\omega)} J_p(\varphi)\,.$$

Then for n large enough, we obtain

$$0 \geq J_p(u_n) = \frac{1}{p}\int_\Omega |\nabla u_n|^p\,\omega\,dx - \int_\Omega f\,u_n\,dx - \int_\Omega \langle G, \nabla u_n\rangle\,dx,$$

and we get

$$\begin{aligned}
&\|\,|\nabla u_n|\,\|^p_{L^p(\Omega,\omega)} \leq p\left(\int_\Omega f\,u_n\,dx + \int_\Omega \langle G, \nabla u_n\rangle\,dx\right)\\
&\leq p\,(\,\|f/\omega\|_{L^{p'}(\Omega,\omega)}\,\|u_n\|_{L^p(\Omega,\omega)}\\
&+\||G|/\omega\|_{L^{p'}(\Omega,\omega)}\|\,|\nabla u_n|\,\|_{L^p(\Omega,\omega)}\,)\\
&\leq p\,(C_\Omega\,\|f/\omega\|_{L^{p'}(\Omega,\omega)} + \||G|/\omega\|_{L^{p'}(\Omega,\omega)}\,)\|\,|\nabla u_n|\,\|_{L^p(\Omega,\omega)}.
\end{aligned}$$

Hence

$$\|\,|\nabla u_n|\,\|_{L^p(\Omega,\omega)} \leq [p\,(\,C_\Omega\,\|f/\omega\|_{L^{p'}(\Omega,\omega)} + \||G|/\omega\|_{L^{p'}(\Omega,\omega)})]^{1/(p-1)}.$$

Therefore $\{u_n\}$ is bounded in $W^{1,p}_0(\Omega,\omega)$. Since $W^{1,p}_0(\Omega,\omega)$ is reflexive, there exists $u \in W^{1,p}_0(\Omega,\omega)$ such that $u_n \rightharpoonup$ in $W^{1,p}_0(\Omega,\omega)$. Since

$$\begin{aligned}
&W^{1,p}_0(\Omega,\omega) \ni \varphi \mapsto \int_\Omega f\,\varphi\,dx + \int_\Omega \langle G, \nabla\varphi\rangle\,dx \text{ and}\\
&\varphi \mapsto \|\,|\nabla\varphi|\,\|_{L^p(\Omega,\omega)}
\end{aligned}$$

are continuous then J_p is continuous. Moreover since $1 < p < \infty$ we have that J_p is convex and thus lower semi-continuous for the weak convergence. It follows that

$$J_p(u) \leq \liminf_n J_p(u_n) = \inf_{\varphi \in W^{1,p}_0(\Omega,\omega)} J_p(\varphi),$$

and thus u is a minimizer of J_p on $W^{1,p}_0(\Omega,\omega)$. For any $\varphi \in W^{1,p}_0(\Omega,\omega)$ the function

$$\lambda \mapsto \frac{1}{p}\int_\Omega |\nabla(u+\lambda\varphi)|^p\,\omega\,dx - \int_\Omega (u+\lambda\,\varphi)\,f\,dx - \int_\Omega \langle G, \nabla(u+\lambda\,\varphi)\rangle\,dx$$

has a minimum at $\lambda = 0$. Hence

$$\frac{d}{d\lambda}\Big(J_p(u+\lambda\,\varphi)\Big)\Big|_{\lambda=0} = 0,\ \ \forall\,\varphi\in W_0^{1,p}(\Omega,\omega).$$

We have

$$\frac{d}{d\,\lambda}\Big(|\,\nabla(u+\lambda\,\varphi|^p\,\omega\Big) = p\,\{|\nabla(u+\lambda\,\varphi)|^{p-2}(\langle\nabla u,\nabla\varphi\rangle+\lambda\,|\nabla\varphi|^2)\}\,\omega,$$

and we obtain

$$\begin{aligned}
0 &= \frac{d}{d\lambda}\Big(J_p(u+\lambda\,\varphi)\Big)\Big|_{\lambda=0}\\
&= \Bigg[\frac{1}{p}\Big(p\int_\Omega|\nabla(u+\lambda\,\varphi)|^{p-2}(\langle\nabla u,\nabla\varphi\rangle+\lambda\,|\nabla\varphi|^2)\omega\,dx\Big)\\
&\qquad -\int_\Omega\varphi\,f\,dx-\int_\Omega\langle G,\nabla\varphi\rangle\,dx\Bigg]\Bigg|_{\lambda=0}\\
&= \int_\Omega|\nabla u|^{p-2}\,\langle\nabla u,\nabla\varphi\rangle\,\omega\,dx-\int_\Omega f\,\varphi\,dx-\int_\Omega\langle G,\nabla\varphi\rangle\,dx.
\end{aligned}$$

Therefore

$$\int_\Omega|\nabla u|^{p-2}\,\langle\nabla u,\nabla\varphi\rangle\,\omega\,dx = \int_\Omega f\,\varphi\,dx+\int_\Omega\langle G,\nabla\varphi\rangle\,dx,$$

that is, $u\in W_0^{1,p}(\Omega,\omega)$ is a solution of problem $(P15)$.
Step 2. *Uniqueness.* If $u_1, u_2\in W_0^{1,p}(\Omega,\omega)$ are two weak solutions of problem $(P15)$, we have

$$\int_\Omega|\nabla u_i|^{p-2}\,\langle\nabla u_i,\nabla\varphi\rangle\,\omega\,dx = \int_\Omega f\,\varphi\,dx+\int_\Omega\langle G,\nabla\varphi\rangle\,dx,\ \ (i=1,2),$$

for all $\varphi\in W_0^{1,p}(\Omega,\omega)$. Hence

$$\int_\Omega\Big(|\nabla u_1|^{p-2}\langle\nabla u_1,\nabla\varphi\rangle-|\nabla u_2|^{p-2}\langle\nabla u_2,\nabla\varphi\rangle\Big)\,\omega\,dx = 0.$$

Taking $\varphi = u_1 - u_2$, and using that for every $x, y \in \mathbb{R}^N$ there exist two positive constants α_p and β_p (see Lemma 6.2.1(b)) such that

$$\begin{aligned}&\beta_p\,(|x|+|y|)^{p-2}|x-y|^2\\ &\le \langle\, |x|^{p-2}x - |y|^{p-2}y, x-y\rangle \le \alpha_p\,(|x|+|y|)^{p-2}|x-y|^2,\end{aligned}$$

we obtain

$$\begin{aligned}0 \quad &= \int_\Omega |\nabla u_1|^{p-2}\langle \nabla u_1, \nabla u_1\rangle\,\omega\,dx\\ &- \int_\Omega \nabla u_2\rangle - |\nabla u_2|^{p-2}\langle \nabla u_2, \nabla u_1 - \nabla u_2\rangle\,\omega\,dx\\ &= \int_\Omega \langle |\nabla u_1|^{p-2}\nabla u_1 - |\nabla u_2|^{p-2}\nabla u_2, \nabla u_1 - \nabla u_2\rangle\,\omega\,dx\\ &\ge \beta_p \int_\Omega \Big(|\nabla u_1| + |\nabla u_2|\Big)^{p-2} |\nabla u_1 - \nabla u_2|^2\,\omega\,dx.\end{aligned}$$

Therefore $\nabla u_1 = \nabla u_2$ μ-a.e. and since $u_1, u_2 \in W_0^{1,p}(\Omega,\omega)$, then $u_1 = u_2$ a.e. (by Lemma 2.1.2). □

Main Result

Now we prove the main result of this section. We need the Lemma 5.2.1.

Theorem 6.2.3. *Let* $\omega \in A_p$, $2 < p < \infty$, $f \in L^1(\Omega)$ *and* $G/\omega \in [L^{p'}(\Omega,\omega)]^N$. *There exists an entropy solutions* u *of problem (P14). Moreover, we have* $u \in \mathcal{M}^{p_1}(\Omega,\omega)$ *(with* $p_1 = \theta\,(p-1)$*) and* $|\nabla u| \in \mathcal{M}^{p_2}(\Omega,\omega)$ *(with* $p_2 = p_1\,p/(p_1+1)$, *where* θ *is the constant in* Theorem 2.1.5*).*

Proof. **Step 1.**Considering a sequence $\{f_n\}$, $f_n \in C_0^\infty(\Omega)$, which

$$f_n \to f \text{ in } L^1(\Omega) \quad\text{and}\quad \|f_n\|_{L^1(\Omega)} \le \|f\|_{L^1(\Omega)},$$

and a sequence $\{G_n\}$, with $G_n/\omega \in [L^{p'}(\Omega,\omega)]^N$ such that $\dfrac{G_n}{\omega} \to \dfrac{G}{\omega}$ in

$[L^{p'}(\Omega,\omega)]^N$ and $\|\,|G_n|/\omega\|_{L^{p'}(\Omega,\omega)} \le \|\,|G|/\omega\|_{L^{p'}(\Omega,\omega)}$. For each n, by Theorem 6.2.2, there exists a solution $u_n \in W_0^{1,p}(\Omega,\omega)$ of the Dirichlet problem

$$(P_n)\begin{cases} -\operatorname{div}[\omega(x)|\nabla u_n|^{p-2}\nabla u_n)] = f_n(x) - \operatorname{div}(G_n(x)) \text{ in } \Omega, \\ u_n(x) = 0 \text{ on } \partial\Omega, \end{cases}$$

that is,

$$\int_\Omega |\nabla u_n|^{p-2}\langle\nabla u_n, \nabla\varphi\rangle\,\omega\,dx = \int_\Omega f_n\varphi\,dx + \int_\Omega \langle G, \nabla\varphi\rangle\,dx \tag{6.14}$$

for all $\varphi \in W_0^{1,p}(\Omega,\omega)$. For $\varphi = T_k(u_n)$ we obtain in (6.14) that

$$\begin{aligned} &\int_\Omega |\nabla u_n|^{p-2}\langle\,\nabla u_n, \nabla T_k(u_n)\rangle\,\omega\,dx \\ &= \int_\Omega f_n\,T_k(u_n)\,dx + \int_\Omega \langle G_n, \nabla T_k(u_n)\rangle\,dx. \end{aligned} \tag{6.15}$$

We have,

$$\begin{aligned} \left|\int_\Omega f_n\,T_k(u_n)\,dx\right| &\le \int_\Omega |f_n||T_k(u_n)|\,dx \\ &\le k\|f_n\|_{L^1(\Omega)} \le k\|f\|_{L^1(\Omega)}, \end{aligned} \tag{6.16}$$

and since $\nabla T_k(u_n) = \chi_{\{|u_n|<k\}}\nabla u_n$, we obtain

$$\begin{aligned} &\int_\Omega |\nabla u_n|^{p-2}\langle\,\nabla u_n, \nabla T_k(u_n)\rangle\,\omega\,dx \\ &= \int_\Omega |\nabla T_k(u_n)|^{p-2}\langle\,\nabla T_k(u_n), \nabla T_k(u_n)\rangle\,\omega\,dx \\ &= \int_\Omega |\nabla T_k(u_n)|^p\omega\,dx. \end{aligned} \tag{6.17}$$

We also have, using Young's inequality, that there exists a constant $C_1 > 0$ (depending only on p) such that

$$\left|\int_\Omega \langle G_n, \nabla T_k(u_n)\rangle\,dx\right| \le \int_\Omega \left|\frac{G_n}{\omega}\right||\nabla T_k(u_n)|\,\omega\,dx$$

$$\leq \left(\int_\Omega |G_n/\omega|^{p'} \omega \, dx \right)^{1/p'} \left(\int_\Omega |\nabla T_k(u_n)|^p \omega \, dx \right)^{1/p}$$
$$\leq \frac{1}{2} \int_\Omega |\nabla T_k(u_n)|^p \omega \, dx + C_1 \int_\Omega |G_n/\omega|^{p'} \omega \, dx$$
$$\leq \frac{1}{2} \int_\Omega |\nabla T_k(u_n)|^p \omega \, dx + C_1 \int_\Omega |G/\omega|^{p'} \omega \, dx. \qquad (6.18)$$

Hence, using (6.16), (6.17) and (6.18), we obtain

$$\int_\Omega |\nabla T_k(u_n)|^p \omega \, dx \leq 2\, k\, \|f\|_{L^1(\Omega)} + 2\, C_1\, \| |G|/\omega \|^{p'}_{L^{p'}(\Omega,\omega)} \leq C_2\, k, \qquad (6.19)$$

where $C_2 = 2\,\|f\|_{L^1(\Omega)} + 2\,C_1\,\||G|/\omega\|^{p'}_{L^{p'}(\Omega,\omega)}$. By Lemma 3.2.5, we have that the sequence $\{u_n\}$ is bounded in $\mathcal{M}^{p_1}(\Omega,\omega)$ (with $p_1 = \theta\,(p-1)$), and $\{|\nabla u_n|\}$ is bounded in $\mathcal{M}^{p_2}(\Omega,\omega)$ (with $p_2 = p_1\, p/(p_1+1)$). Moreover, $\{u_n\}$ is a Cauchy sequence in μ-measure. Consequently, there exists a function u and a subsequence, that we will still denote by $\{u_n\}$, such that

$$u_n \to u \quad \mu - \text{a.e. in } \Omega, \qquad (6.20)$$

and $u_n \to u$ a.e. in Ω (by Lemma 2.1.2). Using (6.19) and (6.20), we have

$$\begin{aligned} T_k(u_n) &\rightharpoonup T_k(u) \text{ weakly in } W^{1,p}_0(\Omega,\omega), \\ T_k(u_n) &\to T_k(u) \text{ strongly in } L^p(\Omega,\omega) \text{ and } \mu - \text{a.e. in } \Omega, \end{aligned} \qquad (6.21)$$

for all $k > 0$. Hence $T_k(u) \in W^{1,p}_0(\Omega,\omega)$.

Furthermore, by the weak lower semicontinuity of the norm $W^{1,p}_0(\Omega,\omega)$, we have that (6.19) still holds for u, that is,

$$\int_\Omega |\nabla T_k(u)|^p \omega \, dx \leq k\, C_2.$$

Applying Lemma 3.2.5, we have that $u \in \mathcal{M}^{p_1}(\Omega,\omega)$ (with $p_1 = \eta\,(p-1)$) and $|\nabla u| \in \mathcal{M}^{p_2}(\Omega,\omega)$ (with $p_2 = p_1\, p/(p_1+1)$).
Step 2. We need to show that $T_k(u_n) \to T_k(u)$ strongly in $W_0^{1,p}(\Omega,\omega)$, for all $k > 0$. Let $h > k$ and applying (6.14) with function

$$\varphi_n = T_{2k}(u_n - T_h(u_n) + T_k(u_n) - T_k(u)),$$

we get

$$\int_\Omega |\nabla u_n|^{p-2}\, \langle \nabla u_n, \nabla \varphi_n\rangle\, \omega\, dx = \int_\Omega f_n\, \varphi_n\, dx + \int_\Omega \langle G, \nabla \varphi_n\rangle\, dx. \tag{6.22}$$

If we set $M = 4k + h$, we have $\nabla \varphi_n = 0$ for $|u_n| > M$. We can write

$$\begin{aligned}&\int_\Omega |\nabla T_M(u_n)|^{p-2}\, \langle \nabla T_M(u_n), \nabla \varphi_n\rangle\, \omega\, dx \\ &= \int_\Omega f_n \varphi_n\, dx + \int_\Omega \langle G, \nabla \varphi_n\rangle\, dx.\end{aligned} \tag{6.23}$$

In the left-hand side of (6.23), we have

$$\begin{aligned}&\int_\Omega |\nabla T_M(u_n)|^{p-2}\, \langle \nabla T_M(u_n), \nabla T_{2k}(u_n - T_h(u_n) \\ &\qquad + T_k(u_n) - T_k(u))\rangle\, \omega dx \\ &= \int_{\{|u_n| \le k\}} |\nabla T_M(u_n)|^{p-2}\, \langle \nabla T_M(u_n)), \nabla T_{2k}(u_n - T_h(u_n) \\ &\qquad + T_k(u_n) - T_k(u))\rangle\, \omega\, dx \\ &+ \int_{\{|u_n| > k\}} |\nabla T_M(u_n)|^{p-2}\, \langle \nabla T_M(u_n), \nabla T_{2k}(u_n - T_h(u_n) \\ &\qquad + T_k(u_n) - T_k(u))\rangle\, \omega\, dx.\end{aligned} \tag{6.24}$$

(a) If $|u_n| \le k$.
Since $h > k$, if $|u_n| \le k < h$, then $T_h(u_n) = T_k(u_n) = u_n$. Hence,

$$u_n - T_h(u_n) + T_k(u_n) - T_k(u) = u_n - T_k(u).$$

We also have that $|u_n - u| \leq 2k$. Then, since $\nabla T_M(u_n) = \nabla T_k(u_n)$ (because $|u_n| \leq k < M$),

$$\begin{aligned}
&\int_{\{|u_n|\leq k\}} |\nabla T_M(u_n)|^{p-2} \langle \nabla T_M(u_n), \\
&\qquad\qquad \nabla T_{2k}(u_n - T_h(u_n) + T_k(u_n) - T_k(u))\rangle \, \omega \, dx \\
&= \int_{\{|u_n|\leq k\}} |\nabla T_k(u_n)|^{p-2} \langle \nabla T_k(u_n), \nabla(T_k(u_n) - T_k(u))\rangle \, \omega \, dx \\
&= \int_\Omega |\nabla T_k(u_n)|^{p-2} \langle \nabla T_k(u_n), \nabla(T_k(u_n) - T_k(u))\rangle \, \omega \, dx.
\end{aligned}$$

(b) If $|u_n| > k$. Since u_n, $T_k(u_n)$ and $T_k(u)$ are in $W_0^{1,p}(\Omega, \omega)$, if

$$|u_n - T_h(u_n) + T_k(u_n) - T_k(u)| \leq 2k,$$

we obtain

$$\begin{aligned}
&\nabla T_{2k}(u_n - T_h(u_n) + T_k(u_n) - T_k(u)) \\
&= \nabla(u_n - T_h(u_n) + T_k(u_n) - T_k(u)) \\
&= \nabla u_n - \nabla T_h(u_n) + \nabla T_k(u_n) - \nabla T_k(u) \\
&= \nabla u_n - \nabla T_h(u_n) - \nabla T_k(u),
\end{aligned}$$

(because $\nabla T_k(u_n) = 0$ if $|u_n| > k$). There are two possible cases as follows:

(i) If $k < |u_n| < h$, we have $\nabla T_h(u_n) = \nabla u_n$. Then

$$\nabla T_{2k}(u_n - T_h(u_n) + T_k(u_n) - T_k(u)) = -\nabla T_k(u);$$

(ii) If $h < |u_n| \leq M$, we have that $\nabla T_h(u_n) = 0$. Then

$$\begin{aligned}
\nabla T_{2k}(u_n - T_h(u_n) + T_k(u_n) - T_k(u)) &= \nabla u_n - \nabla T_k(u) \\
&= \nabla T_M(u_n) - \nabla T_k(u).
\end{aligned}$$

In both cases we obtain

$$\begin{aligned}
&|\nabla T_M(u_n)|^{p-2}\,\langle\,\nabla T_M(u_n),\nabla T_{2k}(u_n-T_h(u_n)+T_k(u_n)-T_k(u))\rangle\\
&\geq -\,|\nabla T_M(u_n)|^{p-2}\,\langle\,\nabla T_M(u_n),\nabla T_k(u)\rangle\\
&\geq -\,|\nabla T_M(u_n)|^{p-2}\,|\nabla T_M(u_n)|\,|\nabla T_k(u)|.
\end{aligned}$$

Therefore we obtain in (6.24)

$$\begin{aligned}
&\int_\Omega |\nabla T_M(u_n)|^{p-2}\,\langle\,\nabla T_M(u_n),\nabla T_{2k}(u_n-T_h(u_n)\\
&\qquad\qquad +T_k(u_n)-T_k(u))\rangle\,\omega\,dx\\
&=\int_{\{|u_n|\leq k\}} |\nabla T_M(u_n)|^{p-2}\,\langle\,\nabla T_M(u_n),\nabla T_{2k}(u_n-T_h(u_n)\\
&\qquad\qquad +T_k(u_n)-T_k(u))\rangle\,\omega\,dx\\
&+\int_{\{|u_n|>k\}} |\nabla T_M(u_n)|^{p-2}\,\langle\,\nabla T_M(u_n),\nabla T_{2k}(u_n-T_h(u_n)\\
&\qquad\qquad +T_k(u_n)-T_k(u))\rangle\,\omega\,dx\\
&\geq \int_\Omega |\nabla T_M(u_n)|^{p-2}\,\langle\,\nabla T_k(u_n),\nabla(T_k(u_n)-T_k(u))\rangle\,\omega\,dx\\
&-\int_{\{|u_n|>k\}} |\nabla T_M(u_n)|^{p-2}|\nabla T_M(u_n)|\,|\nabla T_k(u)|\,\omega\,dx.
\end{aligned}$$

Hence, in (6.23) we obtain

$$\begin{aligned}
&\int_\Omega \langle\,|\nabla T_k(u_n)|^{p-2}\nabla T_k(u_n)-|\nabla T_k(u)|^{p-2}\,\nabla T_k(u),\\
&\qquad\qquad \nabla(T_k(u_n)-T_k(u))\,\rangle\,\omega\,dx\\
&\leq \int_{\{|u_n|>k\}} |\nabla T_M(u_n)|\,|\nabla T_k(u)|\,\omega\,dx\\
&+\int_\Omega f_n T_{2k}(u_n-T_h(u_n)+T_k(u_n)-T_k(u))\,dx\\
&+\int_\Omega \langle G_n,\nabla T_{2k}(u_n-T_h(u_n)+T_k(u_n)-T_k(u))\rangle\,dx
\end{aligned}$$

$$- \int_\Omega |\nabla T_k(u)|^{p-2} \langle \nabla T_k(u), \nabla(T_k(u_n) - T_k(u))\rangle \, \omega \, dx. \tag{6.25}$$

Considering the test function $\psi_n = T_{2k}(u_n - T_h(u_n))$ in (6.14), we have

$$\int_\Omega |\nabla u_n|^{p-2} \langle \nabla u_n, \nabla \psi_n \rangle \, \omega \, dx = \int_\Omega f_n \, \psi_n \, dx + \int_\Omega \langle G_n, \nabla \psi_n \rangle \, dx,$$

and using that

$$\left| \int_\Omega f_n \psi_n \, dx \right| \leq \int_\Omega |f_n||\psi_n| \, dx \leq (2k+1) \|f_n\|_{L^1(\Omega)} \leq (2k+1) \|f\|_{L^1(\Omega)},$$

and

$$\begin{aligned} \int_\Omega |\nabla u_n|^{p-2} \langle \nabla u_n, \nabla \psi_n \rangle \, \omega \, dx &= \int_\Omega |\nabla \psi_n|^{p-2} \langle \nabla \psi_n, \nabla \psi_n \rangle \, \omega \, dx \\ &= \int_\Omega |\nabla \psi_n|^p \omega \, dx \\ &= \int_\Omega |\nabla T_{2k}(u_n - T_h(u_n))|^p \omega \, dx, \end{aligned}$$

we obtain

$$\int_\Omega |\nabla T_{2k}(u_n - T_h(u_n))|^p \omega \, dx \leq (2k+1) \, C_2.$$

Now using that $T_{2k}(u_n - T_h(u_n)) \rightharpoonup T_{2k}(u - T_h(u))$ weakly in $W_0^{1,p}(\Omega, \omega)$ (by (6.21) and Remark 3.2.1(c)), we have

$$\int_\Omega |\nabla T_{2k}(u - T_h(u))|^p \omega \, dx \leq (2k+1) \, C_2. \tag{6.26}$$

We have (by Remark 3.2.1 (a),(c) and (6.26))

$$\int_\Omega |G||\nabla T_{2k}(u - T_h(u))| \, dx = \int_{\{h<|u|<2k+h\}} |G||\nabla u| \, dx$$

$$\leq \left(\int_{\{|u|\geq h\}} |G/\omega|^{p'} \omega \, dx \right)^{1/p'} \left(\int_{\{h<|u|<2k+h\}} |\nabla u|^p \omega dx \right)^{1/p}$$
$$= \left(\int_{\{|u|\geq h\}} |G/\omega|^{p'} \omega \, dx \right)^{1/p'} \left(\int_\Omega |\nabla T_{2k}(u - T_h(u))|^p \omega \, dx \right)^{1/p}$$
$$\leq C_3 \left(\int_{\{|u|\geq h\}} |G/\omega|^{p'} \omega \, dx \right)^{1/p'},$$

where C_3 depends on k but not on h. Therefore, we have

$$\lim_{h\to\infty} \int_\Omega \langle G, \nabla T_{2k}(u - T_h(u)) \rangle \, dx = 0.$$

We also have (by Theorem 2.1.5 and (6.26))

$$\begin{aligned} \int_\Omega |T_{2k}(u - T_h(u))|^p \omega \, dx &\leq C_\Omega \int_\Omega |\nabla T_{2k}(u - T_h(u))|^p \omega \, dx \\ &\leq C_\Omega \, C_2 \, (2k+1). \end{aligned}$$

Moreover, by the Lebesgue's Theorem, we obtain

$$\lim_{h\to\infty} \int_\Omega f \, T_{2k}(u - T_h(u)) \, dx = 0.$$

We can fix a positive real number h_ε sufficiently large to have

$$\int_\Omega f \, T_{2k}(u - T_{h_\varepsilon}) \, dx + \int_\Omega \langle G, \nabla T_{2k}(u - T_{h_\varepsilon}(u)) \rangle \, dx \leq \varepsilon. \quad (6.27)$$

Considering $h = h_\varepsilon$ in (6.25) (and $M = M_\varepsilon = 4k + h_\varepsilon$), by (6.19), we have

$$\int_\Omega | \, |\nabla T_M(u_n)|^{p-2} \, \nabla T_M(u_n)|^{p'} \omega \, dx$$
$$= \int_\Omega |\nabla T_M(u_n)|^{(p-2)\,p'} |\nabla T_M(u_n)|^{p'} \omega \, dx$$

$$= \int_\Omega |\nabla T_M(u_n)|^p \, \omega \, dx$$
$$\leq M\, C_2,$$

that is, $|\,|\nabla T_M(u_n)|^{p-2}\,\nabla T_M(u_n)|$ is bounded in $L^{p'}(\Omega,\omega)$. Moreover, we have $\chi_{\{|u_n|>k\}}|\nabla T_k(u)| \to 0$ in $L^p(\Omega,\omega)$ as $n\to\infty$. Therefore,

$$\lim_{n\to\infty} \int_{\{|u_n|>k\}} |\,|\nabla T_M(u_n)|^{p-2}\,\nabla T_M(u_n)||\nabla T_k(u)|\,\omega\,dx = 0. \tag{6.28}$$

Furthermore, we have that

$$T_{2k}(u_n - T_h(u_n) + T_k(u_n) - T_k(u)) \rightharpoonup T_{2k}(u - T_h(u))$$

weakly in $W_0^{1,p}(\Omega,\omega)$, as $n\to\infty$.
Hence, by (6.21), (6.27) and (6.28), passing to the limit in (6.25), we have

$$\begin{aligned}
&\lim_{n\to\infty} \int_\Omega \langle\, |\nabla T_k(u_n)|^{p-2}\,\nabla T_k(u_n) - |\nabla T_k(u)|^{p-2}\,\nabla T_k(u),\\
&\qquad\qquad \nabla(T_k(u_n) - T_k(u))\rangle\,\omega\,dx\\
&\leq \int_\Omega f\,T_{2k}(u - T_{h_\varepsilon})\,dx + \int_\Omega \langle G, \nabla T_{2k}(u - T_{h_\varepsilon}(u))\rangle\,dx\\
&\leq \varepsilon,
\end{aligned}$$

for all $\varepsilon > 0$, that is,

$$\begin{aligned}
&\lim_{n\to\infty} \int_\Omega \langle\, |\nabla T_k(u_n)|^{p-2}\,\nabla T_k(u_n) - |\nabla T_k(u)|^{p-2}\,\nabla T_k(u),\\
&\qquad\qquad \nabla(T_k(u_n) - T_k(u))\rangle\,\omega\,dx = 0.
\end{aligned}$$

Applying Lemma 5.2.1 we get

$$T_k(u_n) \to T_k(u) \tag{6.29}$$

strongly in $W_0^{1,p}(\Omega,\omega)$ for every $k > 0$.

This convergence implies that for every fixed $k > 0$

$$|\nabla T_k(u_n)|^{p-2}\,\nabla T_k(u_n) \to |\nabla T_k(u)|^{p-2}\,\nabla T_k(u) \tag{6.30}$$

in $(L^{p'}(\Omega,\omega))^N = L^{p'}(\Omega,\omega)\times ...\times L^{p'}(\Omega,\omega)$.

Step 3. Finally, we need to show that u is an entropy solution to the Dirichlet problem (P14). Let us take $\psi_n = T_k(u_n-\varphi)$ as test function in (6.14), with $\varphi\in W_0^{1,p}(\Omega,\omega)\cap L^\infty(\Omega)$. We obtain,

$$\int_\Omega |\nabla u_n|^{p-2}\,\langle\,\nabla u_n,\nabla\psi_n\rangle\,\omega\,dx = \int_\Omega f_n\psi_n\,dx + \int_\Omega \langle G_n,\nabla\psi_n\rangle\,dx. \tag{6.31}$$

If $M = k+\|\varphi\|_{L^\infty(\Omega)}$ and $n > M$, we have

$$\int_\Omega |\nabla u_n|^{p-2}\,\langle\,\nabla u_n,\nabla T_k(u_n-\varphi)\rangle\,\omega\,dx$$
$$= \int_\Omega |\nabla T_M(u_n)|^{p-2}\,\langle\,\nabla T_M(u_n),\nabla T_k(u_n-\varphi)\rangle\,\omega\,dx.$$

Hence, in (6.31) we obtain

$$\int_\Omega |\nabla T_M(u_n)|^{p-2}\,\langle\,\nabla T_M(u_n),\nabla T_k(u_n-\varphi)\rangle\,\omega\,dx$$
$$= \int_\Omega f_n\,T_k(u_n-\varphi)\,dx + \int_\Omega \langle G,\nabla T_k(u-\varphi)\rangle\,dx. \tag{6.32}$$

Therefore, by (6.21) and (6.30), passing to the limit as $n\to\infty$ in (6.32), we obtain

$$\int_\Omega |\nabla u|^{p-2}\,\langle\,\nabla u,\nabla T_k(u-\varphi)\rangle\,\omega\,dx = \int_\Omega f\,T_k(u-\varphi)\,dx$$

for all $\varphi\in W_0^{1,p}(\Omega,\omega)\cap L^\infty(\Omega)$ and for each $k > 0$.

Therefore u is an entropy solutions of problem $(P14)$. □

Example 6.2.1. *Let* $\Omega = \{(x,y)\in\mathbb{R}^2 \ :\ x^2+y^2 < 1\}$, *and consider the weight* $\omega(x,y) = (x^2+y^2)^{-1/6}$ ($\omega\in A_3$, $p=3$), *and* *the functions*

$$f(x,y) = \frac{\sin(xy)}{(x^2+y^2)^{1/3}} \quad (f \in L^1(\Omega))$$
$$G(x,y) = ((x^2+y^2)\sin(x\,y), (x^2+y^2)^{-1/3}\cos(x\,y)).$$

By Theorem 6.2.3, the problem

$$\begin{cases} -\operatorname{div}[(x^2+y^2)^{-1/6}|\nabla u|\nabla u] \\ = \dfrac{\sin(xy)}{(x^2+y^2)^{1/3}} - \operatorname{div}(G(x,y)), \quad \text{in } \Omega, \\ u(x,y) = 0 \quad \text{on } \partial\Omega, \end{cases}$$

has an entropy solution.

6.3 The Degenerated p - Laplacian and p-Biharmonic operators

In this section we prove the existence and uniqueness of (weak) solutions in the weighted Sobolev space $W^{2,p}(\Omega,\omega) \cap W_0^{1,p}(\Omega,\omega)$ (see Definition 6.3.1) for the Navier problem

$$(P16)\begin{cases} Lu(x) &= f_0(x) - \sum_{j=1}^{N} D_j f_j(x) \;\text{ in } \Omega, \\ u(x) &= 0 \;\text{ on } \partial\Omega, \\ \Delta u &= 0, \;\text{ on } \partial\Omega, \end{cases}$$

where L is the partial differential operator

$$Lu = \Delta(\omega\,|\Delta u|^{p-2}\Delta u) - \sum_{j=1}^{N} D_j\left[\omega \mathcal{A}_j(x,u,\nabla u)\right]$$

where $D_j = \partial/\partial x_j$, Ω is a bounded open set in $\mathbb{R}^N$, ω is a weight function, Δ is the Laplacian operator, $1 < p < \infty$ and the functions $\mathcal{A}_j : \Omega\times\mathbb{R}\times\mathbb{R}^N\to\mathbb{R}$ $(j = 1, ..., N)$ satisfies certain conditions

In the non-degenerate case (i.e. with $\omega(x) \equiv 1$), for all $f\in L^p(\Omega)$ the Poisson equation associated with the Dirichlet problem

$$\begin{cases} -\Delta u = f \text{ in } \Omega, \\ u = 0 \text{ on } \partial\Omega, \end{cases}$$

is uniquely solvable in $W^{2,p}(\Omega)\cap W_0^{1,p}(\Omega)$ (see [39]), and the nonlinear Dirichlet problem

$$\begin{cases} -\Delta_p u = f \text{ in } \Omega, \\ u = 0 \text{ on } \partial\Omega, \end{cases}$$

is uniquely solvable in $W_0^{1,p}(\Omega)$ (see [26]), where

$$\Delta_p u = \operatorname{div}(|\nabla u|^{p-2}\nabla u)$$

is the p-Laplacian operator. In the degenerate case, the weighted p-Biharmonic operator have been studied by many authors (see [54] and the references therein), and the degenerated p-Laplacian has been studied in [32].

We denote by $X = W^{2,p}(\Omega,\omega)\cap W_0^{1,p}(\Omega,\omega)$ with the norm

$$\|u\|_X = \left(\int_\Omega |\nabla u|^p\,\omega\,dx + \int_\Omega |\Delta u|^p\,\omega\,dx\right)^{1/p}.$$

Definition 6.3.1. *We say that an element $u\in X$ is a (weak) solution of problem (P16) if, for all $\varphi\in X$,*

$$\int_\Omega |\Delta u|^{p-2}\,\Delta u\,\Delta\varphi\,\omega\,dx + \sum_{j=1}^N \int_\Omega \mathcal{A}_j(x,u,\nabla u)D_j\varphi\,\omega\,dx$$

$$= \int_\Omega f_0\,\varphi\,dx + \sum_{j=1}^N \int_\Omega f_j\,D_j\varphi\,dx,$$

Theorem 6.3.1. *Assume that the functions* $\mathcal{A}_j : \Omega\times\mathbb{R}\times\mathbb{R}^N\to\mathbb{R}$ $(j = 1, ..., N)$ *satisfies the following conditions*
(H1) $x\mapsto\mathcal{A}_j(x,\eta,\xi)$ *is measurable on* Ω *for all* $(\eta,\xi)\in\mathbb{R}\times\mathbb{R}^N$,
$(\eta,\xi)\mapsto\mathcal{A}_j(x,\eta,\xi)$ *is continuous on* $\mathbb{R}\times\mathbb{R}^N$ *for almost all* $x\in\Omega$.
(H2) *there exist a constant* $\theta_1 > 0$ *such that*

$$\langle\mathcal{A}(x,\eta,\xi) - \mathcal{A}(x,\eta',\xi'),(\xi-\xi'\rangle \geq \theta_1\,|\xi-\xi'|^p,$$

whenever $\xi,\xi'\in\mathbb{R}^N$, $\xi\neq\xi'$, *where*

$$\mathcal{A}(x,\eta,\xi) = (\mathcal{A}_1(x,\eta,\xi), ..., \mathcal{A}_N(x,\eta,\xi))$$

(where $\langle.,.\rangle$ *denote here the Euclidean scalar product in* $\mathbb{R}^N$*).*
(H3) $\langle\mathcal{A}(x,\eta,\xi),\xi\rangle \geq \lambda_1|\xi|^p$, *where* λ_1 *is a positive constant.*
(H4) $|\mathcal{A}(x,\eta,\xi)| \leq K_1(x)+h_1(x)|\eta|^{p/p'}+h_2(x)|\xi|^{p/p'}$, *where* K_1, h_1 *and* h_2 *are positive functions, with* h_1 *and* $h_2\in L^\infty(\Omega)$, *and* $K_1\in L^{p'}(\Omega,\omega)$ *(with* $1/p+1/p' = 1$*).*

If $\omega\in A_p$ *(with* $1<p<\infty$*),* $f_j/\omega\in L^{p'}(\Omega,\omega)$ $(j = 0, 1, ..., N$ *and* $1/p+1/p' = 1)$ *then the problem (P16) has a unique solution* $u\in X = W^{2,p}(\Omega,\omega)\cap W^{1,p}_0(\Omega,\omega)$. *Moreover, we have*

$$\|u\|_X \leq \frac{1}{\gamma_2^{p'/p}}\left(C_\Omega\|f_0/\omega\|_{L^{p'}(\Omega,\omega)} + \sum_{j=1}^N \|f_j/\omega\|_{L^{p'}(\Omega,\omega)}\right)^{p'/p},$$

where $\gamma_2 = \min\{\lambda_1, 1\}$ *and* C_Ω *is the constant in Theorem 2.1.5..*

Proof. The basic idea is to reduce the problem (P16) to an operator equation $Au = T$ and apply the Theorem 5.1.1. To prove the existence of solutions, we define $B, B_1, B_2 : X\times X\to\mathbb{R}$ and $T : X\to\mathbb{R}$ by

$$\begin{aligned} B(u,\varphi) &= B_1(u,\varphi) + B_2(u,\varphi) \\ B_1(u,\varphi) &= \sum_{j=1}^N \int_\Omega \mathcal{A}_j(x,u,\nabla u)D_j\varphi\,\omega\,dx \end{aligned}$$

$$\begin{aligned} &= \int_\Omega \langle \mathcal{A}(x,u,\nabla u), \nabla \varphi\rangle\, \omega\, dx \\ B_2(u,\varphi) &= \int_\Omega |\Delta u|^{p-2}\Delta u\, \Delta\varphi\, \omega\, dx \\ T(\varphi) &= \int_\Omega f_0\, \varphi\, dx + \sum_{j=1}^N \int_\Omega f_j\, D_j\varphi\, dx. \end{aligned}$$

Then $u \in X$ is a (weak) solution to problem (P16) if $B(u,\varphi) = B_1(u,\varphi) + B_2(u,\varphi) = T(\varphi)$, for all $\varphi \in X$. For $j = 1, ..., N$ we define the operators $F_j, G : X \to L^{p'}(\Omega,\omega)$ by

$$\begin{aligned} (F_j u)(x) &= \mathcal{A}_j(x, u(x), \nabla u(x)) \\ (Gu)(x) &= |\Delta u(x)|^{p-2}\, \Delta u(x). \end{aligned}$$

Step 1. We have that the operator F_j is bounded and continuous. In fact:
(i) Using (H4) we obtain

$$\begin{aligned} \|F_j u\|^{p'}_{L^{p'}(\Omega,\omega)} &= \int_\Omega |F_j u(x)|^{p'}\omega\, dx = \int_\Omega |\mathcal{A}_j(x,u,\nabla u)|^{p'}\omega\, dx \\ &\le \int_\Omega \left(K_1 + h_1|u|^{p/p'} + h_2|\nabla u|^{p/p'}\right)^{p'} \omega\, dx \\ &\le C_p \int_\Omega \left[(K_1^{p'} + h_1^{p'}|u|^p + h_2^{p'}|\nabla u|^p)\omega\right] dx \\ &= C_p\left[\int_\Omega K_1^{p'}\, \omega\, dx + \int_\Omega h_1^{p'}|u|^p\, \omega\, dx + \int_\Omega h_2^{p'}|\nabla u|^p \omega\, dx\right], \end{aligned} \tag{6.33}$$

where the constant C_p depends only on p. We have, by Theorem 2.1.5,

$$\int_\Omega h_1^{p'}|u|^p\, \omega\, dx \le \|h_1\|^{p'}_{L^\infty(\Omega)} \int_\Omega |u|^p\, \omega\, dx$$

$$\leq C_\Omega^p \|h_1\|_{L^\infty(\Omega)}^{p'} \int_\Omega |\nabla u|^p \, \omega \, dx \leq C_\Omega^p \|h_1\|_{L^\infty(\Omega)}^{p'} \|u\|_X^p,$$

and

$$\int_\Omega h_2^{p'} |\nabla u|^p \omega \, dx \leq \|h_2\|_{L^\infty(\Omega)}^{p'} \int_\Omega |\nabla u|^p \, \omega \, dx \leq \|h_2\|_{L^\infty(\Omega)}^{p'} \|u\|_X^p.$$

Therefore, in (6.33) we obtain

$$\|F_j u\|_{L^{p'}(\Omega,\omega)}$$
$$\leq C_p \left(\|K\|_{L^{p'}(\Omega,\omega)} + (C_\Omega^{p/p'} \|h_1\|_{L^\infty(\Omega)} + \|h_2\|_{L^\infty(\Omega)}) \|u\|_X^{p/p'} \right).$$

(ii) Let $u_m \to u$ in X as $m \to \infty$. We need to show that $F_j u_m \to F_j u$ in $L^{p'}(\Omega,\omega)$. If $u_m \to u$ in X, then $u_m \to u$ in $L^p(\Omega,\omega)$ and $|\nabla u_m| \to |\nabla u|$ in $L^p(\Omega,\omega)$. Using Theorem 1.4.11, there exist a subsequence $\{u_{m_k}\}$ and functions Φ_1 and Φ_2 in $L^p(\Omega,\omega)$ such that

$$\begin{aligned}
&u_{m_k}(x) \to u(x), \ \mu - \text{a.e. in } \Omega, \\
&|u_{m_k}(x)| \leq \Phi_1(x), \ \mu - \text{a.e. in } \Omega, \\
&|\nabla u_{m_k}(x)| \to |\nabla u(x)|, \ \mu - \text{a.e. in } \Omega, \\
&|\nabla u_{m_k}(x)| \leq \Phi_2(x), \ \mu - \text{a.e. in } \Omega.
\end{aligned}$$

Hence, using (H4), we obtain

$$\|F_j u_{m_k} - F_j u\|_{L^{p'}(\Omega,\omega)}^{p'} = \int_\Omega |F_j u_{m_k}(x) - F_j u(x)|^{p'} \omega \, dx$$
$$= \int_\Omega |\mathcal{A}_j(x, u_{m_k}, \nabla u_{m_k}) - \mathcal{A}_j(x, u, \nabla u)|^{p'} \omega \, dx$$
$$\leq C_p \int_\Omega \left(|\mathcal{A}_j(x, u_{m_k}, \nabla u_{m_k})|^{p'} + |\mathcal{A}_j(x, u, \nabla u)|^{p'} \right) \omega \, dx$$
$$\leq C_p \left[\int_\Omega \left(K_1 + h_1 |u_{m_k}|^{p/p'} + h_2 |\nabla u_{m_k}|^{p/p'} \right)^{p'} \omega \, dx \right.$$

$$+\int_\Omega \left(K_1 + h_1|u|^{p/p'} + h_2|\nabla u|^{p/p'}\right)^{p'} \omega\, dx\Bigg]$$

$$\leq 2\,C_p \int_\Omega \left(K_1 + h_1\Phi_1^{p/p'} + h_2\Phi_2^{p/p'}\right)^{p'} \omega\, dx$$

$$\leq 2\,C_p \left[\int_\Omega K_1^{p'}\,\omega\, dx + \int_\Omega h_1^{p'}\,\Phi_1^p\,\omega\, dx + \int_\Omega h_2^{p'}\,\Phi_2^p\,\omega\, dx\right]$$

$$\leq 2\,C_p \Bigg[\|K_1\|^{p'}_{L^{p'}(\Omega,\omega)} + \|h_1\|^{p'}_{L^\infty(\Omega)} \int_\Omega \Phi_1^p\omega\, dx$$

$$+\|h_2\|^{p'}_{L^\infty(\Omega)} \int_\Omega \Phi_2^p\,\omega\, dx\Bigg]$$

$$\leq 2\,C_p \Bigg[\|K_1\|^{p'}_{L^{p'}(\Omega,\omega)} + \|h_1\|^{p'}_{L^\infty(\Omega)}\ \|\Phi_1\|^p_{L^p(\Omega,\omega)}$$

$$+\|h_2\|^{p'}_{L^\infty(\Omega)}\|\Phi_2\|^p_{L^p(\Omega,\omega)}\Bigg].$$

By condition (H1), we have

$$F_j u_m(x) = \mathcal{A}_j(x, u_m(x), \nabla u_m(x)) \to \mathcal{A}_j(x, u(x), \nabla u(x)) = F_j u(x),$$

as $m \to +\infty$. Therefore, by the Lebesgue Dominated Convergence Theorem, we obtain $\|F_j u_{m_k} - F_j u\|_{L^{p'}(\Omega,\omega)} \to 0$, that is, $F_j u_{m_k} \to F_j u$ in $L^{p'}(\Omega, \omega)$. By the Convergence Principle in Banach spaces (see Proposition 10.13 in [59]), we have

$$F_j u_m \to F_j u \ \text{ in } \ L^{p'}(\Omega, \omega). \tag{6.34}$$

Step 2. We also have that the operator G is continuous and bounded. In fact:
(i) We have

$$\|Gu\|^{p'}_{L^{p'}(\Omega,\omega)} = \int_\Omega \left|\,|\Delta u|^{p-2}\,\Delta u\right|^{p'} \omega\, dx$$

$$= \int_\Omega |\Delta u|^{(p-2)\,p'} |\Delta u|^{p'}\,\omega\, dx = \int_\Omega |\Delta u|^p\,\omega\, dx \leq \|u\|^p_X.$$

Hence, $\|Gu\|_{L^{p'}(\Omega,\omega)} \leq \|u\|_X^{p/p'}$.
(ii) If $u_m \to u$ in X then $\Delta u_m \to \Delta u$ in $L^p(\Omega,\omega)$.By Theorem 1.4.11, there exist a subsequence $\{u_{m_k}\}$ and a function $\Phi_3 \in L^p(\Omega,\omega)$ such that

$$\Delta u_{m_k}(x) \to \Delta u(x), \;\; \mu - a.e. \text{ in } \Omega$$
$$|\Delta u_{m_k}(x)| \leq \Phi_3(x), \;\; \mu - a.e. \text{ in } \Omega.$$

Hence, using Lemma 6.2.1(a), we obtain, if $p \neq 2$

$$\begin{aligned}
&\|Gu_{m_k} - Gu\|^{p'}_{L^{p'}(\Omega,\omega)} = \int_\Omega |\, Gu_{m_k} - Gu|^{p'}\,\omega\,dx \\
&= \int_\Omega \left|\, |\Delta u_{m_k}|^{p-2}\,\Delta u_{m_k} - |\,\Delta u|^{p-2}\,\Delta u\right|^{p'}\,\omega\,dx \\
&\leq \int_\Omega \left[\alpha_p\,|\Delta u_{m_k} - \Delta u|\,(\,|\Delta u_{m_k}| + |\Delta u|)^{(p-2)}\right]^{p'}\,\omega\,dx \\
&\leq \alpha_p^{p'} \int_\Omega |\,\Delta u_{m_k} - \Delta u|^{p'}\,(2\,\Phi_3)^{(p-2)\,p'}\,\omega\,dx \\
&\leq \alpha_p^{p'}\,2^{(p-2)p'}\left(\int_\Omega |\Delta u_{m_k} - \Delta u|^p\,\omega\,dx\right)^{p'/p} \times \\
&\times \left(\int_\Omega \Phi_3^{(p-2)\,p\,p'/(p-p')}\,\omega\,dx\right)^{(p-p')/p} \\
&\leq \alpha_p^{p'}\,2^{(p-2)\,p'}\,\|u_{m_k} - u\|_X^{p'}\,\|\Phi\|^{p-p'}_{L^p(\Omega,\omega)},
\end{aligned}$$

since $(p-2)\,p\,p'/(p-p') = p$ if $p \neq 2$. If $p = 2$,we have

$$\|Gu_{m_k} - Gu\|^2_{L^2(\Omega,\omega)} = \int_\Omega |\Delta u_{m_k} - \Delta u|^2\,\omega\,dx \leq \|u_{m_k} - u\|^2_X.$$

Therefore (for $1 < p < \infty$), by the Lebesgue Dominated Convergence Theorem, we obtain $\|Gu_{m_k} - Gu\|_X \to 0$, that is, $Gu_{m_k} \to Gu$ in $L^{p'}(\Omega,\omega)$. By the Convergence Principle in Banach spaces (see

Proposition 10.13 in [59]), we have

$$Gu_m \to Gu \text{ in } L^{p'}(\Omega, \omega). \tag{6.35}$$

Step 3. We have, by Theorem 2.1.5 (with $\theta = 1$),

$$\begin{aligned}
|T(\varphi)| &\leq \int_\Omega |f_0||\varphi|\,dx + \sum_{j=1}^N \int_\Omega |f_j||D_j\varphi|\,dx \\
&= \int_\Omega \frac{|f_0|}{\omega}|\varphi|\omega\,dx + \sum_{j=1}^N \int_\Omega \frac{|f_j|}{\omega}|D_j\varphi|\,\omega\,dx \\
&\leq \|f_0/\omega\|_{L^{p'}(\Omega,\omega)}\|\varphi\|_{L^p(\Omega,\omega)} \\
&+ \sum_{j=1}^N \|f_j/\omega\|_{L^{p'}(\Omega,\omega)}\|D_j\varphi\|_{L^p(\Omega,\omega)} \\
&\leq \left(C_\Omega\,\|f_0/\omega\|_{L^{p'}(\Omega,\omega)} + \sum_{j=1}^N \|f_j/\omega\|_{L^{p'}(\Omega,\omega)}\right)\|\varphi\|_X.
\end{aligned}$$

Moreover, using (H4) and the Hölder inequality, we also have

$$\begin{aligned}
&|B(u,\varphi)| \leq |B_1(u,\varphi)| + |B_2(u,\varphi)| \\
&\leq \sum_{j=1}^N \int_\Omega |\mathcal{A}_j(x,u,\nabla u)||D_j\varphi|\,\omega\,dx \\
&+ \int_\Omega |\,\Delta u|^{p-2}\,|\,\Delta u|\,|\,\Delta\varphi|\,\omega\,dx. \qquad (6.36)
\end{aligned}$$

In (6.36) we have

$$\begin{aligned}
&\int_\Omega |\mathcal{A}(x,u,\nabla u)|\,|\nabla\varphi|\,\omega\,dx \\
&\leq \int_\Omega \left(K_1 + h_1|u|^{p/p'} + h_2|\nabla u|^{p/p'}\right)|\nabla\varphi|\,\omega\,dx
\end{aligned}$$

$$\leq \|K_1\|_{L^{p'}(\Omega,\omega)}\| \, |\nabla\varphi| \, \|_{L^p(\Omega,\omega)} + \|h_1\|_{L^\infty(\Omega)}\|u\|^{p/p'}_{L^p(\Omega,\omega)}\| \, |\nabla\varphi| \, \|_{L^p(\Omega,\omega)}$$
$$+ \|h_2\|_{L^\infty(\Omega)}\| \, |\nabla u| \, \|^{p/p'}_{L^p(\Omega,\omega)}\| \, |\nabla\varphi| \, \|_{L^p(\Omega,\omega)}$$
$$\leq \left(\|K_1\|_{L^{p'}(\Omega,\omega)} + (C_\Omega^{p/p'} \|h_1\|_{L^\infty(\Omega)} + \|h_2\|_{L^\infty(\Omega)})\|u\|_X^{p/p'} \right) \|\varphi\|_X,$$

and

$$\int_\Omega |\,\Delta u|^{p-2}\,|\,\Delta u|\,|\,\Delta\,\varphi|\,\omega\,dx = \int_\Omega |\,\Delta u|^{p-1}\,|\,\Delta\,\varphi|\,\omega\,dx$$
$$\leq \left(\int_\Omega |\,\Delta u|^p\,\omega\,dx\right)^{1/p'} \left(\int_\Omega |\,\Delta\varphi|^p\,\omega\,dx\right)^{1/p}$$
$$\leq \|u\|_X^{p/p'}\,\|\varphi\|_X.$$

Therefore, in (6.36) we obtain, for all $u, \varphi \in X$

$$|B(u,\varphi)| \leq \left[\|K_1\|_{L^{p'}(\Omega,\omega)} + C_\Omega^{p/p'}\,\|h_1\|_{L^\infty(\Omega)}\|u\|_X^{p/p'} + \|h_2\|_{L^\infty(\Omega,\omega)}\|u\|_X^{p/p'} + \|u\|_X^{p/p'}\right] \|\varphi\|_X.$$

Since $B(u,.)$ is linear, for each $u \in X$, there exists a linear and continuous functional on X denoted by Au such that $(Au|\varphi) = B(u,\varphi)$ for all $\varphi \in X$ (recall that $(f|x)$ denotes the value of the linear functional f at the point x). Moreover

$$\|Au\|_* \leq \|K_1\|_{L^{p'}(\Omega,\omega)} + C_\Omega^{p/p'}\,\|h_1\|_{L^\infty(\Omega)}\|u\|_X^{p/p'} + \|h_2\|_{L^\infty(\Omega,\omega)}\|u\|_X^{p/p'} + \|u\|_X^{p/p'}.$$

Hence, we obtain the operator $A : X \to X^*$, $u \mapsto Au$. Consequently, problem (P16) is equivalent to the operator equation $Au = T$, $u \in X$.

Step 4. Using condition (H2) and Lemma 6.2.1(b), we have

$$(Au_1 - Au_2|u_1 - u_2) = B(u_1, u_1 - u_2) - B(u_2, u_1 - u_2)$$

$$
\begin{aligned}
&= \int_\Omega \langle \mathcal{A}(x,u_1,\nabla u_1), \nabla(u_1-u_2)\rangle \,\omega\, dx \\
&+ \int_\Omega |\,\Delta u_1|^{p-2}\,\Delta u_1\,\Delta(u_1-u_2)\,\omega\, dx \\
&- \int_\Omega \langle \mathcal{A}(x,u_2,\nabla u_2), \nabla(u_1-u_2)\rangle \,\omega\, dx \\
&- \int_\Omega |\,\Delta u_2|^{p-2}\,\Delta u_2\,\Delta(u_1-u_2)\,\omega\, dx \\
&= \int_\Omega \langle \mathcal{A}(x,u_1,\nabla u_1) - \mathcal{A}(x,u_2,\nabla u_2), \nabla(u_1-u_2)\rangle \,\omega\, dx \\
&+ \int_\Omega (|\,\Delta u_1|^{p-2}\,\Delta u_1 - |\,\Delta u_2|^{p-2}\,\Delta u_2)\,\Delta(u_1-u_2)\,\omega\, dx \\
&\geq \theta_1 \int_\Omega |\nabla(u_1-u_2)|^p\,\omega\, dx \\
&+ \beta_p \int_\Omega (|\,\Delta u_1| + |\,\Delta u_2|)^{p-2}\,|\Delta u_1 - \Delta u_2|^2\,\omega\, dx \\
&\geq \theta_1 \int_\Omega |\nabla(u_1-u_2)|^p\,\omega\, dx \\
&+ \beta_p \int_\Omega (|\,\Delta u_1 - \Delta u_2|)^{p-2}\,|\Delta u_1 - \Delta u_2|^2\,\omega\, dx \\
&= \theta_1 \int_\Omega |\nabla(u_1-u_2)|^p\,\omega\, dx + \beta_p \int_\Omega |\Delta u_1 - \Delta u_2|^p\,\omega\, dx \\
&\geq \gamma_1\,\|u_1-u_2\|_X^p,
\end{aligned}
$$

where $\gamma_1 = \min\{\theta_1, \beta_p\}$. Therefore, the operator A is strictly monotone. Moreover, using (H3), we obtain

$$
\begin{aligned}
(Au|u) &= B(u,u) = B_1(u,u) + B_2(u,u) \\
&= \int_\Omega \langle \mathcal{A}(x,u,\nabla u), \nabla u\rangle\,\omega\, dx + \int_\Omega |\,\Delta u|^{p-2}\,\Delta u\,\Delta u\,\omega\, dx \\
&\geq \int_\Omega \lambda_1 |\nabla u|^p\,\omega\, dx + \int_\Omega |\,\Delta u|^p\,\omega\, dx
\end{aligned}
$$

$$\geq \gamma_2 \|u\|_X^p,$$

where $\gamma_2 = \min\{\lambda_1, 1\}$. Hence, since $p > 1$, we have

$$\frac{(Au|u)}{\|u\|_X} \to +\infty, \text{ as } \|u\|_X \to +\infty,$$

that is, A is coercive.

Step 5. We show that the operator A is continuous which, in particular means that A is hemicontinuous. Let $u_m \to u$ in X as $m \to \infty$. We have,

$$\begin{aligned}
&|B_1(u_m, \varphi) - B_1(u, \varphi)| \\
&\leq \sum_{j=1}^{N} \int_\Omega |\mathcal{A}_j(x, u_m, \nabla u_m) - \mathcal{A}_j(x, u, \nabla u)||D_j\varphi|\,\omega\,dx \\
&= \sum_{j=1}^{N} \int_\Omega |F_j u_m - F_j u||D_j\varphi|\,\omega\,dx \\
&\leq \sum_{j=1}^{N} \|F_j u_m - F_j u\|_{L^{p'}(\Omega,\omega)} \|D_j\varphi\|_{L^p(\Omega,\omega)} \\
&\leq \sum_{j=1}^{N} \|F_j u_m - F_j u\|_{L^{p'}(\Omega,\omega)} \|\varphi\|_X,
\end{aligned}$$

and

$$\begin{aligned}
&|B_2(u_m, \varphi) - B_2(u, \varphi)| \\
&= \left| \int_\Omega |\Delta u_m|^{p-2} \Delta u_m \, \Delta\varphi\,\omega\,dx - \int_\Omega |\Delta u|^{p-2} \Delta u\,\Delta\varphi\,\omega\,dx \right| \\
&\leq \int_\Omega \left| |\Delta u_m|^{p-2}\Delta u_m - |\Delta u|^{p-2}\Delta u \right| |\Delta\varphi|\,\omega\,dx \\
&= \int_\Omega |Gu_m - Gu|\,|\Delta\varphi|\,\omega\,dx
\end{aligned}$$

$$\leq \|Gu_m - Gu\|_{L^{p'}(\Omega,\omega)} \|\varphi\|_X.$$

for all $\varphi \in X$. Hence,

$$\begin{aligned}
&|B(u_m,\varphi) - B(u,\varphi)| \\
&\leq |B_1(u_m,\varphi) - B_1(u,\varphi)| + |B_2(u_m,\varphi) - B_2(u,\varphi)| \\
&\leq \left[\sum_{j=1}^{N} \|F_j u_m - F_j u\|_{L^{p'}(\Omega,\omega)} + \|Gu_m - Gu\|_{L^{p'}(\Omega,\omega)}\right] \|\varphi\|_X.
\end{aligned}$$

Then we obtain

$$\|Au_m - Au\|_* \leq \sum_{j=1}^{N} \|F_j u_m - F_j u\|_{L^{p'}(\Omega,\omega)} + \|Gu_m - Gu\|_{L^{p'}(\Omega,\omega)}.$$

Therefore, using (6.34) and (6.35) we have $\|Au_m - Au\|_* \to 0$ as $m \to +\infty$, that is, A is continuous.

Therefore, by Theorem 5.1.1, the operator equation $Au = T$ has a unique solution $u \in X$ and it is the unique solution for problem (P16).

Step 6. In particular, by setting $\varphi = u$ in Definition 6.3.1, we have

$$B(u,u) = B_1(u,u) + B_2(u,u) = T(u). \qquad (6.37)$$

Hence, using (H3) and $\gamma_2 = \min\{\lambda_1, 1\}$, we obtain

$$\begin{aligned}
B_1(u,u) + B_2(u,u) &= \int_\Omega \langle \mathcal{A}(x,u,\nabla u), \nabla u\rangle \,\omega\, dx \\
&+ \int_\Omega |\Delta u|^{p-2} \Delta u\, \Delta u\, \omega\, dx \\
&\geq \int_\Omega \lambda_1 |\nabla u|^p \,\omega\, dx + \int_\Omega |\Delta u|^p \,\omega\, dx \\
&\geq \gamma_2 \|u\|_X^p,
\end{aligned}$$

and

$$T(u) = \int_\Omega f_0\, u\, dx + \sum_{j=1}^{N} \int_\Omega f_j\, D_j u\, dx$$

$$\leq \|f_0/\omega\|_{L^{p'}(\Omega,\omega)}\|u\|_{L^p(\Omega,\omega)} + \sum_{j=1}^{N} \|f_j/\omega\|_{L^{p'}(\Omega,\omega)}\|D_ju\|_{L^p(\Omega,\omega)}$$

$$\leq \left(C_\Omega \|f_0/\omega\|_{L^{p'}(\Omega,\omega)} + \sum_{j=1}^{N} \|f_j/\omega\|_{L^{p'}(\Omega,\omega)}\right) \|u\|_X.$$

Therefore, in (6.37), we have

$$\gamma_2 \|u\|_X^p \leq \left(C_\Omega \|f_0/\omega\|_{L^{p'}(\Omega,\omega)} + \sum_{j=1}^{N} \|f_j/\omega\|_{L^{p'}(\Omega,\omega)}\right) \|u\|_X,$$

and we obtain

$$\|u\|_X \leq \frac{1}{\gamma_2^{p'/p}} \left(C_\Omega \|f_0/\omega\|_{L^{p'}(\Omega,\omega)} + \sum_{j=1}^{N} \|f_j/\omega\|_{L^{p'}(\Omega,\omega)}\right)^{p'/p}.$$

□

Example 6.3.1. *Let* $\Omega = \{(x,y)\in\mathbb{R}^2 : x^2+y^2<1\}$, *the weight function* $\omega(x,y) = (x^2+y^2)^{-1/2}$ ($\omega \in A_3$, $p=3$), *and the function*

$$\mathcal{A} : \Omega \times \times \mathbb{R}^2 \to \mathbb{R}^2$$
$$\mathcal{A}((x,y),\eta,\xi) = h_2(x,y)\,|\xi|\,\xi,$$

where $h(x,y) = 2\,\mathrm{e}^{(x^2+y^2)}$. *Let us consider the partial differential operator*

$$\begin{aligned} Lu(x,y) &= \Delta((x^2+y^2)^{-1/2}|\Delta u|\,\Delta u) \\ &- \operatorname{div}((x^2+y^2)^{-1/2}\,\mathcal{A}((x,y),u,\nabla u)). \end{aligned}$$

Therefore, by Theorem 6.3.1, the problem

$$\begin{cases} Lu(x) &= \dfrac{\cos(xy)}{(x^2+y^2)} - \dfrac{\partial}{\partial x}\left(\dfrac{\sin(xy)}{(x^2+y^2)}\right) \\ & -\dfrac{\partial}{\partial y}\left(\dfrac{\sin(xy)}{(x^2+y^2)}\right) \ \text{ in } \ \Omega, \\ u(x) &= 0 \ \text{ on } \ \partial\Omega, \\ \Delta u &= 0 \ \text{ on } \ \partial\Omega, \end{cases}$$

has a unique solution $u \in X = W^{2,3}(\Omega, \omega) \cap W_0^{1,3}(\Omega, \omega)$.

6.4 L^∞-solutions for some degenerate nonlinear elliptic equations

In this section we prove the existence of (weak) solutions in the weighted Sobolev spaces $\mathrm{W}_0^{1,p}(\Omega, \omega)$ for the Dirichlet problem

$$(P17)\begin{cases} Lu(x) &= f(x), \quad \text{on } \Omega, \\ u(x) &= 0, \quad \text{on } \partial\Omega, \end{cases}$$

where L is the partial differential operator

$$\begin{aligned} Lu(x) \quad &= -\mathrm{div}\big[\omega(x)\mathcal{A}(x,u,\nabla u)\big] + g(x,u,\nabla u)\,\omega(x) \\ &+ H(x,u,\nabla u)\,\omega(x) \end{aligned} \tag{6.38}$$

where Ω is a bounded open set in $\mathbb{R}^N$ $(N \geq 2)$, ω is a weight function, and the functions $\mathcal{A} : \Omega \times \mathbb{R} \times \mathbb{R}^N \to \mathbb{R}^N$, $g : \Omega \times \mathbb{R} \times \mathbb{R}^N \to \mathbb{R}$ and $H : \Omega \times \mathbb{R} \times \mathbb{R}^N \to \mathbb{R}$ are Carathéodory functions.

Equations like (6.38) have been studied by many authors in the non-degenerate case (i.e. with $\omega(x) \equiv 1$) (see e.g. [7] and the references therein). The degenerate case with different conditions haven been studied by many authors. In [31] Drabek, Kufner and Mustonen proved that under certain condition, the Dirichlet problem associated with the equation

$$-\mathrm{div}(a(x,u,\nabla u)) = h,$$

$h \in [W_0^{1,p}(\Omega, \omega)]^*$ has at least one solution $u \in W_0^{1,p}(\Omega, \omega)$.

Firstly, we prove an L^∞ estimate for the bounded solutions of $(P17)$: we assume that $f/\omega \in L^q(\Omega, \omega)$, with $\theta/(\theta - 1) < q < \infty$ (where $\theta > 1$ as in Theorem 2.1.5) and we prove that any $u \in W_0^{1,p}(\Omega, \omega) \cap L^\infty(\Omega)$ that solves $(P17)$ satisfies $\|u\|_{L^\infty(\Omega)} \leq C$, where C depends only of the data, i.e., Ω, $N, p, q, \alpha_1, \alpha_2, C_0, C_1$

and $\|f/\omega\|_{L^q(\Omega,\omega)}$. After that, we prove the existence of solution for problem $(P17)$ if $f/\omega \in L^q(\Omega,\omega)$, with $p'\theta/(\theta-1) < q < \infty$.

Note that, in the proof of our main result, many ideas have been adapted from [7], [6] and [49].

Definition 6.4.1. *We say that $u \in W_0^{1,p}(\Omega,\omega) \cap L^\infty(\Omega)$ is a (weak) solution of problem (P17) if*

$$\int_\Omega \langle \mathcal{A}(x,u,\nabla u), \nabla\varphi\rangle\, \omega\, dx + \int_\Omega g(x,u,\nabla u)\varphi\,\omega\, dx$$
$$+ \int_\Omega H(x,u,\nabla u)\,\varphi\,\omega\, dx = \int_\Omega f\,\varphi\, dx, \tag{6.39}$$

for all $\varphi \in W_0^{1,p}(\Omega,\omega) \cap L^\infty(\Omega)$.

Theorem 6.4.1. *Let ω be an A_p-weight, $1 < p < \infty$. Suppose that*

(H1) *$x \mapsto \mathcal{A}(x,\eta,\xi)$ is measurable in Ω for all $(\eta,\xi) \in \mathbb{R}\times\mathbb{R}^N$,*
$(\eta,\xi) \mapsto \mathcal{A}(x,\eta,\xi)$ is continuous in $\mathbb{R}\times\mathbb{R}^N$ for almost all $x \in \Omega$.

(H2) *$\langle \mathcal{A}(x,\eta,\xi) - \mathcal{A}(x,\eta',\xi'), \xi-\xi'\rangle > 0$, whenever $\xi,\xi' \in \mathbb{R}^N$, $\xi \neq \xi'$;*

(H3) *$\langle \mathcal{A}(x,\eta,\xi),\xi\rangle \geq \alpha_1|\xi|^p$, with $1 < p < \infty$, where $\alpha_1 > 0$;*

(H4) *$|\mathcal{A}(x,\eta,\xi)| \leq K_1(x) + h_1(x)|\eta|^{p/p'} + h_2(x)|\xi|^{p/p'}$, where K_1, h_1 and h_2 are positive functions, with h_1 and $h_2 \in L^\infty(\Omega)$, and $K_1 \in L^{p'}(\Omega,\omega)$ $(1/p + 1/p' = 1)$.*

(H5) *$x \mapsto g(x,\eta,\xi)$ is measurable in Ω for all $(\eta,\xi) \in \mathbb{R}\times\mathbb{R}^N$,*
$(\eta,\xi) \mapsto g(x,\eta,\xi)$ is continuous in $\mathbb{R}\times\mathbb{R}^N$ for almost all $x \in \Omega$.

(H6) *$|g(x,\eta,\xi)| \leq K_2(x) + h_3(x)|\eta|^{p/p'} + h_4(x)|\xi|^{p/p'}$, where K_2, h_3 and h_4 are positive functions, with $h_3, h_4 \in L^\infty(\Omega)$ and $K_2 \in L^{p'}(\Omega,\omega)$.*

(H7) *$g(x,\eta,\xi)\eta \geq \alpha_0|\eta|^p$, for all $\eta \in \mathbb{R}$, where $\alpha_0 > 0$.*

(H8) *$x \mapsto H(x,\eta,\xi)$ is measurable in Ω for all $(\eta,\xi) \in \mathbb{R}\times\mathbb{R}^N$*

$(\eta,\xi)\mapsto H(x,\eta,\xi)$ is continuous in $\mathbb{R}\times\mathbb{R}^N$ for almost all $x\in\Omega$.

(H9) *$|H(x,\eta,\xi)|\le C_0 + C_1|\xi|^p$, where C_0 and C_1 are positive constants.*

(H10) *$f/\omega\in L^q(\Omega,\omega)$, with $\theta/(\theta-1) < q < \infty$ (where $\theta > 1$ as in Theorem 2.1.5).*

Let $u\in W_0^{1,p}(\Omega,\omega)\cap L^\infty(\Omega)$ be a solution of problem $(P17)$. Then there

exists a constant $C > 0$, which depends only on $\Omega, n, p, \alpha_1, \alpha_0, C_0, C_1$ and $\|f/\omega\|_{L^q(\Omega,\omega)}$, such that $\|u\|_{L^\infty(\Omega)}\le C$.

Proof. Set $\lambda = \dfrac{C_1}{\alpha_1} + 1$ and define for $k > 0$ the functions $\phi\in C^1(\mathbb{R})$ and $G_k\in W^{1,\infty}(\mathbb{R})$ by

$$\phi(s) = \begin{cases} e^{\lambda s} - 1, & \text{if } s\ge 0,\\ -e^{-\lambda s} + 1, & \text{if } s\le 0,\end{cases}$$

$$G_k(s) = \begin{cases} s-k, & \text{if } s\ge k,\\ 0, & \text{if } -k\le s\le k,\\ s+k, & \text{if } s\le -k.\end{cases}$$

If $u\in W_0^{1,p}(\Omega,\omega)\cap L^\infty(\Omega)$ is a solution of problem (P17), define the set $A(k) = \{x\in\Omega : |u(x)| > k\}$.

We will use the test functions $v(x) = \phi(G_k(u(x)))$. We have

$$v(x) = \phi((|u|-k)^+)\,\chi_{A(k)}\mathrm{sign}(u),$$
$$\nabla v = \phi'((|u|-k)^+)\chi_{A(k)}\nabla u,$$

where $\chi_{A(k)}$ is the characteristic function of the set $A(k)$.
Since $u\in W_0^{1,p}(\Omega,\omega)\cap L^\infty(\Omega)$, we have that $v\in W_0^{1,p}(\Omega,\omega)\cap L^\infty(\Omega)$.
Using the function v in (6.39) we obtain

$$\int_\Omega \langle\mathcal{A}(x,u,\nabla u),\nabla v\rangle\,\omega\,dx + \int_\Omega g(x,u,\nabla u)v\,\omega\,dx$$

$$+\int_\Omega H(x,u,\nabla u)\, v\, \omega\, dx$$
$$= \int_\Omega f\, v\, dx. \tag{6.40}$$

We have the following estimates.
(i) By (H3) we obtain

$$\begin{aligned}
&\int_\Omega \langle \mathcal{A}(x,u,\nabla u), \nabla v\rangle\, \omega\, dx\\
&= \int_{A(k)} \phi'((|u|-k)^+)\mathcal{A}(x,u,\nabla u).\nabla u\, \omega\, dx\\
&\geq \alpha_1 \int_{A(k)} |\nabla u|^p \phi'((|u|-k)^+)\, \omega\, dx.
\end{aligned}$$

(ii) By (H7) we obtain

$$\begin{aligned}
&\int_\Omega g(x,u,\nabla u)\, v\, \omega\, dx\\
&= \int_{A(k)} g(x,u,\nabla u)\, \phi((|u|-k)^+)\, \omega\, dx\\
&\geq \alpha_0 \int_{A(k)} |u|^{p-1} \phi((|u|-k)^+)\, \omega\, dx.
\end{aligned}$$

(iii) Using (H9) we obtain

$$\begin{aligned}
&\left|\int_\Omega H(x,u,\nabla u)\, v\, \omega\, dx\right|\\
&\leq \int_\Omega |H(x,u,\nabla u)||v|\, \omega\, dx\\
&\leq \int_{A(k)} (C_0 + C_1|\nabla u|^p)\phi((|u|-k)^+)\, \omega\, dx.
\end{aligned}$$

And we also have

$$\left| \int_\Omega f\, v\, dx \right| \leq \int_{A(k)} |f|\, \phi((|u|-k)^+)\, dx.$$

Hence in (6.40) we obtain

$$\begin{aligned}
&\alpha_1 \int_{A(k)} |\nabla u|^p \phi'((|u|-k)^+)\, \omega\, dx \\
&+ \alpha_0 \int_{A(k)} |u|^{p-1} \phi((|u|-k)^+)\, \omega\, dx \\
&\leq \int_{A(k)} (C_0 + C_1 |\nabla u|^p) \phi((|u|-k)^+)\, \omega\, dx \\
&+ \int_{A(k)} |f| \phi((|u|-k)^+)\, dx. \qquad (6.41)
\end{aligned}$$

Since $\lambda = \dfrac{C_1}{\alpha_1} + 1$, we have for $s \geq 0$

$$\begin{aligned}
&\alpha_1 \phi'(s) - C_1 \phi(s) = \alpha_1 \lambda \mathrm{e}^{\lambda s} - C_1 (\mathrm{e}^{\lambda s} - 1) \\
&= (\alpha_1 \lambda - C_1) \mathrm{e}^{\lambda s} + C_1 = \alpha_1 \mathrm{e}^{\lambda s} + C_1 \\
&\geq \alpha_1 \mathrm{e}^{\lambda s} = \frac{\alpha_1}{\lambda^p} [\lambda \mathrm{e}^{\lambda s/p}]^p \\
&= \frac{\alpha_1}{\lambda^p} [\phi'(s/p)]^p. \qquad (6.42)
\end{aligned}$$

Hence in (6.41) we obtain

$$\begin{aligned}
&\int_{A(k)} \Big[\alpha_1 |\nabla u|^p\, \phi'((|u|-k)^+) - C_1 |\nabla u|^p\, \phi((|u|-k)^+) \Big]\, \omega\, dx \\
&+ \alpha_0 \int_{A(k)} |u|^{p-1} \phi((|u|-k)^+)\, \omega\, dx \\
&\leq \int_{A(k)} (|f| + C_0\, \omega) \phi((|u|-k)^+)\, dx. \qquad (6.43)
\end{aligned}$$

Using (6.42) and $k < |u(x)|$ if $x \in A(k)$, we obtain

$$\begin{aligned} &\frac{\alpha_1}{\lambda^p} \int_{A(k)} \left| \phi'\left(\frac{(|u|-k)^+}{p}\right) \nabla u \right|^p \omega\, dx \\ &+ \alpha_0 k^{p-1} \int_{A(k)} \phi((|u|-k)^+)\, \omega\, dx \\ &\leq \int_{A(k)} (|f| + C_0\, \omega)\, \phi((|u|-k)^+)\, dx. \end{aligned} \tag{6.44}$$

Let us define the function ψ_k by $\psi_k(x) = \phi\left(\dfrac{(|u(x)|-k)^+}{p}\right)$. We have that $\psi_k \in W_0^{1,p}(\Omega, \omega)$ and

$$\nabla \psi_k = \frac{1}{p}\, \phi'\left(\frac{(|u|-k)^+}{p}\right) \chi_{A(k)} \mathrm{sign}(u)\, \nabla u. \tag{6.45}$$

We have that
(a) For all $s \geq 0$, $\mathrm{e}^{\lambda s} - 1 \geq (\mathrm{e}^{\lambda s/p} - 1)^p$;
(b) There exist a constant $C_2 > 0$ ($C_2 = C_2(\lambda, p)$) such that for all $s \geq 1$

$$\mathrm{e}^{\lambda s} - 1 \leq C_2 (\mathrm{e}^{\lambda s/p} - 1)^p \text{ and } \lambda \mathrm{e}^{\lambda s} \leq C_2\, \lambda (\mathrm{e}^{\lambda s/p} - 1)^p.$$

This implies
(I1) $\phi((|u|-k)^+) = \mathrm{e}^{\lambda(|u|-k)^+} - 1 \geq (\mathrm{e}^{\lambda(|u|-k)^+/p} - 1)^p = |\,\psi_k|^p$ a.e. on Ω;
(I2) If $x \in \mathrm{A}(k+1)$ then
$\phi((|u|-k)^+) = \mathrm{e}^{\lambda(|u|-k)^+} - 1 \leq C_2 (\mathrm{e}^{\lambda(|u|-k)^+/p} - 1)^p = C_2 |\psi_k|^p$ and
$\phi'((|u|-k)^+) = \lambda\, \mathrm{e}^{\lambda(|u|-k)^+} \leq C_2\, \lambda\, (\mathrm{e}^{\lambda(|u|-k)^+/p} - 1)^p = C_2\, \lambda\, |\psi_k|^p$.
Combining (I1) and (I2) with (6.44) and (6.45) we obtain

$$\frac{\alpha_1 p^p}{\lambda^p} \int_\Omega |\nabla \psi_k|^p \omega\, dx + \alpha_0 k^{p-1} \int_\Omega |\psi_k|^p\, \omega\, dx$$

$$
\begin{aligned}
&\le \int_{A(k)} (|f| + C_0\,\omega)\,\phi((|u| - k)^+)\,dx \\
&\le \int_{A(k+1)} (|f| + C_0\,\omega)\,C_2\,|\psi_k|^p\,dx \\
&+ \int_{A(k)-A(k+1)} (|f| + C_0\,\omega)\phi((|u| - k)^+)\,dx. \qquad (6.46)
\end{aligned}
$$

Define the function $h = |f| + C_0\,\omega$. Since $f/\omega \in L^q(\Omega, \omega)$ and $\mu(\Omega) < \infty$, we have that $h/\omega \in L^q(\Omega, \omega)$. Hence

$$
\begin{aligned}
\int_\Omega h\,|\psi_k|^p\,dx &= \int_\Omega \frac{h}{\omega}\,|\psi_k|^p\,\omega^{1/q}\,\omega^{1/q'}\,dx \\
&\le \|h/\omega\|_{L^q(\Omega,\omega)}\|\psi_k\|^p_{L^{p\,q'}(\Omega,\omega)}. \qquad (6.47)
\end{aligned}
$$

If $x \in A(k) - A(k+1)$, we have $k < |u| < k + 1$. Hence

$$
\phi((|u| - k)^+) = \mathrm{e}^{\lambda(|u|-k)^+} - 1 \le \mathrm{e}^\lambda - 1
$$

and we obtain

$$
\begin{aligned}
&\int_{A(k)-A(k+1)} (|f| + C_0\,\omega)\,\phi((|u| - k)^+)\,dx \\
&\le \int_{A(k)-A(k+1)} (\mathrm{e}^\lambda - 1)\,h\,dx \\
&\le \mathrm{e}^\lambda \int_{A(k)} h\,dx. \qquad (6.48)
\end{aligned}
$$

By Theorem 2.1.5 (with $p^* = p\,\theta$), (6.46) and (6.48) we have

$$
\begin{aligned}
&\frac{\alpha_1\,p^p}{\lambda^p}\,\frac{1}{C_\Omega^p}\left(\int_\Omega |\psi_k|^{p^*}\,\omega\,dx\right)^{p/p^*} + \alpha_0\,k_0^{p-1}\int_\Omega |\psi_k|^p\,\omega\,dx \\
&\le C_2 \int_\Omega h\,|\psi_k|^p\,dx + \mathrm{e}^\lambda \int_{A(k)} h\,dx
\end{aligned}
$$

Therefore, there exist positive constants C_3 and C_4 (depending only on Ω, α_1, p,λ and C_2) such that

$$C_3\bigg(\int_\Omega |\psi_k|^{p^*}\omega\,dx\bigg)^{p/p^*} + C_4\,\alpha_0\,k_0^{p-1}\int_\Omega |\psi_k|^p\,\omega\,dx \\ \leq \int_\Omega h\,|\psi_k|^p\,dx + \int_{A(k)} h\,dx \tag{6.49}$$

Since $\theta/(\theta-1) < q$ then $q' < \theta$ and $p < p\,q' < p^*$. For $0 < r < 1$ such that $\dfrac{1}{p\,q'} = \dfrac{r}{p} + \dfrac{1-r}{p^*}$, using an interpolation inequality, Young's inequality (with $0 < \gamma < \infty$) and Hölder's inequality with exponents q and q' we thus obtain

$$\begin{aligned} \int_\Omega h\,|\psi_k|^p\,dx &\leq \|h/\omega\|_{L^q(\Omega,\omega)}\|\psi_k\|^p_{L^{p\,q'}(\Omega,\omega)} \\ &\leq \|h/\omega\|_{L^q(\Omega,\omega)}\|\psi_k\|^{r\,p}_{L^p(\Omega,\omega)}\|\psi_k\|^{(1-r)p}_{L^{p^*}(\Omega,\omega)} \\ &\leq (1-r)\gamma^{1/(1-r)}\|\psi_k\|^p_{L^{p^*}(\Omega,\omega)} \\ &+ r\,\gamma^{-1/r}\|h/\omega\|^{1/r}_{L^q(\Omega,\omega)}\|\psi_k\|^p_{L^p(\Omega,\omega)}. \end{aligned} \tag{6.50}$$

Hence in (6.49) we obtain

$$C_3\bigg(\int_\Omega |\psi_k|^{p^*}\omega\,dx\bigg)^{p/p^*} + C_4\,\alpha_0\,k_0^{p-1}\int_\Omega |\psi_k|^p\,\omega\,dx \\ \leq (1-r)\gamma^{1/(1-r)}\|\psi_k\|^p_{L^{p^*}(\Omega,\omega)} + r\,\gamma^{-1/r}\|h/\omega\|^{1/r}_{L^q(\Omega,\omega)}\|\psi_k\|^p_{L^p(\Omega,\omega)} \\ + \int_{A(k)} h\,dx. \tag{6.51}$$

Now, we can choose γ in order to have $(1-r)\,\gamma^{1/(1-r)} = C_3/2$ and k_0 such that

$$C_4\,\alpha_0\,k_0^{p-1} = r\,\gamma^{-1/r}\,\|h/\omega\|^{1/r}_{L^q(\Omega,\omega)}.$$

We obtain, from (6.51), that for every $k \geq k_0$ it results

$$\frac{C_3}{2}\left(\int_\Omega |\psi_k|^{p^*}\,\omega\,dx\right)^{p/p^*} \leq \int_{A(k)} h\,dx \leq \|h/\omega\|_{L^q(\Omega,\omega)}\,[\mu(A_k)]^{1/q'}.$$

Hence for all $k \geq k_0$ we have

$$\begin{aligned}\int_\Omega |\psi_k|^{p^*}\,\omega\,dx &\leq \left(\frac{2}{C_3}\|h/\omega\|_{L^q(\Omega,\omega)}[\mu(A(k))]^{1/q'}\right)^{p^*/p}\\ &= \left(\frac{2}{C_3}\right)^{p^*/p}\|h/\omega\|^{p^*/p}_{L^q(\Omega,\omega)}[\mu(A(k))]^{p^*/p\,q'}\\ &= C_5[\mu(A(k))]^{p^*/p\,q'}.\end{aligned}$$

Let us now take $l > k \geq k_0$ we have

$$\begin{aligned}\mu(A(l))\left[\lambda\left(\frac{l-k}{p}\right)\right]^{p^*} &\leq \mu(A(k))|\phi((l-k)/p)|^{p^*}\\ &\leq \int_{A(k)} |\psi_k|^{p^*}\omega\,dx\\ &\leq \int_\Omega |\psi_k|^{p^*}\omega\,dx.\end{aligned}$$

Therefore for all $l > k \geq k_0$ we obtain

$$\begin{aligned}(l-k)^{p^*}\mu(A(l)) &\leq \frac{p^{p^*}}{\lambda^{p^*}}\,C_5[\mu(A(k))]^{p^*/p\,q'}\\ &= C_6[\mu(A(k))]^{p^*/p\,q'}.\end{aligned}$$

that is, $\mu(A(l)) \leq \dfrac{C_6}{(l-k)^{p^*}}[\mu(A(k))]^{p^*/p\,q'}$.

Let $\varphi(k) = \mu(A(k))$. Since $\beta = p^*/p\,q' > 1$, by Lemma 4.1.7 there exists a constant $C_7 > 0$ such that

$$\mu(A(k)) = 0,\ \forall\, k \geq C_7.$$

Using Lemma 2.1.2 we have $|A(k)| = 0$ for all $k \geq C_7$. Therefore any solution u of problem $(P17)$ satisfies the estimate $\|u\|_{L^\infty(\Omega)} \leq C_7$. □

The main result of this section is given in the next theorem.

Theorem 6.4.2. *Let us assume that* (H1) – (H9) *(in Theorem 6.4.1) hold true and suppose that*
(H11) $f/\omega \in L^q(\Omega, \omega)$, *with* $p'\theta/(\theta-1) < q < \infty$ *(where* $\theta > 1$ *is in Theorem 2.1.5);*
(H12) $H(x,\eta,\xi)\,\eta \geq 0$, *for all* $\eta \in \mathbb{R}$.
Then there exists at least one solution $u \in W_0^{1,p}(\Omega,\omega) \cap L^\infty(\Omega)$ *of the problem* $(P17)$.

Theorem 6.4.2 will be proved by approximation problem $(P17)$ with the following problems

$$(P_m)\begin{cases} -\operatorname{div}\big[\omega\,\mathcal{A}(x,u,\nabla u)\big] + g(x,u,\nabla u)\,\omega \\ +\,H_m(x,u,\nabla u)\,\omega = f, \text{ on } \Omega \\ u(x) = 0, \text{ on } \partial\Omega \end{cases}$$

where $H_m(x,\eta,\xi) = \dfrac{H(x,\eta,\xi)}{1+\dfrac{1}{m}\,|H(x,\eta,\xi)|}$, for $m \in \mathbb{N}$. Note that $|H_m| \leq |H|$ and that $|H_m| \leq m$.

Proof. **Step 1.** Let us define for $m \in \mathbb{N}$ the approximation

$$H_m(x,\eta,\xi) = \frac{H(x,\eta,\xi)}{1+\dfrac{1}{m}\,|H(x,\eta,\xi)|}.$$

We have that

$$|H_m(x,\eta,\xi)| \leq |H(x,\eta,\xi)|\;,\;\; |H_m(x,\eta,\xi)| < m$$

and $H_m(x,\eta,\xi)$ satisfies the conditions (H9) and (H12). We consider the approximate problem

$$(P_m)\begin{cases} -\operatorname{div}\big[\omega\,\mathcal{A}(x,u,\nabla u)\big]+g(x,u,\nabla u)\,\omega \\ +H_m(x,u,\nabla u)\,\omega = f \ \text{ in } \Omega, \\ u(x)=0 \text{ on } \partial\Omega, \end{cases}$$

We say that $u\in W_0^{1,p}(\Omega,\omega)$ is (weak) solution of problem (P_m) if

$$\begin{aligned} &\int_\Omega \langle \mathcal{A}(x,u,\nabla u),\nabla\varphi\rangle\,\omega\,dx+\int_\Omega g(x,u,\nabla u)\,\varphi\,\omega\,dx \\ &+\int_\Omega H_m(x,u,\nabla u)\,\varphi\,\omega\,dx=\int_\Omega f\,\varphi\,dx, \end{aligned} \qquad (6.52)$$

for all $\varphi\in W_0^{1,p}(\Omega,\omega)$. We will prove that there exists at least one solution u_m of the problem (P_m).
For $u,v,\varphi\in W_0^{1,p}(\Omega,\omega)$ we define

$$\begin{aligned} B(u,v,\varphi) &= \int_\Omega \langle \mathcal{A}(x,u,\nabla v),\nabla\varphi\rangle\,\omega\,dx, \\ B_m(u,\varphi) &= \int_\Omega g(x,u,\nabla u)\,\varphi\,\omega\,dx+\int_\Omega H_m(x,u,\nabla u)\,\varphi\,\omega\,dx, \\ T(\varphi) &= \int_\Omega f\,\varphi\,dx. \end{aligned}$$

Then $u\in W_0^{1,p}(\Omega,\omega)$ is a (weak) solution of problem (P_m) if

$$B(u,u,\varphi)+B_m(u,\varphi)=T(\varphi), \quad \text{for all } \varphi\in W_0^{1,p}(\Omega,\omega).$$

Let $a(u,v,\varphi)=B(u,v,\varphi)+B_m(u,\varphi)$.
(i) Using (H4) we obtain

$$|B(u,v,\varphi)| \ \leq \ \Big(\|K_1\|_{L^{p'}(\Omega,\omega)}+\|h_1\|_{L^\infty(\Omega)}\,\|u\|^{p/p'}_{W_0^{1,p}(\Omega,\omega)}$$

$$+\|h_2\|_{L^\infty(\Omega)}\|v\|^{p/p'}_{W_0^{1,p}(\Omega,\omega)}\Big)\|\varphi\|_{W_0^{1,p}(\Omega,\omega)}.$$

(ii) Using (H6) and $|H_m(x,\eta,\xi)|\leq m$, we obtain

$$\begin{aligned}|B_m(u,\varphi)| &\leq \Big(\|K_2\|_{L^{p'}(\Omega,\omega)}+\|h_3\|_{L^\infty(\Omega)}\|u\|^{p/p'}_{W_0^{1,p}(\Omega,\omega)}\\ &+ \|h_4\|_{L^\infty(\Omega)}\|u\|^{p/p'}_{W_0^{1,p}(\Omega,\omega)}+m\,[\mu(\Omega)]^{1/p'}\Big)\|\varphi\|_{W_0^{1,p}(\Omega,\omega)}\end{aligned}$$

Hence,

$$\begin{aligned}&|\,a(u,v,\varphi)|\\ &\leq\Big(\|K_1\|_{L^{p'}(\Omega,\omega)}+(\|h_1\|_{L^\infty(\Omega)}+\|h_3\|_{L^\infty(\Omega)}\\ &+\|h_4\|_{L^\infty(\Omega)})\,\|u\|^{p/p'}_{W_0^{1,p}(\Omega,\omega)}+\|K_2\|_{L^{p'}(\Omega,\omega)}\\ &+\|h_2\|_{L^\infty(\Omega)}\|v\|^{p/p'}_{W_0^{1,p}(\Omega,\omega)}+m\,[\mu(\Omega)]^{1/p'}\Big)\|\varphi\|_{W_0^{1,p}(\Omega,\omega)}.\end{aligned}$$

Since $a(u,v,.)$ is linear and continuous for each

$$(u,v)\in W_0^{1,p}(\Omega,\omega)\times W_0^{1,p}(\Omega,\omega)$$

there exists a linear and continuous operator

$$A(u,v): W_0^{1,p}(\Omega,\omega)\to[W_0^{1,p}(\Omega,\omega)]^*$$

such that

$$(A(u,v)|\varphi)=a(u,v,\varphi).$$

We set $\tilde{A}(u)=A(u,u)$ for all $u\in W_0^{1,p}(\Omega,\omega)$. The operator

$$\tilde{A}: W_0^{1,p}(\Omega,\omega)\to\ [W_0^{1,p}(\Omega,\omega)]^*$$

is semimonotone, that is, by similar arguments as in the proof of Theorem 2 in [49] we have:

(i) $(A(u,u) - A(u,v)|u - v) \geq 0$ for all $u, v \in W_0^{1,p}(\Omega,\omega)$;
(ii) For each $u \in W_0^{1,p}(\Omega,\omega)$, the operator $v \mapsto A(u,v)$ is hemicontinuous and bounded from $W_0^{1,p}(\Omega,\omega)$ to $[W_0^{1,p}(\Omega,\omega)]^*$ and for each
$v \in W_0^{1,p}(\Omega,\omega)$ the operator $u \mapsto A(u,v)$ is hemicontinuous and bounded from $W_0^{1,p}(\Omega,\omega)$ to $[W_0^{1,p}(\Omega,\omega)]^*$;
(iii) If $u_n \rightharpoonup u$ in $W_0^{1,p}(\Omega,\omega)$ and $(A(u_n,u_n) - A(u_n,u)|u_n - u) \to 0$, then
$A(u_n,u) \rightharpoonup A(u,v)$ in $[W_0^{1,p}(\Omega,\omega)]^*$ as $n \to \infty$ for all $v \in W_0^{1,p}(\Omega,\omega)$;
(iv) If $v \in W_0^{1,p}(\Omega,\omega)$, $u_n \rightharpoonup u$ in $W_0^{1,p}(\Omega,\omega)$ and

$$A(u_n,v) \rightharpoonup \tilde{v} \text{ in } [W_0^{1,p}(\Omega,\omega)]^*$$

then $(A(u_n,v)|u_n) \to (\tilde{v}|u)$ as $n \to \infty$;
(v) The operator $\tilde{A} : W_0^{1,p}(\Omega,\omega) \to [W_0^{1,p}(\Omega,\omega)]^*$ is bounded. Hence the operator $\tilde{A} : W_0^{1,p}(\Omega,\omega) \to [W_0^{1,p}(\Omega,\omega)]^*$ is pseudomonotone (see [61]).
(vi) By (H3), (H7) and (H12) we have

$$(\tilde{A}(u)|u) \geq \alpha_1 \int_\Omega |\nabla u|^p \omega \, dx + \alpha_0 \int_\Omega |u|^p \omega \, dx \geq \tilde{\alpha} \|u\|^p_{W_0^{1,p}(\Omega,\omega)}.$$

where $\tilde{\alpha} = \min\{\alpha_0, \alpha_1\}$. Since $p > 1$, we have

$$\frac{(\tilde{A}(u)|u)}{\|u\|_{W_0^{1,p}(\Omega,\omega)}} \to \infty \text{ as } \|u\|_{W_0^{1,p}(\Omega,\omega)} \to \infty,$$

that is, the operator $\tilde{A}$ is coercive.
Then, by Theorem 6.1.1, for each $T \in [W_0^{1,p}(\Omega,\omega)]^*$, the equation $\tilde{A}u = T$, $u \in W_0^{1,p}(\Omega,\omega)$ has a solution. Therefore, the problem (P_m) has a solution $u_m \in W_0^{1,p}(\Omega,\omega)$.
Step 2. We will show that $u_m \in L^\infty(\Omega)$ and $\|u_m\|_{L^\infty(\Omega)} \leq C$, where C is independent of m. If $u \in W_0^{1,p}(\Omega,\omega)$ is a solution of

problem (P_m) we define

$$u_n(x) = \begin{cases} u(x), & \text{if } |u(x)| \le n, \\ n, & \text{if } u(x) > n, \\ -n, & \text{if } u(x) < -n. \end{cases}$$

We have $D_i u_n = D_i u$ if $|u(x)| \le n$. For $k > 0$, let us define the function $\psi_n(x) = \text{sign}(u_n(x)) \max\{|u_n(x)| - k, 0\}$. We have

$$\psi_n \in W_0^{1,p}(\Omega, \omega) \cap L^\infty(\Omega).$$

Now consider the function

$$\Phi(t) = \begin{cases} t + k, & \text{if } t \le -k, \\ 0, & \text{if } |t| \le k, \\ t - k, & \text{if } t \ge k. \end{cases}$$

We have that Φ is a Lipschitz function and $\Phi(0) = 0$. Then $\Phi(\psi_n) \in W_0^{1,p}(\Omega, \omega)$. Moreover $D_i \Phi(\psi_n) = \Phi'(\psi_n) D_i \psi_n$ and

$$\Phi'(u_n) \nabla u_n \to \Phi'(u) \nabla u, \quad \mu - a.e. \text{ in } \Omega.$$

We also have, for all measurable subset $E \subset \Omega$

$$\int_E |\Phi'(u_n) \nabla u_n|^p \, \omega \, dx \le \int_E |\nabla u_n|^p \, \omega \, dx.$$

By applying the Vitali's Convergence Theorem, with $\psi = \Phi(u)$, we obtain

$$\nabla \psi_n \to \nabla \psi \quad \text{in} \quad L^p(\Omega, \omega). \tag{6.53}$$

Since $u \in W_0^{1,p}(\Omega, \omega)$ is a solution of problem (P_m) and the functions $\psi_n \in W_0^{1,p}(\Omega, \omega) \cap L^\infty(\Omega)$, we have

$$\int_\Omega \langle \mathcal{A}(x, u, \nabla u), \nabla \psi_n \rangle \, \omega \, dx + \int_\Omega g(x, u, \nabla u) \, \psi_n \, \omega \, dx$$
$$+ \int_\Omega H_m(x, u, \nabla u) \, \psi_n \, \omega \, dx = \int_\Omega f \, \psi_n \, dx. \tag{6.54}$$

Using (H4), (H6), $|H_m(x,\eta,\xi)|\le m$ and (6.53), we obtain in (6.54) as $n\to\infty$

$$\int_\Omega \langle \mathcal{A}(x,u,\nabla u),\nabla\psi\rangle\,\omega\,dx+\int_\Omega g(x,u,\nabla u)\,\psi\,\omega\,dx$$
$$+\int_\Omega H_m(x,u,\nabla u)\,\psi\,\omega\,dx=\int_\Omega f\,\psi\,dx.$$

Using $\varphi=\psi\,\chi_{A(k)}$ in (6.52) (where $A(k)=\{x\in\Omega\,:\,|u(x)|>k\}$) we obtain

$$\int_{A(k)} \langle \mathcal{A}(x,u,\nabla u),\nabla\psi\rangle\,\omega\,dx+\int_{A(k)} g(x,u,\nabla u)\,\psi\,\omega\,dx$$
$$+\int_{A(k)} H_m(x,u,\nabla u)\,\psi\,\omega\,dx=\int_{A(k)} f\,\psi\,dx. \tag{6.55}$$

Since

$$\psi=\Phi(u)=\begin{cases} u+k, & \text{if } u\le -k,\\ 0, & \text{if } |u|\le k,\\ u-k, & \text{if } u\ge k,\end{cases}$$

we obtain:
(i) By (H7) we have $g(x,\eta,\xi)\,\eta\ge 0$ for all $\eta\in\mathbb{R}$, and

$$\begin{aligned}\int_{A(k)} g(x,u,\nabla u)\,\psi\,\omega\,dx &= \int_{\{u\le -k\}} g(x,u,\nabla u)(u+k)\,\omega\,dx\\ &+ \int_{\{u\ge k\}} g(x,u,\nabla u)(u-k)\,\omega\,dx\ge 0;\end{aligned}$$

(ii) Using (H12) we have $H_m(x,\eta,\xi)\,\eta\ge 0$ for all $\eta\in\mathbb{R}$, and

$$\int_{A(k)} H_m(x,u,\nabla u)\,\psi\,\omega\,dx$$
$$=\int_{\{u\le -k\}} H_m(x,u,\nabla u)(u+k)\,\omega\,dx$$

$$+\int_{\{u\geq k\}} H_m(x,u,\nabla u)(u-k)\,\omega\,dx \geq 0.$$

We have $\nabla\psi = \nabla u$ in $A(k)$. Using (H3), (i) and (ii) we obtain in (6.55)

$$\alpha_1 \int_{A(k)} |\nabla u|^p\,\omega\,dx \leq \int_{A(k)} f\,\psi\,dx. \tag{6.56}$$

By Theorem 2.1.5 (with $\theta = 1$) there is a positive constant C_Ω such that

$$\int_\Omega |\psi|^p\omega\,dx \leq C_\Omega^p \int_\Omega |\nabla\psi|^p\omega\,dx.$$

Then we obtain

$$\begin{aligned}\int_{A(k)} f\,\psi\,dx &\leq \left(\int_{A(k)} \left|\frac{f}{\omega}\right|^{p'}\omega\right)^{1/p'}\left(\int_{A(k)} |\,\psi|^p\omega\,dx\right)^{1/p}\\ &\leq C_\Omega\left(\int_{A(k)} \left|\frac{f}{\omega}\right|^{p'}\omega\right)^{1/p'}\left(\int_{A(k)} |\,\nabla\psi|^p\omega\,dx\right)^{1/p}\\ &= C_\Omega\left(\int_{A(k)} \left|\frac{f}{\omega}\right|^{p'}\omega\right)^{1/p'}\left(\int_{A(k)} |\,\nabla u|^p\omega\,dx\right)^{1/p}.\end{aligned} \tag{6.57}$$

Using (6.57) and Young's inequality we obtain in (6.56) (for all $\varepsilon > 0$)

$$\begin{aligned}&\alpha_1 \int_{A(k)} |\,\nabla u|^p\omega\,dx\\ &\leq C_\Omega\left(\int_{A(k)} \left|\frac{f}{\omega}\right|^{p'}\omega\right)^{1/p'}\left(\int_{A(k)} |\,\nabla u|^p\omega\,dx\right)^{1/p}\\ &\leq C_\Omega\left[\varepsilon\int_{A(k)} |\,\nabla u|^p\,\omega\,dx + C(\varepsilon)\int_{A(k)} \left|\frac{f}{\omega}\right|^{p'}\omega\,dx\right],\end{aligned}$$

where $C(\varepsilon) = (\varepsilon p)^{-p'/p}/p'$. We can choose $\varepsilon > 0$ so that $C_\Omega\,\varepsilon = \alpha_1/2$, and there exists a constant C_8 such that

$$\int_{A(k)} |\nabla u|^p \omega\, dx \le C_8 \int_{A(k)} \left|\frac{f}{\omega}\right|^{p'} \omega\, dx. \tag{6.58}$$

Using the Sobolev's inequality (Theorem 2.1.5) and Hölder's inequality with exponents q and q' we obtain (since $q > p'\theta/(\theta - 1) > p'$ and $p^* = p\,\theta$)

$$\begin{aligned}
\left(\int_{A(k)} (|\,u| - k)^{p^*} \omega\, dx\right)^{p/p^*} &= \left(\int_{A(k)} |\psi|^{p^*}\, \omega\, dx\right)^{p/p^*}\\
\le & C_\Omega \int_{A(k)} |\nabla\psi|^p \omega\, dx = C_\Omega \int_{A(k)} |\nabla u|^p \omega\, dx\\
\le & C_\Omega\, C_8 \int_{A(k)} \left|\frac{f}{\omega}\right|^{p'} \omega\, dx \le C_9 \left(\int_\Omega \left|\frac{f}{\omega}\right|^q \omega\, dx\right)^{p'/q} [\mu(\Omega)]^{1 - \dfrac{p'}{q}}.
\end{aligned}$$

Let us now take $l > k > 0$, and observe that $A(l) \subset A(k)$. Then, from the previous inequality, it follows that

$$\begin{aligned}
&\mu(A(l))\,(l - k)^{p^*}\\
&= \int_{A(l)} (l - k)^{p^*} \omega\, dx\\
&\le \int_{A(l)} (|\,u| - k)^{p^*} \omega\, dx\\
&\le \int_{A(k)} (|\,u| - k)^{p^*} \omega\, dx\\
&\le C_9 \left(\int_\Omega \left|\frac{f}{\omega}\right|^q \omega\, dx\right)^{\dfrac{p'\,p^*}{q\,p}} [\mu(A(k))]^{(1 - \dfrac{p'}{q})\dfrac{p^*}{p}}.
\end{aligned}$$

Hence we obtain

$$\mu(A(l)) \leq \frac{C_9\left(\int_\Omega \left|\frac{f}{\omega}\right|^q \omega\, dx\right)^{\dfrac{p'\,p^*}{q\,p}}}{(l-k)^{p^*}}\,[\mu(A(k))]^{\left(1-\dfrac{p'}{q}\right)\dfrac{p^*}{p}}.$$

Since $\left(1-\dfrac{p'}{q}\right)\dfrac{p^*}{p} > 1$, by Lemma 4.1.7 there exists a constant $C_{10} > 0$ such that $\mu(A(k)) = 0$ for all $k \geq C_{10}$, and using Lemma 2.1.2 we obtain $|A(k)| = 0$. Therefore if u_m is a solution of problem (P_m) we have $\|u_m\|_{L^\infty(\Omega)} \leq C_{10}$ and C_{10} is independent of m.

Step 3. Since $u_m \in W_0^{1,p}(\Omega,\omega) \cap L^\infty(\Omega)$ and $\|u_m\|_{L^\infty(\Omega)} \leq C_{10}$, then the sequence $\{u_m\}$ is relative compact in the strong topology of $W_0^{1,p}(\Omega,\omega)$ (by apply the analogous results of [6] and Lemma 5.2.1). Then, by extracting a subsequence $\{u_{m_k}\}$ which strongly converges in $W_0^{1,p}(\Omega,\omega)$ (there exists $u \in W_0^{1,p}(\Omega,\omega)$ such that $u_{m_k} \to u$ in $W_0^{1,p}(\Omega,\omega)$), we have for all $\varphi \in W_0^{1,p}(\Omega,\omega) \cap L^\infty(\Omega)$

$$\begin{aligned}
&\int_\Omega \langle \mathcal{A}(x,u_{m_k},\nabla u_{m_k}),\nabla\varphi\rangle\,\omega\,dx + \int_\Omega g(x,u_{m_k},\nabla u_{m_k})\,\varphi\,\omega\,dx\\
&+ \int_\Omega H_{m_k}(x,u_{m_k},\nabla u_{m_k})\,\varphi\,\omega\,dx\\
&\to \int_\Omega \langle \mathcal{A}(x,u,\nabla u),\nabla\varphi\rangle\,\omega\,dx + \int_\Omega g(x,u,\nabla u)\,\varphi\,\omega\,dx\\
&+ \int_\Omega H(x,u,\nabla u)\,\varphi\,\omega\,dx.
\end{aligned}$$

Therefore $u \in W_0^{1,p}(\Omega,\omega) \cap L^\infty(\Omega)$ is the solution of problem $(P17)$.

□

Example 6.4.1. *Let* $\Omega = \{(x,y) \in \mathbb{R}^2 \;:\; x^2+y^2 < 1\}$, *and consider the weight function* $\omega(x,y) = (x^2+y^2)^{-1/2}$ $(\omega \in A_2)$, *the functions* $\mathcal{A} : \Omega\times\mathbb{R}\times\mathbb{R}^2 \to \mathbb{R}^2$, $g : \Omega\times\mathbb{R}\times\mathbb{R}^2 \to \mathbb{R}$ *and*

$H:\Omega\times\mathbb{R}\times\mathbb{R}^2\to\mathbb{R}$ *defined by*

$$\mathcal{A}((x,y),\eta,\xi)=h_2(x,y)\,\xi,$$
$$g((x,y),\eta,\xi)=\eta\,(\cos^2(xy)+1),$$
$$H((x,y),\eta,\xi)=|\xi|^2\sin^2(xy).\arctan(\eta),$$

where $h_2(x,y)=2e^{x^2+y^2}$. *Let us consider the partial differential operator*

$$\begin{aligned} Lu(x,y) &= -\operatorname{div}\Big[\omega(x,y)\mathcal{A}((x,y),u,\nabla u)\Big] \\ &\quad +\omega(x,y)g((x,y),u,\nabla u) \\ &+ \ \omega(x,y)\,H((x,y),\eta,\xi), \end{aligned}$$

and $f(x,y)=(x^2+y^2)^{-1/3q}\cos(1/(x^2+y^2))$, *with* $q>2r/(r-1)>2$. *Therefore, by Theorem 6.4.2, the problem*

$$\begin{cases} Lu(x,y) & = f(x,y) \text{ on } \Omega \\ u(x,y) & = 0, \text{ on } \partial\Omega \end{cases}$$

has a solution $u\in W_0^{1,2}(\Omega,\omega)\cap L^\infty(\Omega)$.

Now we will prove the existence of a solution for a nonlinear degenerate elliptic equations having natural growth terms, that is, we prove the existence of (weak) solutions in the weighted Sobolev space $W_0^{1,p}(\Omega,\omega)$ for the nonlinear degenerate elliptic problem with Dirichlet boundary conditions

$$(P18)\begin{cases} Lu(x)+H(x,u,\nabla u)\,\omega=T, \\ H(x,u,\nabla u)\in L^1(\Omega,\omega), \\ H(x,u,\nabla u)\,u\in L^1(\Omega,\omega), \\ u\in W_0^{1,p}(\Omega,\omega), \end{cases}$$

where L is the partial differential operator

$$Lu(x)=-\operatorname{div}(\omega\,\mathcal{A}(x,u,\nabla u))$$

and $H(x,u,\nabla u)$ is a non linear lower order term having natural growth (of order p) with respect to $|\nabla u|$ (with respect to $|u|$ we do not assume any growth restrictions, but we assume the sign-condition $H(x,\eta,\xi)\eta\geq 0$), Ω is a bounded open set in $\mathbb{R}^N$, ω is a weight function, $1<p<\infty$, $T\in[W_0^{1,p}(\Omega,\omega)]^*$ and $\mathcal{A}:\Omega\times\mathbb{R}\times\mathbb{R}^N\to\mathbb{R}^N$ and $H:\Omega\times\mathbb{R}\times\mathbb{R}^N\to\mathbb{R}$ are given functions.

In Theorem 6.4.2 we have shown the existence of solutions for nonlinear degenerate elliptic equations analogous to (P18) but assuming that $T=f$ and $f/\omega\in L^q(\Omega,\omega)$ (with $p'\theta/(\theta-1)<q<\infty$, where $\theta>1$ as in Theorem 2.1.5) and $|H(x,\eta,\xi)|\leq C_0+C_1|\xi|^p$. We mention that L^∞ bounds for similar problems have been recently proved in [40] by different methods.

Note that, in the proof of our main results, many ideas have been adapted from [3], [6] and Theorem 6.4.2.

Definition 6.4.2. *We say that* $u\in W_0^{1,p}(\Omega,\omega)$ *is a (weak) solution of problem (P18) if for any* $\varphi\in W_0^{1,p}(\Omega,\omega)\cap L^\infty(\Omega)$ *we have*

$$\int_\Omega \langle\mathcal{A}(x,u,\nabla u),\nabla\varphi\rangle\,\omega\,dx+\int_\Omega H(x,u,\nabla u)\,\varphi\,\omega\,dx$$

$$=(T|\varphi), \tag{6.59}$$

$$H(x,u,\nabla u)\in L^1(\Omega,\omega), \tag{6.60}$$

$$H(x,u,\nabla u)\,u\in L^1(\Omega,\omega). \tag{6.61}$$

Theorem 6.4.3. *Let* $1<p<\infty$ *and* $\omega\in A_p$. *Assume (H1)-(H7).*

(H1) $x\mapsto\mathcal{A}(x,\eta,\xi)$ *is measurable on* Ω *for all* $(\eta,\xi)\in\mathbb{R}\times\mathbb{R}^N$, $(\eta,\xi)\mapsto\mathcal{A}(x,\eta,\xi)$ *is continuous on* $\mathbb{R}\times\mathbb{R}^N$ *for almost all* $x\in\Omega$.

(H2) $|\mathcal{A}(x,\eta,\xi)|\leq K(x)+h_1(x)\,|\eta|^{p-1}+h_2(x)\,|\xi|^{p-1}$, $K\in L^{p'}(\Omega,\omega)$, *with* $K\geq 0$ *(with* $1/p+1/p'=1$*),* h_1, $h_2\in L^\infty(\Omega)$, $h_1,h_2\geq 0$.

(H3) *$\langle \mathcal{A}(x,\eta,\xi) - \mathcal{A}(x,\eta',\xi'), \xi - \xi' \rangle \geq 0$, whenever $\xi,\xi' \in \mathbb{R}^N$, $\xi \neq \xi'$, where $\mathcal{A}(x,\eta,\xi) = (\mathcal{A}_1(x,\eta,\xi), ..., \mathcal{A}_N(x,\eta,\xi))$ (where $\langle .,. \rangle$ denotes here the Euclidean scalar product in $\mathbb{R}^N$).*
(H4) *$\langle \mathcal{A}(x,\eta,\xi), \xi \rangle \geq \alpha\,|\xi|^p$, where α is a positive constant.*
(H5) *$x \mapsto H(x,\eta,\xi)$ is measurable on Ω for all $(\eta,\xi) \in \mathbb{R}\times\mathbb{R}^N$*
$(\eta,\xi) \mapsto H(x,\eta,\xi)$ is continuous on $\mathbb{R}\times\mathbb{R}^N$ for almost all $x \in \Omega$.
(H6) *$|H(x,\eta,\xi)| \leq b(\eta)\big(|\xi|^p + h(x)\big)$, where $h \in L^1(\Omega,\omega)$, $h \geq 0$ and $0 \leq b(\eta) \leq \beta$ for all $\eta \in \mathbb{R}$.*
(H7) *$H(x,\eta,\xi)\,\eta \geq 0$.*

Then there exist a (weak) solution $u \in W_0^{1,p}(\Omega,\omega)$ of problem (P18).

Proof. **Step 1.** Let us define for $\varepsilon > 0$ the approximation

$$H_\varepsilon(x,\eta,\xi) = \frac{H(x,\eta,\xi)}{1 + \varepsilon\,|H(x,\eta,\xi)|}.$$

We consider the approximate problem

$$(P_\varepsilon) \begin{cases} Lu_\varepsilon(x) + H_\varepsilon(x,u_\varepsilon,\nabla u_\varepsilon)\,\omega = T, \\ u_\varepsilon \in W_0^{1,p}(\Omega,\omega) \end{cases}$$

which has a solution $u_\varepsilon \in W_0^{1,p}(\Omega,\omega)$ (see Theorem 6.4.2 Step 1), that is,

$$\int_\Omega \langle \mathcal{A}(x,u_\varepsilon,\nabla u_\varepsilon), \nabla\varphi \rangle\,\omega\,dx + \int_\Omega H_\varepsilon(x,u_\varepsilon\,\nabla u_\varepsilon)\,\varphi\,\omega\,dx = (T|\varphi) \tag{6.62}$$

for all $\varphi \in W_0^{1,p}(\Omega,\omega) \cap L^\infty(\Omega)$, and we also have $\|u_\varepsilon\|_{L^\infty(\Omega)} \leq C_1$ (where C_1 is independent of ε, see Theorem 6.4.2 Step 2). Using the function $\varphi = u_\varepsilon$ in (6.62), and by (H4) and (H7), we obtain

$$\begin{aligned} \alpha \int_\Omega |\nabla u_\varepsilon|^p\,\omega\,dx &\leq \int_\Omega \langle \mathcal{A}(x,u_\varepsilon\,\nabla u_\varepsilon), \nabla u_\varepsilon \rangle\,\omega\,dx \\ &\leq \int_\Omega \langle \mathcal{A}(x,u_\varepsilon\,\nabla u_\varepsilon), \nabla u_\varepsilon \rangle\,\omega\,dx + \int_\Omega H_\varepsilon(x,u_\varepsilon,\nabla u_\varepsilon)\,u_\varepsilon\,\omega\,dx \end{aligned}$$

$$= (T|u_\varepsilon) \leq \|T\|_* \|u_\varepsilon\|_{W_0^{1,p}(\Omega,\omega)}.$$

Hence, by Theorem 2.1.5 (with $\theta = 1$) we have

$$\begin{aligned} \|u_\varepsilon\|^p_{W_0^{1,p}(\Omega,\omega)} &= \int_\Omega |u_\varepsilon|^p \,\omega\, dx + \int_\Omega |\nabla u_\varepsilon|^p \,\omega\, dx \\ &\leq (C_\Omega^p + 1) \int_\Omega |\nabla u_\varepsilon|^p \,\omega\, dx \\ &\leq \frac{(C_\Omega^p + 1)}{\alpha} \|T\|_* \|u_\varepsilon\|_{W_0^{1,p}(\Omega,\omega)}, \end{aligned}$$

that is,

$$\|u_\varepsilon\|_{W_0^{1,p}(\Omega,\omega)} \leq C_2 \tag{6.63}$$

where C_2 is independent of ε. Therefore, there exists a subsequence (still denoted by $\{u_\varepsilon\}$) and $u \in W_0^{1,p}(\Omega, \omega)$ such that

$$u_\varepsilon \rightharpoonup u \text{ in } W_0^{1,p}(\Omega, \omega), \tag{6.64}$$

$$u_\varepsilon \to u \quad \mu - a.e., \tag{6.65}$$

and also $u_\varepsilon \to u$ a.e. by Lemma 2.1.2(a).

Step 2. We will prove that $u_\varepsilon^+ \to u^+$ in $W_0^{1,p}(\Omega, \omega)$.
For $k \in \mathbb{R}$, $k > 0$, we define $u_k^+ = \min\{u^+, k\}$. For k fixed, we define $g_\varepsilon = u_\varepsilon^+ - u_k^+$. We will study the behavior of g_ε^+ and of g_ε^-.
(I) Behavior of $g_\varepsilon^+ = (u_\varepsilon^+ - u_k^+)^+$.
Note that $g_\varepsilon \in W_0^{1,p}(\Omega, \omega)$, then $g_\varepsilon^+ \in W_0^{1,p}(\Omega, \omega)$ and

$$\|g_\varepsilon^+\|_{L^\infty(\Omega)} \leq \|u_\varepsilon\|_{L^\infty(\Omega)} \leq C_1$$

(C_1 independent of ε). Hence, using g_ε^+ as test function in (6.62), we obtain

$$\int_\Omega \langle \mathcal{A}(x, u_\varepsilon, \nabla u_\varepsilon), \nabla g_\varepsilon^+ \rangle \,\omega\, dx + \int_\Omega H_\varepsilon(x, u_\varepsilon \nabla u_\varepsilon)\, g_\varepsilon^+ \,\omega\, dx = (T|g_\varepsilon^+). \tag{6.66}$$

Note that where $g_\varepsilon^+ = (u_\varepsilon^+ - u_k^+)^+ > 0$, then $u_\varepsilon^+ > 0$, and we have $u_\varepsilon > 0$ and from (H7) $H_\varepsilon(x, u_\varepsilon, \nabla u_\varepsilon) \geq 0$. Therefore, in (6.66) we get

$$\int_\Omega \langle \mathcal{A}(x, u_\varepsilon \nabla u_\varepsilon) \nabla g_\varepsilon^+ \rangle \, \omega \, dx \leq (T | g_\varepsilon^+).$$

Since $u_\varepsilon = u_\varepsilon^+$ on the set $\{x \in \Omega : g_\varepsilon^+(x) > 0\}$, we can also write

$$\int_\Omega \langle \mathcal{A}(x, u_\varepsilon, \nabla u_\varepsilon^+), \nabla g_\varepsilon^+ \rangle \, \omega \, dx \leq (T | g_\varepsilon^+). \tag{6.67}$$

Hence, by (6.67), we obtain

$$\begin{aligned} &\int_\Omega \langle \mathcal{A}(x, u_\varepsilon, \nabla u_\varepsilon^+) - \mathcal{A}(x, u_\varepsilon, \nabla u_k^+), \nabla (u_\varepsilon^+ - u_k^+)^+ \rangle \, \omega \, dx \\ &= \int_\Omega \langle \mathcal{A}(x, u_\varepsilon, \nabla u_\varepsilon^+), \nabla (u_\varepsilon^+ - u_k^+)^+ \rangle \, \omega \, dx \\ &- \int_\Omega \langle \mathcal{A}(x, u_\varepsilon, \nabla u_k^+), \nabla (u_\varepsilon^+ - u_k^+)^+ \rangle \, \omega \, dx \\ &\leq - \int_\Omega \langle \mathcal{A}(x, u_\varepsilon \nabla u_k^+), \nabla (u_\varepsilon^+ - u_k^+)^+ \rangle \, \omega \, dx + (T | g_\varepsilon^+). \end{aligned} \tag{6.68}$$

As $\varepsilon \to 0$, we have $g_\varepsilon^+ = (u_\varepsilon^+ - u_k^+)^+ \to (u^+ - u_k^+)^+$ a.e. and μ-a.e. (by Lemma 2.1.2(a)). Furthermore, since g_ε^+ is bounded in $W_0^{1,p}(\Omega, \omega)$ (by (6.63), we have $g_\varepsilon^+ \rightharpoonup (u^+ - u_k^+)^+$ in $W_0^{1,p}(\Omega, \omega)$, as $\varepsilon \to 0$ and k fixed. We define

$$R_k = - \int_\Omega \langle \mathcal{A}(x, u, \nabla u_k^+), \nabla (u^+ - u_k^+)^+ \rangle \, \omega \, dx + (T | (u^+ - u_k^+)^+). \tag{6.69}$$

We have, by (H2),

$$\begin{aligned} |R_k| \leq &\int_\Omega |\mathcal{A}(x, u, \nabla u_k^+)| |\nabla (u^+ - u_k^+)^+| \, \omega \, dx \\ &+ |(T | (u^+ - u_k^+)^+)| \end{aligned}$$

$$\le \Big(\int_\Omega |\mathcal{A}(x,u,\nabla u_k^+)|^{p'}\omega\,dx\Big)^{1/p'}\Big(\int_\Omega |\nabla(u^+-u_k^+)^+|^p\omega\,dx\Big)^{1/p}$$
$$+\,\|T\|_*\|(u^+-u_k^+)\|_{W_0^{1,p}(\Omega,\omega)}$$
$$\le C\Big(\int_\Omega (K^{p'}+h_1^{p'}|u|^p+h_2^{p'}|\nabla u_k^+|^p)\omega\,dx\Big)\|(u^+-u_k^+)^+\|_{W_0^{1,p}(\Omega,\omega)}$$
$$+\,\|T\|_*\|(u^+-u_k^+)^+\|_{W_0^{1,p}(\Omega,\omega)}$$
$$\le C\Big(\|K\|^{p'}_{L^{p'}(\Omega,\omega)}+\|h_1\|^{p'}_{L^\infty(\Omega)}\|u\|^p_{L^p(\Omega,\omega)}$$
$$+\,\|h_2\|^{p'}_{L^\infty(\Omega)}\|\nabla u_k^+\|^p_{L^p(\Omega,\omega)}\Big)\|(u^+-u_k^+)^+\|_{W_0^{1,p}(\Omega,\omega)}$$
$$+\,\|T\|_*\|(u^+-u_k^+)^+\|_{W_0^{1,p}(\Omega,\omega)}.$$

Since $(u^+-u_k^+)^+\to 0$ in $W_0^{1,p}(\Omega,\omega)$ as $k\to\infty$, we obtain $R_k\to 0$ as $k\to\infty$. Now, passing to the limit in $\varepsilon\to 0$ (with k fixed) in (6.68) we obtain

$$\overline{\lim_{\varepsilon\to 0}}\int_\Omega \langle \mathcal{A}(x,u_\varepsilon,\nabla u_\varepsilon^+)-\mathcal{A}(x,u_\varepsilon,\nabla u_k^+),\nabla(u_\varepsilon^+-u_k^+)^+\rangle\,\omega\,dx$$
$$\le R_k. \tag{6.70}$$

(II) Behavior of $g_\varepsilon^-=(u_\varepsilon^+-u_k^+)^-$.
We shall use as a test function in (6.62) the function $v_\varepsilon=\varphi_\lambda(g_\varepsilon^-)$, where $\varphi_\lambda(s)=s\,e^{\lambda s^2}$, and $\lambda\in\mathbb{R}$ will be chosen later. We have $0\le g_\varepsilon^-\le k$. Hence, $g_\varepsilon^-\in L^\infty(\Omega)$, and since $g_\varepsilon\in W_0^{1,p}(\Omega,\omega)$, then we also have $v_\varepsilon\in W_0^{1,p}(\Omega,\omega)\cap L^\infty(\Omega)$. Therefore, using v_ε in (6.62) we get

$$\int_\Omega \langle \mathcal{A}(x,u_\varepsilon,\nabla u_\varepsilon),\nabla v_\varepsilon\rangle\,\omega\,dx+\int_\Omega H_\varepsilon(x,u_\varepsilon\,\nabla u_\varepsilon)\,v_\varepsilon\,\omega\,dx=(T|v_\varepsilon),$$

that is,

$$\int_\Omega \langle \mathcal{A}(x,u_\varepsilon,\nabla u_\varepsilon),\nabla g_\varepsilon^-\,\varphi'_\lambda(g_\varepsilon^-)\rangle\,\omega\,dx$$

$$+\int_\Omega H_\varepsilon(x,u_\varepsilon,\nabla u_\varepsilon)\,\varphi_\lambda(g_\varepsilon^-)\,\omega\,dx=(T|\varphi_\lambda(g_\varepsilon^-)). \quad (6.71)$$

Considering the sets

$$E_\varepsilon=\{x\in\Omega: u_\varepsilon^+(x)\le u_k^+(x)\}$$
$$F_\varepsilon=\{x\in\Omega: 0\le u_\varepsilon(x)\le u_k^+(x)\}.$$

If $x\notin E_\varepsilon$, then $g_\varepsilon^-=(u_\varepsilon^+-u_k^+)^-=0$ and $\varphi_\lambda(0)=0$. Then we have

$$\int_\Omega H_\varepsilon(x,u_\varepsilon,\nabla u_\varepsilon)\,\varphi_\lambda(g_\varepsilon^-)\,\omega\,dx$$
$$=\int_{E_\varepsilon} H_\varepsilon(x,u_\varepsilon,\nabla u_\varepsilon)\,\varphi_\lambda(g_\varepsilon^-)\,\omega\,dx. \quad (6.72)$$

If $u_\varepsilon\le 0$, then $g_\varepsilon=u_\varepsilon^+-u_k^+\le 0$, and $g_\varepsilon^-=(u_\varepsilon^+-u_k^+)^-\ge 0$. Hence, $\varphi_\lambda(g_\varepsilon^-)\ge 0$, and since $H_\varepsilon(x,u_\varepsilon,\nabla u_\varepsilon)\,u_\varepsilon\ge 0$, then $H_\varepsilon(x,u_\varepsilon\,\nabla u_\varepsilon)\le 0$. Using (H3) and (H6), we obtain

$$\int_{E_\varepsilon} H_\varepsilon(x,u_\varepsilon,\nabla u_\varepsilon)\,\varphi_\lambda(g_\varepsilon^-)\,\omega\,dx$$
$$\le\int_{F_\varepsilon} H_\varepsilon(x,u_\varepsilon,\nabla u_\varepsilon)\,\varphi_\lambda(g_\varepsilon^-)\,\omega\,dx$$
$$\le\int_{F_\varepsilon} |H_\varepsilon(x,u_\varepsilon,\nabla u_\varepsilon)|\,\varphi_\lambda(g_\varepsilon^-)\,\omega\,dx$$
$$\le\int_{F_\varepsilon} b(u_\varepsilon)\Big(|\nabla u_\varepsilon|^p+h(x)\Big)\varphi_\lambda(g_\varepsilon^-)\,\omega\,dx$$
$$\le\beta\int_{F_\varepsilon}|\nabla u_\varepsilon|^p\varphi_\lambda(g_\varepsilon^-)\,\omega\,dx+\beta\int_\Omega h(x)\varphi_\lambda(g_\varepsilon^-)\,\omega\,dx$$
$$\le\frac{\beta}{\alpha}\int_{F_\varepsilon}\big(\mathcal{A}(x,u_\varepsilon,\nabla u_\varepsilon).\nabla u_\varepsilon\big)\,\varphi_\lambda(g_\varepsilon^-)\,\omega\,dx$$
$$+\beta\int_\Omega h(x)\,\varphi_\lambda(g_\varepsilon^-)\,\omega\,dx. \quad (6.73)$$

Since $u_\varepsilon \le 0$ implies $g_\varepsilon^- = (u_\varepsilon^+ - u_k^+)^- = u_k^+$, we have

$$\begin{aligned}
&\int_\Omega - \langle \mathcal{A}(x, u_\varepsilon, \nabla u_\varepsilon^+), \nabla g_\varepsilon^- \, \varphi_\lambda'(g_\varepsilon^-)\rangle \, \omega \, dx \\
&= \int_\Omega - \langle \mathcal{A}(x, u_\varepsilon, \nabla u_\varepsilon), \nabla g_\varepsilon^- \, \varphi_\lambda'(g_\varepsilon^-)\rangle \, \omega \, dx \\
&+ \int_\Omega \langle \mathcal{A}(x, u_\varepsilon, \nabla u_\varepsilon), \nabla g_\varepsilon^- \, \varphi_\lambda'(g_\varepsilon^-)\rangle \, \omega \, dx \\
&- \int_\Omega \langle \mathcal{A}(x, u_\varepsilon, \nabla u_\varepsilon^+), \nabla g_\varepsilon^- \, \varphi_\lambda'(g_\varepsilon^-)\rangle \, \omega \, dx \\
&= \int_\Omega - \langle \mathcal{A}(x, u_\varepsilon, \nabla u_\varepsilon), \nabla g_\varepsilon^- \, \varphi_\lambda'(g_\varepsilon^-)\rangle \, \omega \, dx \\
&+ \int_\Omega \langle \mathcal{A}(x, u_\varepsilon, \nabla u_\varepsilon) - \mathcal{A}(x, u_\varepsilon, \nabla u_\varepsilon^+), \nabla g_\varepsilon^- \, \varphi_\lambda'(g_\varepsilon^-)\rangle \, \omega \, dx \\
&= - \int_\Omega \langle \mathcal{A}(x, u_\varepsilon, \nabla u_\varepsilon), \nabla g_\varepsilon^- \, \varphi_\lambda(g_\varepsilon^-) \, \rangle \, \omega dx \\
&+ \int_\Omega \langle \mathcal{A}(x, u_\varepsilon, \nabla u_\varepsilon) - \mathcal{A}(x, u_\varepsilon, \nabla u_\varepsilon^+), \nabla u_k^+ \varphi_\lambda'(u_k^+)\rangle \omega \, dx.
\end{aligned} \tag{6.74}$$

Using (6.71) and (6.74) we obtain

$$\begin{aligned}
&\int_\Omega - \langle \mathcal{A}(x, u_\varepsilon, \nabla u_\varepsilon^+) - \mathcal{A}(x, u_\varepsilon, \nabla u_k^+), \nabla (u_\varepsilon^+ - u_k^+)^- \, \varphi_\lambda'(g_\varepsilon^-)\rangle \, \omega \, dx \\
&= \int_\Omega - \langle \mathcal{A}(x, u_\varepsilon, \nabla u_\varepsilon^+), \nabla (u_\varepsilon^+ - u_k^+)^- \, \varphi_\lambda'(g_\varepsilon^-)\rangle \, \omega \, dx \\
&+ \int_\Omega \langle \mathcal{A}(x, u_\varepsilon, \nabla u_k^+), \nabla (u_\varepsilon^+ - u_k^+)^- \, \varphi_\lambda'(g_\varepsilon^-)\rangle \, \omega \, dx \\
&= \int_\Omega - \langle \mathcal{A}(x, u_\varepsilon, \nabla u_\varepsilon), \nabla (u_\varepsilon^+ - u^+)^- \varphi_\lambda'(g_\varepsilon^-)\rangle \, \omega \, dx \\
&+ \int_\Omega \langle \mathcal{A}(x, u_\varepsilon, \nabla u_\varepsilon) - \mathcal{A}(x, u_\varepsilon, \nabla u_\varepsilon^+), \nabla u_k^+ \, \varphi_\lambda'(u_k^+)\rangle \, \omega \, dx \\
&+ \int_\Omega \langle \mathcal{A}(x, u_\varepsilon, \nabla u_k^+), \nabla (u_\varepsilon^+ - u_k^+)^- \, \varphi_\lambda'(g_\varepsilon^-)\rangle \, \omega \, dx
\end{aligned}$$

$$= \int_\Omega H_\varepsilon(x, u_\varepsilon, \nabla u_\varepsilon)\, \varphi_\lambda(g_\varepsilon^-)\, \omega\, dx + (-T|\varphi_\lambda(g_\varepsilon^-))$$
$$+ \int_\Omega \langle \mathcal{A}(x, u_\varepsilon, \nabla u_\varepsilon) - \mathcal{A}(x, u_\varepsilon, \nabla u_\varepsilon^+), \nabla u_k^+\, \varphi_\lambda'(u_k^+)\rangle\, \omega\, dx$$
$$+ \int_\Omega \langle \mathcal{A}(x, u_\varepsilon, \nabla u_k^+), \nabla (u_\varepsilon^+ - u_k^+)^+\, \varphi_\lambda'(g_\varepsilon^-)\rangle\, \omega\, dx = I.$$

Now, by (6.72) and (6.73) we obtain

$$\begin{aligned}
I \le & \int_\Omega \langle \mathcal{A}(x, u_\varepsilon, \nabla u_\varepsilon) - \mathcal{A}(x, u_\varepsilon, \nabla u_\varepsilon^+), \nabla u_k^+\, \varphi_\lambda'(u_k^+)\rangle\, \omega\, dx \\
& + (-T|\varphi_\lambda(g_\varepsilon^-)) \\
& + \int_\Omega \langle \mathcal{A}(x, u_\varepsilon, \nabla u_k^+), \nabla (u_\varepsilon^+ - u_k^+)^-\varphi_\lambda(g_\varepsilon^-)\rangle\, \omega\, dx \\
& + \frac{\beta}{\alpha} \int_{F_\varepsilon} \langle \mathcal{A}(x, u_\varepsilon, \nabla u_\varepsilon), \nabla u_\varepsilon\, \varphi_\lambda(g_\varepsilon^-)\rangle\, \omega\, dx \\
& + \beta \int_\Omega h(x)\, \varphi_\lambda(g_\varepsilon^-)\, \omega\, dx \\
= & \int_\Omega \langle \mathcal{A}(x, u_\varepsilon, \nabla u_\varepsilon) - \mathcal{A}(x, u_\varepsilon, \nabla u_\varepsilon^+), \nabla u_k^+\, \varphi_\lambda'(u_k^+)\rangle\, \omega\, dx \\
& + (-T|\varphi_\lambda(g_\varepsilon^-)) \\
& + \int_\Omega \langle \mathcal{A}(x, u_\varepsilon, \nabla u_k^+), \nabla (u_\varepsilon^+ - u_k^+)^-\varphi_\lambda(g_\varepsilon^-)\rangle\, \omega\, dx \\
& + \frac{\beta}{\alpha} \int_\Omega \langle \mathcal{A}(x, u_\varepsilon, \nabla u_\varepsilon^+) - \mathcal{A}(x, u_\varepsilon, \nabla u_k^+), \nabla (u_\varepsilon^+ - u_k^+)\, \varphi_\lambda(g_\varepsilon^-)\rangle\, \omega\, dx \\
& + \frac{\beta}{\alpha} \int_\Omega \langle \mathcal{A}(x, u_\varepsilon, \nabla u_\varepsilon^+), \nabla u_k^+\, \varphi_\lambda(g_\varepsilon^-)\rangle\, \omega\, dx + \beta \int_\Omega h(x) \varphi_\lambda(g_\varepsilon^-)\, \omega\, dx \\
& + \frac{\beta}{\alpha} \int_\Omega \langle \mathcal{A}(x, u_\varepsilon, \nabla u_k^+), \nabla (u_\varepsilon^+ - u_k^+)\, \varphi_\lambda(g_\varepsilon^-)\rangle\, \omega\, dx.
\end{aligned} \tag{6.75}$$

Hence, by (6.75) we have

$$\int_\Omega - \langle \mathcal{A}(x, u_\varepsilon, \nabla u_\varepsilon^+) - \mathcal{A}(x, u_\varepsilon, \nabla u_k^+), \nabla (u_\varepsilon^+ - u_k^+)^-\, \varphi_\lambda'(g_\varepsilon^-)\rangle\, \omega\, dx$$

$$\leq \int_\Omega \langle \mathcal{A}(x, u_\varepsilon, \nabla u_\varepsilon) - \mathcal{A}(x, u_\varepsilon, \nabla u_\varepsilon^+), \nabla u_k^+ \, \varphi_\lambda'(u_k^+)\rangle \, \omega \, dx$$
$$+(-T|\varphi_\lambda(g_\varepsilon^-)) + \int_\Omega \langle \mathcal{A}(x, u_\varepsilon, \nabla u_k^+), \nabla(u_\varepsilon^+ - u_k^+)^- \varphi_\lambda(g_\varepsilon^-)\rangle \, \omega \, dx$$
$$+\frac{\beta}{\alpha}\int_\Omega \langle\big(\mathcal{A}(x, u_\varepsilon, \nabla u_\varepsilon^+) - \mathcal{A}(x, u_\varepsilon, \nabla u_k^+)\big), \nabla(u_\varepsilon^+ - u_k^+)\, \varphi_\lambda(g_\varepsilon^-)\rangle \, \omega \, dx$$
$$+ \frac{\beta}{\alpha}\int_\Omega \langle \mathcal{A}(x, u_\varepsilon, \nabla u_\varepsilon^+), \nabla u_k^+ \, \varphi_\lambda(g_\varepsilon^-)\rangle \, \omega \, dx$$
$$+ \frac{\beta}{\alpha}\int_\Omega \langle \mathcal{A}(x, u_\varepsilon, \nabla u_k^+), \nabla(u_\varepsilon^+ - u_k^+)\, \varphi_\lambda(g_\varepsilon^-)\rangle \, \omega \, dx$$
$$+ \beta \int_\Omega h(x)\varphi_\lambda(g_\varepsilon^-)\, \omega \, dx. \tag{6.76}$$

Now we choose $\lambda = \dfrac{\beta^2}{4\alpha^2}$, we have $\varphi_\lambda' - \dfrac{\beta}{\alpha}\varphi_\lambda \geq \dfrac{1}{2}$. Thus, in (6.76) we obtain

$$-\frac{1}{2}\int_\Omega \langle \mathcal{A}(x, u_\varepsilon, \nabla u_\varepsilon^+) - \mathcal{A}(x, u_\varepsilon, \nabla u_k^+), \nabla(u_\varepsilon^+ - u_k^+)^-\rangle \, \omega \, dx$$
$$\leq \int_\Omega -\langle \mathcal{A}(x, u_\varepsilon, \nabla u_\varepsilon^+) - \mathcal{A}(x, u_\varepsilon, \nabla u_k^+), \nabla(u_\varepsilon^+ - u_k^+)\, \varphi_\lambda'(g_\varepsilon^-)\rangle \, \omega \, dx$$
$$-\frac{\beta}{\alpha}\int_\Omega \langle \mathcal{A}(x, u_\varepsilon, \nabla u_\varepsilon^+) - \mathcal{A}(x, u_\varepsilon, \nabla u_k^+), \nabla(u_\varepsilon^+ - u_k^+)\, \varphi_\lambda(g_\varepsilon^-)\rangle \, \omega \, dx$$
$$= \int_\Omega \langle \mathcal{A}(x, u_\varepsilon, \nabla u_\varepsilon) - \mathcal{A}(x, u_\varepsilon, \nabla u_\varepsilon^+), \nabla u_k^+ \, \varphi_\lambda'(u_k^+)\rangle \, \omega \, dx$$
$$+ \quad (-T|\varphi_\lambda(g_\varepsilon^-))$$
$$+ \int_\Omega \langle \mathcal{A}(x, u_\varepsilon, \nabla u_k^+), \nabla(u_\varepsilon^+ - u_k^+)^- \, \varphi_\lambda'(g_\varepsilon^-)\rangle \, \omega \, dx$$
$$+ \frac{\beta}{\alpha}\int_\Omega \langle \mathcal{A}(x, u_\varepsilon, \nabla u_\varepsilon^+), \nabla u_k^+ \, \varphi_\lambda(g_\varepsilon^-)\rangle \, \omega \, dx$$
$$+ \frac{\beta}{\alpha}\int_\Omega \langle \mathcal{A}(x, u_\varepsilon, \nabla u_k^+), \nabla(u_\varepsilon^+ - u_k^+)\, \varphi_\lambda(g_\varepsilon^-)\rangle \, \omega \, dx$$

$$+\ \beta\int_\Omega h(x)\,\varphi_\lambda(g_\varepsilon^-)\,\omega\,dx. \tag{6.77}$$

By (H2) and (6.63) we have

$$\begin{aligned}
&\|\mathcal{A}(x,u_\varepsilon,\nabla u_\varepsilon)\|^{p'}_{L^{p'}(\Omega,\omega)} = \int_\Omega |\mathcal{A}(x,u_\varepsilon,\nabla u_\varepsilon)|^{p'}\,\omega\,dx\\
&\leq C_p\bigg[\int_\Omega |K|^{p'}\omega\,dx + \|h_1\|^{p'}_{L^\infty(\Omega)}\int_\Omega |u_\varepsilon|^p\,\omega\,dx\\
&+\|h_2\|^{p'}_{L^\infty(\Omega)}\int_\Omega |\nabla u_\varepsilon|^p\,\omega\,dx\bigg]\\
&\leq C_p\bigg[\|K\|^{p'}_{L^{p'}(\Omega,\omega)} + (\|h_1\|^{p'}_{L^\infty(\Omega)}+\|h_2\|^{p'}_{L^\infty(\Omega)})\|u_\varepsilon\|^p_{W^{1,p}_0(\Omega,\omega)}\bigg]\\
&\leq C_p\bigg[\|K\|^{p'}_{L^{p'}(\Omega,\omega)} + C_2^p\,(\,\|h_1\|^{p'}_{L^\infty(\Omega)}+\|h_2\|^{p'}_{L^\infty(\Omega)})\bigg],
\end{aligned}$$

that is, $\|\mathcal{A}(x,u_\varepsilon,\nabla u_\varepsilon)\|_{L^{p'}(\Omega,\omega)}\leq C_4$ (where C_4 independent of ε). And, analogously, we have $\|\mathcal{A}(x,u_\varepsilon,\nabla u_\varepsilon^+)\|_{L^{p'}(\Omega,\omega)}\leq C_4$. Then, by extracting a subsequence of $\{u_\varepsilon\}$ (which is still denoted by $\{u_\varepsilon\}$) such that

$$\begin{aligned}
\mathcal{A}(x,u_\varepsilon,\nabla u_\varepsilon) &\rightharpoonup \mathcal{F}\ \text{ in }\ (L^{p'}(\Omega,\omega))^N\\
\mathcal{A}(x,u_\varepsilon,\nabla u_\varepsilon^+) &\rightharpoonup \mathcal{G}\ \text{ in }\ (L^{p'}(\Omega,\omega))^N,
\end{aligned}$$

and using the Lebesgue Dominated Convergence Theorem, we can pass the limit $\varepsilon\to 0$ (with k fixed) in the right hand of (6.77), we obtain

$$\begin{aligned}
&\int_\Omega \langle \mathcal{F}(x)-\mathcal{G}(x),\nabla u_k^+\,\varphi'_\lambda(u_k^+)\rangle\,\omega\,dx + (-T|\varphi_\lambda((u^+-u_k^+)^-))\\
&+\int_\Omega \langle \mathcal{A}(x,u,\nabla u_k^+),\nabla(u^+-u_k^+)^-\varphi'_\lambda((u^+-u_k^+)^-)\rangle\,\omega\,dx\\
&+\frac{\beta}{\alpha}\int_\Omega \langle \mathcal{G}(x),\nabla u_k^+\,\varphi_\lambda((u^+-u_k^+)^-)\rangle\,\omega\,dx
\end{aligned}$$

$$+\frac{\beta}{\alpha}\int_\Omega \langle \mathcal{A}(x,u,\nabla u_k^+), \nabla(u^+ - u_k^+)\,\varphi_\lambda((u^+ - u_k^+)^-)\rangle\,\omega\,dx$$
$$+\beta\int_\Omega h(x)\,\varphi_\lambda((u^+ - u_k^+)^-)\,\omega\,dx$$
$$=\int_\Omega \langle \mathcal{F}(x) - \mathcal{G}(x)), \nabla u_k^+\,\varphi_\lambda'(u_k^+)\rangle\,\omega\,dx,$$

since $(u^+ - u_k^+)^- = 0$ and $\varphi_\lambda(0) = 0$. Moreover, since

$$\langle \mathcal{A}(x,u_\varepsilon,\nabla u_\varepsilon) - \mathcal{A}(x,u_\varepsilon,\nabla u_\varepsilon^+), \nabla(u_\varepsilon)_k^+\rangle = 0 \text{ a.e.}$$

where $(u_\varepsilon)_k^+ = \min\{u_\varepsilon, k\}$, then $\langle \mathcal{F}(x) - \mathcal{G}(x)), \nabla u_k^+\rangle = 0$ a.e and μ-a.e.. Hence,

$$\int_\Omega \langle \mathcal{F}(x) - \mathcal{G}(x), \nabla u_k^+\,\varphi_\lambda'(u_k^+)\rangle\,\omega\,dx = 0.$$

Thus, passing to the limit $\varepsilon \to 0$ (with k fixed) in (6.77) we obtain

$$\overline{\lim_{\varepsilon\to 0}} - \int_\Omega \langle \mathcal{A}(x,u_\varepsilon,\nabla u_\varepsilon^+) - \mathcal{A}(x,u_\varepsilon,\nabla u_k^+), \nabla(u_\varepsilon^+ - u_k^+)^-\rangle\,\omega\,dx$$
$$\leq 0. \tag{6.78}$$

Using (6.70) and (6.78) we have

$$\overline{\lim_{\varepsilon\to 0}}\int_\Omega \langle \mathcal{A}(x,u_\varepsilon,\nabla u_\varepsilon^+) - \mathcal{A}(x,u_\varepsilon,\nabla u^+), \nabla(u_\varepsilon^+ - u^+)\rangle\,\omega\,dx$$
$$\leq \overline{\lim_{\varepsilon\to 0}}\int_\Omega \langle \mathcal{A}(x,u_\varepsilon,\nabla u_\varepsilon^+) - \mathcal{A}(x,u_\varepsilon,\nabla u_k^+), \nabla(u_\varepsilon^+ - u_k^+)^+\rangle\,\omega\,dx$$
$$+\overline{\lim_{\varepsilon\to 0}} - \int_\Omega \langle \mathcal{A}(x,u_\varepsilon,\nabla u_\varepsilon^+) - \mathcal{A}(x,u_\varepsilon,\nabla u_k^+), \nabla(u_\varepsilon^+ - u^+)^-\rangle\,\omega\,dx$$
$$+\overline{\lim_{\varepsilon\to 0}}\int_\Omega \langle \mathcal{A}(x,u_\varepsilon,\nabla u_\varepsilon^+) - \mathcal{A}(x,u_\varepsilon,\nabla u_k^+), \nabla(u_k^+ - u^+)\rangle\,\omega\,dx$$
$$+\overline{\lim_{\varepsilon\to 0}}\int_\Omega \langle \mathcal{A}(x,u_\varepsilon,\nabla u_k^+) - \mathcal{A}(x,u_\varepsilon,\nabla u^+), \nabla(u_\varepsilon^+ - u^+)\rangle\,\omega\,dx$$

$$\leq R_k + \int_\Omega \langle \mathcal{G}(x) - \mathcal{A}(x, u, \nabla u^+), \nabla (u_k^+ - u^+) \rangle \, \omega \, dx. \tag{6.79}$$

Hence, using (H2) and passing to the limit $k \to \infty$ we obtain

$$R_k + \int_\Omega \langle \mathcal{G}(x) - \mathcal{A}(x, u, \nabla u^+), \nabla (u_k^+ - u^+) \rangle \, \omega \, dx \to 0.$$

Therefore,

$$\int_\Omega \langle \mathcal{A}(x, u_\varepsilon, \nabla u_\varepsilon^+) - \mathcal{A}(x, u_\varepsilon, \nabla u^+), \nabla (u_\varepsilon^+ - u^+) \rangle \, \omega \, dx \to 0.$$

Thus, by (6.64), (6.65) and Lemma 5.2.1, we have

$$u_\varepsilon^+ \to u^+ \;\text{ in }\; W_0^{1,p}(\Omega, \omega). \tag{6.80}$$

Step 3. We will prove that $u_\varepsilon^- \to u^-$ in $W_0^{1,p}(\Omega, \omega)$.
We define $u_k^- = \min\{u^-, k\}$ and $f_\varepsilon = u_\varepsilon^- - u_k^-$. We have

$$f_\varepsilon^+ \in W_0^{1,p}(\Omega, \omega) \cap L^\infty(\Omega).$$

Hence, using f_ε^+ as a test function in (6.62) we obtain

$$\int_\Omega \langle \mathcal{A}(x, u_\varepsilon, \nabla u_\varepsilon), \nabla f_\varepsilon^+ \rangle \, \omega \, dx + \int_\Omega H_\varepsilon(x, u_\varepsilon, \nabla u_\varepsilon) \, f_\varepsilon^+ \, \omega \, dx = (T | f_\varepsilon^+).$$

Note that $H_\varepsilon(x, u_\varepsilon, \nabla u_\varepsilon) \, f_\varepsilon^+ \leq 0$ a.e. and μ-a.e..Thus, analogously to (6.70) we obtain (with k fixed)

$$\overline{\lim_{\varepsilon \to 0}} \int_\Omega - \langle \mathcal{A}(x, u_\varepsilon, -\nabla u_\varepsilon^-) - \mathcal{A}(x, u_\varepsilon, -\nabla u_k^-), \nabla (u_\varepsilon^- - u_k^-)^+ \rangle \, \omega \, dx$$
$$\leq S_k, \tag{6.81}$$

where $\lim_{k \to \infty} S_k = 0$. Now, considering the test function $v_\varepsilon = \varphi_\lambda(f_\varepsilon^-)$, we obtain as in (6.78) that

$$\overline{\lim_{\varepsilon \to 0}} \int_\Omega \langle \mathcal{A}(x, u_\varepsilon, -\nabla u_\varepsilon^-) - \mathcal{A}(x, u_\varepsilon, -\nabla u_k^-), \nabla (u_\varepsilon^- - u_k^-)^- \rangle \, \omega \, dx \leq 0. \tag{6.82}$$

Therefore, using (6.81) and (6.82), we have as in (6.80) that

$$u_\varepsilon^- \to u^- \quad \text{em} \quad W_0^{1,p}(\Omega,\omega). \tag{6.83}$$

Step 4. Convergence.
Using (6.82) and (6.83) we obtain a subsequence of $\{u_\varepsilon\}$ satisfying

$$u_\varepsilon \to u \quad \text{in} \quad W_0^{1,p}(\Omega,\omega), \;\; \mu - a.e. \text{ and } a.e. \tag{6.84}$$

$$\nabla u_\varepsilon \to \nabla u \quad \mu - a.e. \text{ and } a.e. \tag{6.85}$$

Now, using (H5) we obtain

$$\begin{aligned} &H_\varepsilon(x,u_\varepsilon,\nabla u_\varepsilon) \to H(x,u,\nabla u) \;\; a.e. \text{ and } \mu - a.e., \\ &H_\varepsilon(x,u_\varepsilon,\nabla u_\varepsilon)\, u_\varepsilon \to H(x,u,\nabla u)\, u \;\; a.e. \text{ and } \;\mu - a.e., \end{aligned} \tag{6.86}$$

and using (H4), we have

$$\int_\Omega \langle \mathcal{A}(x,u_\varepsilon,\nabla u_\varepsilon), \nabla u_\varepsilon\rangle\, \omega\, dx \geq \alpha\, \|\nabla u_\varepsilon\|^p_{L^p(\Omega,\omega)} \geq 0.$$

We obtain, as in (6.63),

$$\begin{aligned} 0 &\leq \int_\Omega H_\varepsilon(x,u_\varepsilon,\nabla u_\varepsilon)\, u_\varepsilon\, \omega\, dx \\ &\leq \int_\Omega \langle \mathcal{A}(x,u_\varepsilon,\nabla u_\varepsilon), \nabla u_\varepsilon\rangle\, \omega\, dx + \int_\Omega H_\varepsilon(x,u_\varepsilon,\nabla u_\varepsilon)\, u_\varepsilon\, \omega\, dx \\ &= (T|u_\varepsilon) \\ &\leq \|T\|_* \|u_\varepsilon\|_{W_0^{1,p}(\Omega,\omega)} \\ &\leq \|T\|_* C_2 = C_3. \end{aligned} \tag{6.87}$$

For $m > 0$, we define

$$X_m^\varepsilon = \{x \in \Omega : |u_\varepsilon(x)| \leq m\} \text{ and } Y_m = \{x \in \Omega : |u_\varepsilon(x)| > m\}.$$

For any measurable subset $E \subset \Omega$, by (H6), we have

$$
\begin{aligned}
&\int_E |H_\varepsilon(x, u_\varepsilon, \nabla u_\varepsilon)| \, \omega \, dx \\
&= \int_{E \cap X_m^\varepsilon} |H_\varepsilon(x, u_\varepsilon \nabla u_\varepsilon)| \, \omega \, dx + \int_{E \cap Y_m^\varepsilon} |H_\varepsilon(x, u_\varepsilon \nabla u_\varepsilon)| \, \omega \, dx \\
&\le \int_{E \cap X_m^\varepsilon} |H_\varepsilon(x, u_\varepsilon \nabla u_\varepsilon)| \, \omega \, dx + \frac{1}{m} \int_{E \cap Y_m^\varepsilon} H_\varepsilon(x, u_\varepsilon \nabla u_\varepsilon) \, u_\varepsilon \, \omega \, dx \\
&\le \int_E |H_\varepsilon(x, u_\varepsilon, \nabla u_\varepsilon)| \, \omega \, dx + \frac{C_3}{m} \\
&\le \beta \int_\Omega \Big(|\nabla u_\varepsilon|^p + h \Big) \, \omega \, dx + \frac{C_3}{m}. \qquad (6.88)
\end{aligned}
$$

Since the sequence $\{\nabla u_\varepsilon\}$ strongly converges in $(L^p(\Omega, \omega))^N$, then (6.88) implies the equi-integrability of $H_\varepsilon(x, u_\varepsilon, \nabla u_\varepsilon)$. Using (6.86) and the Vitali's Theorem (see [61]), we obtain

$$
H_\varepsilon(x, u_\varepsilon, \nabla u_\varepsilon) \to H(x, u, \nabla u) \quad \text{in} \quad L^1(\Omega, \omega). \qquad (6.89)
$$

Now, using (6.84) and (6.86), and passing to the limit $\varepsilon \to 0$ in

$$
\int_\Omega \langle \mathcal{A}(x, u_\varepsilon, \nabla u_\varepsilon), \nabla \varphi \rangle \, \omega \, dx + \int_\Omega H_\varepsilon(x, u_\varepsilon, \nabla u_\varepsilon) \, \varphi \, \omega \, dx = (T|\varphi),
$$

for all $\varphi \in W_0^{1,p}(\Omega, \omega) \cap L^\infty(\Omega)$, we obtain

$$
\begin{aligned}
&\int_\Omega \langle \mathcal{A}(x, u, \nabla u), \nabla \varphi \rangle \, \omega \, dx + \int_\Omega H(x, u, \nabla u) \, \varphi \, \omega \, dx \\
&= (T|\varphi). \qquad (6.90)
\end{aligned}
$$

Moreover, since $H_\varepsilon(x, u_\varepsilon, \nabla u_\varepsilon) \, u_\varepsilon \ge 0$ a.e. and μ-a.e., using (6.86), (6.87) and Fatou's Lemma we have

$$
H(x, u, \nabla u) \, u \in L^1(\Omega, \omega). \qquad (6.91)
$$

Therefore, by (6.89), (6.90) and (6.91), u is a solution of problem (P18). □

Example 6.4.2. *Let* $\Omega = \{(x,y)\in\mathbb{R}^2 \; : x^2+y^2 < 1\}$, *and consider the weight function* $\omega(x,y) = (x^2+y^2)^{-1/2}$ $(\omega\in A_2)$, *the functions* $\mathcal{A}:\Omega\times\mathbb{R}\times\mathbb{R}^2\to\mathbb{R}^2$, *and* $H:\Omega\times\mathbb{R}\times\mathbb{R}^2\to\mathbb{R}$ *defined by*

$$\begin{aligned}\mathcal{A}((x,y),\eta,\xi) &= g(x,y)\,\xi,\\ H((x,y),\eta,\xi) &= |\xi|^2\, sin^2(xy)\,\frac{\eta^2}{\eta^2+1}\, arctan(\eta)\\ &\quad + h(x,y)\,\frac{\eta^2}{\eta^2+1}\, arctan(\eta),\end{aligned}$$

where $g(x,y) = e^{x^2+y^2}$ *and* $h(x,y) = (x^2+y^2)^{1/2} cos^2(xy)$. *Let us consider the partial differential operator*

$$Lu(x,y) = -\operatorname{div}\Big[\omega(x,y)\mathcal{A}((x,y),u,\nabla u)\Big]$$

and $T = f_0(x,y) = (x^2+y^2)^{-1/2}\cos(1/(x^2+y^2))$. *Therefore, by Theorem 6.4.3, the problem*

$$\left\{\begin{array}{l} Lu(x,y) + H(x,u,\nabla u)\,\omega = T,\\ H(x,u,\nabla u)\in L^1(\Omega,\omega),\\ H(x,u,\nabla u)\,u\in L^1(\Omega,\omega),\end{array}\right.$$

has a solution $u\in W_0^{1,2}(\Omega,\omega)$.

6.5 On a class of Nonhomogeneous Navier problems with degenerate operators

In this section we prove the existence of (weak) solutions in the weighted Sobolev space $W_0^{2,p}(\Omega,\omega_1,\omega_3)$ for the Navier problem

$$(P19)\begin{cases} Lu(x) = f_0(x) - \displaystyle\sum_{j=1}^{n} D_j f_j(x) \ \text{ in } \ \Omega, \\ u(x) = \Delta u(x) = 0 \ \text{ on } \ \partial\Omega, \end{cases}$$

where L is the partial differential operator

$$\begin{aligned} Lu(x) &= \Delta\big[\omega_1\,|\Delta u|^{p-2}\Delta u + \omega_2|\Delta u|^{q-2}\Delta u\big] \\ &- \operatorname{div}\big[\mathcal{A}(x,u,\nabla u)\,\omega_3 + \mathcal{B}(x,u,\nabla u)\,\omega_4\big] \\ &+ \mathcal{H}(x,u,\nabla u)\,\omega_5, \end{aligned} \qquad (6.92)$$

where $D_j = \partial/\partial x_j$, Ω is a bounded open set in $\mathbb{R}^n$, ω_1, ω_2, ω_3, ω_4 and ω_5 are five weight functions (which represent the degeneration or singularity in the equation (6.92)), $2 \le q < p < \infty$, $1 < s, z < p < \infty$ and the functions $\mathcal{A}_j : \Omega\times\mathbb{R}\times\mathbb{R}^n \to \mathbb{R}$, $\mathcal{B}_j : \Omega\times\mathbb{R}\times\mathbb{R}^n \to \mathbb{R}$ $(j = 1, ..., n)$ and $\mathcal{H} : \Omega\times\mathbb{R}\times\mathbb{R}^n \to \mathbb{R}$ satisfying the following conditions:

(H1) $x \mapsto \mathcal{A}_j(x,\eta,\xi)$ is measurable on Ω for all $(\eta,\xi) \in \mathbb{R}\times\mathbb{R}^n$, $(\eta,\xi) \mapsto \mathcal{A}_j(x,\eta,\xi)$ is continuous on $\mathbb{R}\times\mathbb{R}^n$ for almost all $x\in\Omega$;

(H2) There exists a constant $\theta_1 > 0$ such that

$$\langle \mathcal{A}(x,\eta,\xi) - \mathcal{A}(x,\eta',\xi'), (\xi - \xi')\rangle \ge \theta_1\,|\xi - \xi'|^p,$$

whenever $\xi, \xi' \in \mathbb{R}^n$, $\xi \neq \xi'$, and $\mathcal{A}(x,\eta,\xi) = (\mathcal{A}_1(x,\eta,\xi), ..., \mathcal{A}_n(x,\eta,\xi))$ (where $\langle .,. \rangle$ denotes here the Euclidian scalar product in $\mathbb{R}^n$);

(H3) $\langle \mathcal{A}(x,\eta,\xi), \xi\rangle \ge \lambda_1 |\xi|^p$, where λ_1 is a positive constant;

(H4) $|\mathcal{A}(x,\eta,\xi)| \le K_1(x) + h_1(x)|\eta|^{p/p'} + h_2(x)|\xi|^{p/p'}$, where K_1, h_1 and h_2 are nonnegative functions, with $h_1, h_2 \in L^\infty(\Omega)$ and $K_1 \in L^{p'}(\Omega,\omega_3)$ (with $1/p + 1/p' = 1$);

(H5) $x\mapsto\mathcal{B}_j(x,\eta,\xi)$ is measurable on Ω for all $(\eta,\xi)\in\mathbb{R}\times\mathbb{R}^n$,
$(\eta,\xi)\mapsto\mathcal{B}_j(x,\eta,\xi)$ is continuous on $\mathbb{R}\times\mathbb{R}^n$ for almost all $x\in\Omega$;
(H6) There exists a constant $\theta_2>0$ such that

$$\langle\mathcal{B}(x,\eta,\xi)-\mathcal{B}(x,\eta',\xi'),(\xi-\xi')\rangle\geq\theta_2\,|\xi-\xi'|^s,$$

whenever $\xi,\xi'\in\mathbb{R}^n$, $\xi\neq\xi'$,
where $\mathcal{B}(x,\eta,\xi)=(\mathcal{B}_1(x,\eta,\xi),...,\mathcal{B}_n(x,\eta,\xi))$;
(H7) $\langle\mathcal{B}(x,\eta,\xi),\xi\rangle\geq\lambda_2|\xi|^s+\Lambda_2|\eta|^s$, where $\lambda_2>0$ and $\Lambda_2\geq 0$ are constants;
(H8) $|\mathcal{B}(x,\eta,\xi)|\leq K_2(x)+g_1(x)|\eta|^{s/s'}+g_2(x)|\xi|^{s/s'}$, where K_2, g_1 and g_2 are nonnegative functions, with g_1 and $g_2\in L^\infty(\Omega)$, and $K_2\in L^{s'}(\Omega,\omega_4)$ (with $1/q+1/q'=1$).
(H9) $x\mapsto\mathcal{H}(x,\eta,\xi)$ is measurable on Ω for all $(\eta,\xi)\in\mathbb{R}\times\mathbb{R}^n$
$(\eta,\xi)\mapsto\mathcal{H}(x,\eta,\xi)$ is continuous on $\mathbb{R}\times\mathbb{R}^n$ for almost all $x\in\Omega$.
(H10) $[\mathcal{H}(x,\eta,\xi)-\mathcal{H}(x,\eta',\xi')](\eta-\eta')>0$, whenever $\eta,\eta'\in\mathbb{R}$, $\eta\neq\eta'$.
(H11) $\mathcal{H}(x,\eta,\xi)\eta\geq\lambda_3|\xi|^z+\Lambda_3|\eta|^z$, λ_3 and Λ_3 are nonnegative constants.
(H12) $|\mathcal{H}(x,\eta,\xi)|\leq K_3(x)+h_3(x)|\eta|^{z/z'}+h_4(x)|\xi|^{z/z'}$, where K_3,h_3 and h_4 are nonnegative functions, with $K_3\in L^{z'}(\Omega,\omega_5)$ (with $1/z+1/z'=1$), h_3 and $h_4\in L^\infty(\Omega)$.

Definition 6.5.1. *Let $\Omega\subset\mathbb{R}^n$ be an open bounded set, $k\in\mathbb{N}$ and $1\leq p<\infty$. Let ω be a given family of weight functions ω_α ($|\alpha|\leq k$):*
$\omega=\{\omega_\alpha(x):\,x\in\Omega,\ |\alpha|\leq k\}$. We denote by $W^{k,p}(\Omega,\omega)$ the set of all functions $u\in L^p(\Omega,\omega_0)$ for which the weak derivatives $D^\alpha u$ (with $|\alpha|\leq k$) belong to $L^p(\Omega,\omega_\alpha)$. The weighted Sobolev space $W^{k,p}(\Omega,\omega)$ is a normed linear space if equipped with the norm

$$\|u\|_{W^{k,p}(\Omega,\omega)}=\left(\sum_{|\alpha|\leq k}\|D^\alpha u\|^p_{L^p(\Omega,\omega_\alpha)}\right)^{1/p}. \tag{6.93}$$

Theorem 6.5.1. *(i) Let $1 < p < \infty$ and suppose that the weight functions ω_α satisfy $\omega_\alpha^{-1/(p-1)} \in L^1_{\text{loc}}(\Omega)$ $(|\alpha| \le k)$. Then $W^{k,p}(\Omega, \omega)$ is a uniformly convex (and hence reflexive) Banach space.*
(ii) If we additionally suppose that also $\omega_\alpha \in L^1_{\text{loc}}(\Omega)$, $|\alpha| \le k$, then $C_0^\infty(\Omega)$ is a subset of $W^{k,p}(\Omega, \omega)$ and we can introduce the space $W_0^{k,p}(\Omega, \omega)$ as the closure of $C_0^\infty(\Omega)$ with respect to the norm $\|.\|_{W^{k,p}(\Omega,\omega)}$.

Proof. See Theorem 1.3 in [32] or [48]. □

Note that if $w_\alpha \in A_p$ ($1 < p < \infty$ and $|\alpha| \le k$) then conditions (i) and (ii) of Theorem 6.5.1 are satisfied.

Lemma 6.5.2. *Let $1 < p < \infty$.*
(a) There exists a constant $\alpha_p > 0$ such that

$$\left| |x|^{p-2}x - |y|^{p-2}y \right| \le \alpha_p\, |x-y|(|x|+|y|)^{p-2}, \forall x, y \in \mathbb{R}^n;$$

(b) There exist two positive constants β_p, γ_p such that for every $x, y \in \mathbb{R}^n$

$$\begin{aligned} \beta_p\, (|x|+|y|)^{p-2}|x-y|^2 &\le \langle |x|^{p-2}x - |y|^{p-2}y, x-y\rangle \\ &\le \gamma_p\, (|x|+|y|)^{p-2}|x-y|^2, \end{aligned}$$

where $\langle ., . \rangle$ denotes the Euclidian scalar product in $\mathbb{R}^n$.

Proof. See [22], Proposition 17.2 and Proposition 17.3. □

Definition 6.5.2. *Let $\Omega \subset \mathbb{R}^n$ be a bounded open set, $\omega_1, \omega_3 \in A_p$, $2 < p < \infty$. We define the space $W^{2,p}(\Omega, \omega_1, \omega_3)$ with the norm*

$$\begin{aligned} \|u\|_{W^{k,p}(\Omega,\omega_1,\omega_3)} &= \left(\int_\Omega |u|^p \omega_3\, dx + \int_\Omega |\nabla u|^p\, \omega_3 dx \right. \\ &\quad \left. + \int_\Omega |\Delta u|^p\, \omega_1\, dx \right)^{1/p}, \end{aligned} \tag{6.94}$$

and $W_0^{2,p}(\Omega,\omega_1,\omega_3)$ *as the closure of* $C_0^\infty(\Omega)$ *with respect to the norm* $\|.\|_{W^{2,p}(\Omega,\omega_1,\omega_3)}$.

Remark 6.5.1. *By Theorem 2.1.5 with* $k=1$ *we obtain*

$$\begin{aligned}\|u\|^p_{W^{2,p}(\Omega,\omega_1,\omega_3)} &= \int_\Omega |u|^p\omega_3\,dx + \int_\Omega |\nabla u|^p\omega_3\,dx + \int_\Omega |\Delta u|^p\,\omega_1\,dx\\ &\leq (C_\Omega^p+1)\int_\Omega |\nabla u|^p\,\omega_3\,dx + \int_\Omega |\Delta u|^p\,\omega_1\,dx\\ &\leq C_1\bigg(\int_\Omega |\nabla u|^p\,\omega_3\,dx + \int_\Omega |\Delta u|^p\,\omega_1\,dx\bigg),\end{aligned}$$

where $C_1=(C_\Omega^p+1)$. *Hence*

$$\|u\|_{W^{2,p}(\Omega,\omega_1,\omega_3)} \leq C_1^{1/p}\bigg(\int_\Omega |\nabla u|^p\,\omega_3\,dx + \int_\Omega |\Delta u|^p\,\omega_1\,dx\bigg)^{1/p}.$$

Since $\bigg(\int_\Omega |\nabla u|^p\,\omega_3\,dx+\int_\Omega |\Delta u|^p\,\omega_1\,dx\bigg)^{1/p} \leq \|u\|_{W^{2,p}(\Omega,\omega_1,\omega_3)}$, *then the norms*

$$\|u\|_{W^{2,p}(\Omega,\omega_1,\omega_3)} \text{ and } \|u\|_X = \bigg(\int_\Omega |\nabla u|^p\,\omega_3\,dx+\int_\Omega |\Delta u|^p\,\omega_1\,dx\bigg)^{1/p}$$

are equivalents.

Definition 6.5.3. *We denote by* $X=W_0^{2,p}(\Omega,\omega_1,\omega_3)$ *with the norm*

$$\|u\|_X = \bigg(\int_\Omega |\nabla u|^p\,\omega_3\,dx + \int_\Omega |\Delta u|^p\,\omega_1\,dx\bigg)^{1/p}.$$

Definition 6.5.4. *We say that an element* $u\in X=W_0^{2,p}(\Omega,\omega_1,\omega_2)$ *is a (weak) solution of problem (P19) if*

$$\int_\Omega |\Delta u|^{p-2}\Delta u\Delta\varphi\,\omega_1\,dx + \int_\Omega |\Delta u|^{q-2}\Delta u\Delta\varphi\,\omega_2\,dx$$

$$+\int_\Omega \langle \mathcal{A}(x,u,\nabla u),\nabla\varphi\rangle\,\omega_3\,dx+\int_\Omega \langle \mathcal{B}(x,u,\nabla u),\nabla\varphi\rangle\,\omega_4\,dx$$
$$+\int_\Omega \mathcal{H}(x,u,\nabla u)\,\varphi\,\omega_5\,dx$$
$$=\int_\Omega f_0\,\varphi\,dx+\sum_{j=1}^n\int_\Omega f_j\,D_j\varphi\,dx,$$

for all $\varphi\in X$.

Remark 6.5.2. *(a) If* $\dfrac{\omega_2}{\omega_1}\in L^{p/(p-q)}(\Omega,\omega_1)$ $(2\le q<p<\infty)$ *then there exits a constant* $M_1>0$ *such that*

$$\|u\|_{L^q(\Omega,\omega_2)}\le M_1\,\|u\|_{L^p(\Omega,\omega_1)},$$

where $M_1=\|\omega_2/\omega_1\|^{1/q}_{L^{p/(p-q)}(\Omega,\omega_1)}$. *In fact, by Hölder's inequality,*

$$\|u\|^q_{L^q(\Omega,\omega_2)}$$
$$=\int_\Omega |u|^q\,\omega_2\,dx=\int_\Omega |u|^q\,\frac{\omega_2}{\omega_1}\,\omega_1\,dx$$
$$\le\left(\int_\Omega |u|^{qp/q}\omega_1\,dx\right)^{q/p}\left(\int_\Omega\left(\frac{\omega_2}{\omega_1}\right)^{p/(p-q)}\omega_1\,dx\right)^{(p-q)/p}$$
$$=\|u\|^q_{L^p(\Omega,\omega_1)}\,\|\omega_2/\omega_1\|_{L^{p/(p-q)}(\Omega,\omega_1)}.$$

(b) Analogously, if $\dfrac{\omega_4}{\omega_3}\in L^{p/(p-s)}(\Omega,\omega_3)$ *and* $\dfrac{\omega_5}{\omega_3}\in L^{p/(p-z)}(\Omega,\omega_3)$ *(with* $1<s,z<p<\infty$*), there exist constants* $M_2>0$ *and* $M_3>0$ *(respectively),* $M_2=\|\omega_4/\omega_3\|^{1/s}_{L^{p/(p-s)}(\Omega,\omega_3)}$ *and* $M_3=\|\omega_5/\omega_3\|^{1/z}_{L^{p/(p-z)}(\Omega,\omega_3)}$ *such that*

$$\|u\|_{L^s(\Omega,\omega_4)}\le M_2\,\|u\|_{L^p(\Omega,\omega_3)},$$

$$\|u\|_{L^z(\Omega,\omega_5)}\le M_3\,\|u\|_{L^p(\Omega,\omega_3)}.$$

Theorem 6.5.3. *Let* $2 \leq q < p < \infty$, $1 < s, z < p < \infty$ *and assume (H1)-(H12). If*
(i) $\omega_1 \in A_p$, $\omega_2 \in \mathcal{W}(\Omega)$, *and* $\dfrac{\omega_2}{\omega_1} \in L^{p/(p-q)}(\Omega, \omega_1)$;
(ii) $\omega_3 \in A_p$, $\omega_4 \in A_s$ *and* $\dfrac{\omega_4}{\omega_3} \in L^{p/(p-s)}(\Omega, \omega_3)$;
(iii) $\omega_5 \in \mathcal{W}(\Omega)$ *and* $\dfrac{\omega_5}{\omega_3} \in L^{p/(p-z)}(\Omega, \omega_3)$.
(iv) $f_j/\omega_3 \in L^{p'}(\Omega, \omega_3)$ $(j = 0, 1, ..., n)$.
Then the problem (P19) has a unique solution $u \in X = W_0^{2,p}(\Omega, \omega_1, \omega_3)$. *Moreover, we have*

$$\|u\|_X \leq \left(\frac{M}{\gamma}\right)^{1/(p-1)}$$

where

$$\gamma = \min\{1, \lambda_1\}, \; M = C_\Omega \|f_0/\omega_3\|_{L^{p'}(\Omega,\omega_3)} + \sum_{j=1}^{n} \|f_j/\omega_3\|_{L^{p'}(\Omega,\omega_3)}$$

and C_Ω *is the constant in Theorem 2.1.5.*

Proof. We define $\mathbf{B}, \mathbf{B_1}, \mathbf{B_2}, \mathbf{B_3}, \mathbf{B_4}, \mathbf{B_5} : X \times X \to \mathbb{R}$ and $\mathbf{T} : X \to \mathbb{R}$ by

$$\begin{aligned}
\mathbf{B}(u,\varphi) &= \mathbf{B_1}(u,\varphi) + \mathbf{B_2}(u,\varphi) + \mathbf{B_3}(u,\varphi) + \mathbf{B_4}(u,\varphi) \\
&+ \mathbf{B_5}(u,\varphi), \\
\mathbf{B_1}(u,\varphi) &= \int_\Omega |\Delta u|^{p-2} \Delta u \Delta\varphi \, \omega_1 \, dx, \\
\mathbf{B_2}(u,\varphi) &= \int_\Omega |\Delta u|^{q-2} \Delta u \Delta\varphi \, \omega_2 \, dx, \\
\mathbf{B_3}(u,\varphi) &= \int_\Omega \langle \mathcal{A}(x,u,\nabla u), \nabla\varphi \rangle \, \omega_3 \, dx, \\
\mathbf{B_4}(u,\varphi) &= \int_\Omega \langle \mathcal{B}(x,u,\nabla u), \nabla\varphi \rangle \, \omega_4 \, dx, \\
\mathbf{B_5}(u,\varphi) &= \int_\Omega \mathcal{H}(x,u,\nabla u) \, \varphi \, \omega_5 \, dx,
\end{aligned}$$

$$\mathbf{T}(\varphi) = \int_\Omega f_0 \, \varphi \, dx + \sum_{j=1}^{n} \int_\Omega f_j \, D_j\varphi \, dx.$$

Then $u \in X$ is a (weak) solution to problem (P19) if

$$\begin{aligned}\mathbf{B}(u,\varphi) &= \mathbf{B_1}(u,\varphi) + \mathbf{B_2}(u,\varphi) + \mathbf{B_3}(u,\varphi) + \mathbf{B_4}(u,\varphi) \\ &+ \mathbf{B_5}(u,\varphi) = \mathbf{T}(\varphi),\end{aligned}$$

for all $\varphi \in X$.

Step 1. For $j = 1, ..., n$ we define the operator $F_j : X \to L^{p'}(\Omega, \omega_3)$ as

$$(F_j u)(x) = \mathcal{A}_j(x, u(x), \nabla u(x)).$$

We now show that the operator F_j is bounded and continuous.

(i) Using (H4) and Theorem 2.1.5 (with $k = 1$, since $\omega_3 \in A_p$) we obtain

$$\begin{aligned}&\|F_j u\|^{p'}_{L^{p'}(\Omega,\omega_3)} \\ &= \int_\Omega |F_j u(x)|^{p'} \omega_3 \, dx \\ &= \int_\Omega |\mathcal{A}_j(x,u,\nabla u)|^{p'} \omega_3 \, dx \\ &\leq \int_\Omega \left(K_1 + h_1 |u|^{p/p'} + h_2 |\nabla u|^{p/p'} \right)^{p'} \omega_3 \, dx \\ &\leq C_p \int_\Omega \left[(K_1^{p'} + h_1^{p'} |u|^p + h_2^{p'} |\nabla u|^p) \omega_3 \right] dx \\ &= C_p \left[\int_\Omega K_1^{p'} \, \omega_3 \, dx + \int_\Omega h_1^{p'} |u|^p \, \omega_3 \, dx + \int_\Omega h_2^{p'} |\nabla u|^p \omega_3 \, dx \right] \\ &\leq C_p \Bigg(\|K_1\|^{p'}_{L^{p'}(\Omega,\omega_3)} + \|h_1\|^{p'}_{L^\infty(\Omega)} \|u\|^p_{L^p(\Omega,\omega_3)} \\ &+ \|h_2\|^{p'}_{L^\infty(\Omega)} \| \, |\nabla u| \, \|^p_{L^p(\Omega,\omega_3)} \Bigg)\end{aligned}$$

$$\leq C_p \Bigg(\|K_1\|^{p'}_{L^{p'}(\Omega,\omega_3)} + \|h_1\|^{p'}_{L^\infty(\Omega)} C^p_\Omega \| \, |\nabla u| \, \|^p_{L^p(\Omega,\omega_3)}$$
$$+ \|h_2\|^{p'}_{L^\infty(\Omega)} \| \, |\nabla u| \, \|^p_{L^p(\Omega,\omega_3)} \Bigg)$$
$$\leq C_p \Bigg(\|K_1\|^{p'}_{L^{p'}(\Omega,\omega_3)} + (C^p_\Omega \, \|h_1\|^{p'}_{L^\infty(\Omega)} + \|h_2\|^{p'}_{L^\infty(\Omega)}) \|u\|^p_X \Bigg), \tag{6.95}$$

where the constant C_p depends only on p. Therefore, in (6.95) we obtain

$$\|F_j u\|_{L^{p'}(\Omega,\omega_3)}$$
$$\leq C_p^{1/p'} \left(\|K_1\|_{L^{p'}(\Omega,\omega_3)} + (C^{p-1}_\Omega \|h_1\|_{L^\infty(\Omega)} + \|h_2\|_{L^\infty(\Omega)}) \, \|u\|^{p-1}_X \right).$$

(ii) Let $u_m \to u$ in X as $m \to \infty$. We need to show that $F_j u_m \to F_j u$ in $L^{p'}(\Omega, \omega_3)$. We will apply the Lebesgue Dominated Convergence Theorem. If $u_m \to u$ in X, then $|\nabla u_m| \to |\nabla u|$ in $L^p(\Omega, \omega_3)$. Using Theorem 1.4.11 (since $\omega_3 \in A_p$), there exist a subsequence $\{u_{m_k}\}$ and a function $\Phi_1 \in L^p(\Omega, \omega_3)$ such that

$$\begin{aligned} &u_{m_k}(x) \to u(x) \text{ a.e. in } \Omega, \\ &D_j u_{m_k}(x) \to D_j u(x) \text{ a.e. in } \Omega, \\ &|\nabla u_{m_k}(x)| \leq \Phi_1(x) \text{ a.e. in } \Omega. \end{aligned} \tag{6.96}$$

By Theorem 2.1.5 (with $k = 1$, since $\omega_3 \in A_p$) we have

$$\|u_{m_k}\|_{L^p(\Omega,\omega_3)} \leq C_\Omega \| \, |\nabla u_{m_k}| \, \|_{L^p(\Omega,\omega_3)} \leq C_\Omega \|\Phi_1\|_{L^p(\Omega,\omega_3)}, \tag{6.97}$$

$$\|u\|_{L^p(\Omega,\omega_3)} \leq C_\Omega \| \, |\nabla u| \, \|_{L^p(\Omega,\omega_3)} \leq C_\Omega \|\Phi_1\|_{L^p(\Omega,\omega_3)}. \tag{6.98}$$

Next, applying (H4), (6.96), (6.97) and (6.98) we obtain

$$\|F_j u_{m_k} - F_j u\|^{p'}_{L^{p'}(\Omega,\omega_3)}$$

$$
\begin{aligned}
&= \int_\Omega |F_j u_{m_k}(x) - F_j u(x)|^{p'} \omega_3 \, dx \\
&= \int_\Omega |\mathcal{A}_j(x, u_{m_k}, \nabla u_{m_k}) - \mathcal{A}_j(x, u, \nabla u)|^{p'} \omega_3 \, dx \\
&\le C_p \int_\Omega \left(|\mathcal{A}_j(x, u_{m_k}, \nabla u_{m_k})|^{p'} + |\mathcal{A}_j(x, u, \nabla u)|^{p'} \right) \omega_3 \, dx \\
&\le C_p \left[\int_\Omega \left(K_1 + h_1 |u_{m_k}|^{p/p'} + h_2 |\nabla u_{m_k}|^{p/p'} \right)^{p'} \omega_3 \, dx \right. \\
&\left. + \int_\Omega \left(K_1 + h_1 |u|^{p/p'} + h_2 |\nabla u|^{p/p'} \right)^{p'} \omega_3 \, dx \right] \\
&\le C_p \left[\int_\Omega K_1^{p'} \omega_3 \, dx + \int_\Omega h_1^{p'} |u_{m_k}|^p \omega_3 \, dx + \int_\Omega h_2^{p'} |\nabla u_{m_k}|^p \omega_3 \, dx \right. \\
&\left. + \int_\Omega K_1^{p'} \omega_3 \, dx + \int_\Omega h_1^{p'} |u|^p \, \omega_3 \, dx + \int_\Omega h_2^{p'} |\nabla u|^p \, \omega_3 \, dx \right] \\
&\le 2\, C_p \left[\|K_1\|^{p'}_{L^{p'}(\Omega,\omega_3)} + (C_\Omega^p \|h_1\|^{p'}_{L^\infty(\Omega)} + \|h_2\|^{p'}_{L^\infty(\Omega)}) \|\Phi_1\|^p_{L^p(\Omega,\omega_3)} \right].
\end{aligned}
$$

By condition (H1), we have

$$
\begin{aligned}
F_j u_{m_k}(x) \quad &= \mathcal{A}_j(x, u_{m_k}(x), \nabla u_{m_k}(x)) \\
&\to \mathcal{A}_j(x, u(x), \nabla u(x)) = F_j u(x),
\end{aligned}
$$

as $m_k \to +\infty$. Therefore, by the Lebesgue Dominated Convergence Theorem, we obtain

$$\|F_j u_{m_k} - F_j u\|_{L^{p'}(\Omega,\omega_3)} \to 0,$$

that is,

$$F_j u_{m_k} \to F_j u \ \text{ in } \ L^{p'}(\Omega, \omega_3).$$

We conclude from the Convergence Principle in Banach spaces (see Proposition 10.13 in [59]) that

$$F_j u_m \to F_j u \ \text{ in } \ L^{p'}(\Omega, \omega_3). \tag{6.99}$$

Step 2. We define the operator $G_j : X \to L^{s'}(\Omega, \omega_4)$ $(j = 1, ..., n)$ by

$$(G_j u)(x) = \mathcal{B}_j(x, u(x), \nabla u(x)).$$

This operator is continuous and bounded. In fact,
(i) Using (H8), Theorem 2.1.5 (since $\omega_3 \in A_p$) and Remark 6.5.2(b) we obtain

$$\begin{aligned}
&\|G_j u\|^{s'}_{L^{s'}(\Omega,\omega_4)} \\
&= \int_\Omega |G_j u(x)|^{s'} \omega_4 \, dx \\
&= \int_\Omega |\mathcal{B}_j(x, u, \nabla u)|^{s'} \omega_4 \, dx \\
&\le \int_\Omega \left(K_2 + g_1 |u|^{s/s'} + g_2 |\nabla u|^{s/s'} \right)^{s'} \omega_4 \, dx \\
&\le C_s \int_\Omega \left[(K_2^{s'} + g_1^{s'} |u|^s + g_2^{s'} |\nabla u|^s) \omega_4 \right] dx \\
&= C_s \left[\int_\Omega K_2^{s'} \, \omega_4 \, dx + \int_\Omega g_1^{s'} |u|^s \, \omega_4 \, dx + \int_\Omega g_2^{s'} |\nabla u|^s \omega_4 \, dx \right] \\
&\le C_s \Big(\|K_2\|^{s'}_{L^{s'}(\Omega,\omega_4)} + \|g_1\|^{s'}_{L^\infty(\Omega)} \|u\|^s_{L^s(\Omega,\omega_4)} \\
&+ \|g_2\|^{s'}_{L^\infty(\Omega)} \| \, |\nabla u| \, \|^s_{L^s(\Omega,\omega_4)} \Big) \\
&\le C_s \Big(\|K_2\|^{s'}_{L^{s'}(\Omega,\omega_4)} + \|g_1\|^{s'}_{L^\infty(\Omega)} \, M_2^s \, \|u\|^s_{L^p(\Omega,\omega_3)} \\
&+ M_2^s \|g_2\|^{s'}_{L^\infty(\Omega)} \| \, |\nabla u| \, \|^s_{L^p(\Omega,\omega_3)} \Big) \\
&\le C_s \Big(\|K_2\|^{s'}_{L^{s'}(\Omega,\omega_4)} + \|g_1\|^{s'}_{L^\infty(\Omega)} \, C_\Omega^s \, M_2^s \, \| \, |\nabla u| \, \|^s_{L^r(\Omega,\omega_3)} \\
&+ M_2^s \|g_2\|^{s'}_{L^\infty(\Omega)} \| \, |\nabla u| \, \|^s_{L^r(\Omega,\omega_3)} \Big)
\end{aligned}$$

$$\leq C_s\left(\|K_2\|^{s'}_{L^{s'}(\Omega,\omega_4)} + (C^s_\Omega\, M^s_2 \|g_1\|^{s'}_{L^\infty(\Omega)} \right.$$

$$\left. + M^s_2 \|g_2\|^{s'}_{L^\infty(\Omega)}) \|u\|^s_X \right), \tag{6.100}$$

where the constant C_s depends only on s. Therefore, in (6.100), we obtain

$$\|G_j u\|_{L^{s'}(\Omega,\omega_4)}$$

$$\leq C_s^{1/s'} \left(\|K_2\|_{L^{s'}(\Omega,\omega_4)} + M_2^{s-1}(C_\Omega^{s-1}\|g_1\|_{L^\infty(\Omega)} \right.$$

$$\left. + \|g_2\|_{L^\infty(\Omega)})\, \|u\|_X^{s-1} \right). \tag{6.101}$$

(ii) Let $u_m \to u$ in X as $m \to \infty$. We need to show that $G_j u_m \to G_j u$ in $L^{s'}(\Omega,\omega_4)$. We will apply the Lebesgue Dominated Theorem. If $u_m \to u$ in X, then $|\nabla u_m| \to |\nabla u|$ in $L^p(\Omega,\omega_3)$. Next, applying (H8), Theorem 2.1.5, Remark 6.5.2(b), (6.96), (6.97) and (6.98) we obtain

$$\|G_j u_{m_k} - G_j u\|^{s'}_{L^{s'}(\Omega,\omega_4)} = \int_\Omega |G_j u_{m_k}(x) - G_j u(x)|^{s'} \omega_4\, dx$$

$$= \int_\Omega |\mathcal{B}_j(x,u_{m_k},\nabla u_{m_k}) - \mathcal{B}_j(x,u,\nabla u)|^{s'}\, \omega_4\, dx$$

$$\leq C_s \int_\Omega \left(|\mathcal{B}_j(x,u_{m_k},\nabla u_{m_k})|^{s'} + |\mathcal{B}_j(x,u,\nabla u)|^{s'} \right) \omega_4\, dx$$

$$\leq C_s \left[\int_\Omega \left(K_2 + g_1 |u_{m_k}|^{s/s'} + g_2 |\nabla u_{m_k}|^{s/s'} \right)^{s'} \omega_4\, dx \right.$$

$$\left. + \int_\Omega \left(K_2 + g_1|u|^{s/s'} + g_2|\nabla u|^{s/s'} \right)^{s'} \omega_4\, dx \right]$$

$$\leq C_s \left[\int_\Omega K_2^{s'} \omega_4\, dx + \|g_1\|^{s'}_{L^\infty(\Omega)} \int_\Omega |u_{m_k}|^s\, \omega_4\, dx \right.$$

$$
\begin{aligned}
&+\|g_2\|_{L^\infty(\Omega)}^{s'} \int_\Omega |\nabla u_{m_k}|^s\,\omega_4\,dx \\
&+ \int_\Omega K_2^{s'}\,\omega_4\,dx + \|g_1\|_{L^\infty(\Omega)}^{s'} \int_\Omega |u|^s\,\omega_4\,dx \\
&+\|g_2\|_{L^\infty(\Omega)}^{s'} \int_\Omega |\nabla u|^s\,\omega_4\,dx\bigg] \\
&\leq C_s\bigg[\int_\Omega K_2^{s'}\omega_4\,dx + \|g_1\|_{L^\infty(\Omega)}^{s'} M_2^s \|u_{m_k}\|_{L^p(\Omega,\omega_3)}^s \\
&+\|g_2\|_{L^\infty(\Omega)}^{s'} M_2^s \|\,|\nabla u_{m_k}|\,\|_{L^p(\Omega,\omega_3)}^s + \int_\Omega K_2^{s'}\omega_4\,dx \\
&+\|g_1\|_{L^\infty(\Omega)}^{s'} M_2^s \|u\|_{L^p(\Omega,\omega_3)}^s + \|g_2\|_{L^\infty(\Omega)}^{s'} M_2^s \|\,|\nabla u|\,\|_{L^p(\Omega,\omega_3)}^s \\
&\leq 2C_s\bigg[\|K_2\|_{L^{s'}(\Omega,\omega_4)}^{s'} + \|g_1\|_{L^\infty(\Omega)}^{s'} C_\Omega^s M_2^s \|\Phi_1\|_{L^p(\Omega,\omega_3)}^s \\
&+\|g_2\|_{L^\infty(\Omega)}^{s'} M_2^s \|\Phi_1\|_{L^p(\Omega,\omega_3)}^s\bigg] \\
&= 2\,C_s\bigg(\|K_2\|_{L^{s'}(\Omega,\omega_4)}^{s'} + M_2^s\big(C_\Omega^s\,\|g_1\|_{L^\infty(\Omega)}^{s'} \\
&+\|g_2\|_{L^\infty(\Omega)}^{s'}\big)\|\Phi_1\|_{L^p(\Omega,\omega_3)}\bigg].
\end{aligned}
$$

By condition (H5), we have

$$
\begin{aligned}
G_j u_{m_k}(x) &= \mathcal{B}_j(x, u_{m_k}(x), \nabla u_{m_k}(x)) \\
&\to \mathcal{B}_j(x, u(x), \nabla u(x)) = G_j u(x),
\end{aligned}
$$

as $m_k \to +\infty$. Therefore, by the Lebesgue Dominated Convergence Theorem, we obtain

$$\|G_j u_{m_k} - G_j u\|_{L^{s'}(\Omega,\omega_4)} \to 0,$$

that is,

$$G_j u_{m_k} \to G_j u \;\text{ in }\; L^{s'}(\Omega, \omega_4).$$

We conclude from the Convergence Principle in Banach spaces (see Proposition 10.13 in [59]) that

$$G_j u_m \to G_j u \quad \text{in} \quad L^{s'}(\Omega, \omega_4). \tag{6.102}$$

Step 3. We define the operator $F : X \to L^{p'}(\Omega, \omega_1)$ by

$$(Fu)(x) = |\Delta u(x)|^{p-2} \, \Delta u(x).$$

We now show that operator F is bounded and continuous.
(i) We have

$$\begin{aligned} \|Fu\|^{p'}_{L^{p'}(\Omega,\omega_1)} &= \int_\Omega |F_1 u(x)|^{p'} \, \omega_1 \, dx \\ &= \int_\Omega \Big| |\Delta u|^{p-2} \, \Delta u \Big|^{p'} \, \omega_1 \, dx \\ &= \int_\Omega |\Delta u|^p \, \omega_1 \, dx \\ &= \|\Delta u\|^p_{L^p(\Omega,\omega_1)} \\ &\leq \|u\|^p_X. \end{aligned} \tag{6.103}$$

Therefore, in (6.103), we obtain

$$\|F\, u\|_{L^{p'}(\Omega,\omega_1)} \leq \|u\|_X^{p-1}, \tag{6.104}$$

and hence the boundedness.
(ii) Let $u_m \to u$ in X as $m \to 0$. We need to show that $Fu_m \to Fu$ in $L^{p'}(\Omega, \omega_1)$. If $u_m \to u$ in X then $\Delta u_m \to \Delta u$ in $L^p(\Omega, \omega_1)$. Using Theorem 1.4.11, there exist a subsequence $\{u_{m_k}\}$ and a function $\Phi_2 \in L^p(\Omega, \omega_1)$ such that

$$\Delta u_{m_k}(x) \to \Delta u(x) \text{ a.e. in } \Omega, \tag{6.105}$$

$$|\Delta u_{m_k}(x)| \leq \Phi_2(x) \text{ a.e. in } \Omega. \tag{6.106}$$

Now, since $p > 2$, using (6.105), (6.106), $a = p/p' = p - 1$ and $a' = (p-1)/(p-2)$, there exists a constant $\alpha_p > 0$ (by Lemma

6.5.2(a)) such that

$$
\begin{aligned}
&\|Fu_{m_k} - Fu\|^{p'}_{L^{p'}(\Omega,\omega_1)}\\
&= \int_\Omega |Fu_{m_k} - Fu|^{p'}\omega_1\,dx\\
&= \int_\Omega \left||\Delta u_{m_k}|^{p-2}\Delta u_{m_k} - |\Delta u|^{p-2}\Delta u\right|^{p'}\omega_1\,dx\\
&\leq \int_\Omega \left[\alpha_p\,|\Delta u_{m_k} - \Delta u|\,(\,|\Delta u_{m_k}| + |\Delta u|\,)^{p-2}\right]^{p'}\omega_1\,dx\\
&\leq \alpha_p^{p'}\int_\Omega |\Delta u_{m_k} - \Delta u|^{p'}(2\Phi_2)^{(p-2)p'}\,\omega_1\,dx\\
&= 2^{(p-2)p'}\alpha_p^{p'}\int_\Omega |\Delta u_{m_k} - \Delta u|^{p'}\,\Phi_2^{(p-2)p'}\,\omega_1\,dx\\
&\leq 2^{(p-2)p'}\alpha_p^{p'}\left(\int_\Omega |\Delta u_{m_k} - \Delta u|^{p'a}\omega_1\,dx\right)^{1/a}\times\\
&\times\left(\int_\Omega \Phi_2^{(p-2)p'a'}\,\omega_1\,dx\right)^{1/a'}\\
&= 2^{(p-2)p'}\alpha_p^{p'}\left(\int_\Omega |\Delta u_{m_k} - \Delta u|^{p}\,\omega_1\,dx\right)^{p'/p}\times\\
&\times\left(\int_\Omega \Phi_2^{p}\,\omega_1\,dx\right)^{(p-2)/(p-1)}\\
&= 2^{(p-2)p'}\alpha_p^{p'}\,\|\Delta u_{m_k} - \Delta u\|^{p'}_{L^p(\Omega,\omega_1)}\|\Phi_2\|^{p'(p-2)}_{L^p(\Omega,\omega_1)}\\
&\leq 2^{(p-2)p'}\alpha_p^{p'}\,\|u_{m_k} - u\|^{p'}_X\,\|\Phi_2\|^{p'(p-2)}_{L^p(\Omega,\omega_1)}.
\end{aligned}
$$

Hence,

$$
\|Fu_{m_k} - Fu\|_{L^{p'}(\Omega,\omega_1)} \leq 2^{p-2}\,\alpha_p\,\|u_{m_k} - u\|_X\,\|\Phi_2\|^{p-2}_{L^p(\Omega,\omega_1)}.
$$

Therefore (since $2 < p < \infty$), we obtain $\|Fu_{m_k} - Fu\|_{L^{p'}(\Omega,\omega_1)} \to 0$, that is,

$$Fu_{m_k} \to Fu \ \text{ in } \ L^{p'}(\Omega, \omega_1).$$

By the Convergence Principle in Banach spaces (see Proposition 10.13 in [59] , we have

$$Fu_m \to Fu \ \text{ in } \ L^{p'}(\Omega, \omega_1). \tag{6.107}$$

Step 4. Define the operator $G : X \to L^{q'}(\Omega, \omega_2)$ by

$$(Gu)(x) = |\Delta u(x)|^{q-2}\Delta u(x).$$

We also have that the operator G is continuous and bounded. In fact:
(i) If $q > 2$, we have by Remark 6.5.2(a)

$$\begin{aligned} \|Gu\|^{q'}_{L^{q'}(\Omega,\omega_2)} &= \int_\Omega \big| |\Delta u|^{q-2}\,\Delta u\big|^{q'}\,\omega_2\,dx = \int_\Omega |\Delta u|^q\,\omega_2\,dx \\ &= \|\Delta u\|^q_{L^q(\Omega,\omega_2)} \\ &\leq M_1^q \|\Delta u\|^q_{L^p(\Omega,\omega_1)} \\ &\leq M_1^q\,\|u\|^q_X. \end{aligned}$$

Hence, $\|Gu\|_{L^{q'}(\Omega,\omega_2)} \leq M_1^{q-1}\|u\|_X^{q-1}$.
(ii) Now using (6.105), (6.106), Remark 6.5.2(a), $b = q/q' = q-1$ and $b' = (q-1)/(q-2)$ (if $q > 2$), there exists a constant $\alpha_q > 0$ (by Lemma 6.5.2(a)) such that

$$\|Gu_{m_k} - Gu\|^{q'}_{L^{q'}(\Omega,\omega_2)} = \int_\Omega |\,Gu_{m_k} - Gu|^{q'}\,\omega_2\,dx$$

$$= \int_\Omega \Big|\,|\Delta u_{m_k}|^{q-2}\,\Delta u_{m_k} - |\,\Delta u|^{q-2}\,\Delta u\Big|^{q'}\,\omega_2\,dx$$

$$\leq \int_\Omega \Big[\alpha_q\,|\Delta u_{m_k} - \Delta u|\,(\,|\Delta u_{m_k}| + |\Delta u|)^{(q-2)}\Big]^{q'}\,\omega_2\,dx$$

$$\leq \alpha_q^{q'}\int_\Omega |\,\Delta u_{m_k} - \Delta u|^{q'}\,(2\,\Phi_2)^{(q-2)\,q'}\,\omega_2\,dx$$

$$\begin{aligned}
&\leq 2^{(q-2)q'}\alpha_q^{q'}\left(\int_\Omega |\Delta u_{m_k}-\Delta u|^{q'b}\omega_2\,dx\right)^{1/b}\times\\
&\times\left(\int_\Omega \Phi_2^{(q-2)q'b'}\omega_2\,dx\right)^{1/b'}\\
&=\alpha_q^{q'}\,2^{(q-2)q'}\left(\int_\Omega |\Delta u_{m_k}-\Delta u|^{q}\,\omega_2\,dx\right)^{q'/q}\times\\
&\times\left(\int_\Omega \Phi_2^{q}\,\omega_2\,dx\right)^{(q-2)/(q-1)}\\
&=\alpha_q^{q'}\,2^{(q-2)q'}\|\Delta u_{m_k}-\Delta u\|^{q'}_{L^q(\Omega,\omega_2)}\|\Phi_2\|^{q'(q-2)}_{L^q(\Omega,\omega_2)}\\
&\leq \alpha_q^{q'}\,2^{(q-2)q'}M_1^{q'}\|\Delta u_{m_k}-\Delta u\|^{q'}_{L^p(\Omega,\omega_1)}M_1^{q'(q-2)}\|\Phi_2\|^{q'(q-2)}_{L^p(\Omega,\omega_1)}\\
&\leq \alpha_q^{q'}\,2^{(q-2)q'}M_1^{q}\|u_{m_k}-u\|^{q'}_{X}\|\Phi_2\|^{q'(q-2)}_{L^p(\Omega,\omega_1)}.
\end{aligned}$$

Hence,

$$\|Gu_{m_k}-Gu\|_{L^{q'}(\Omega,\omega_2)}\leq 2^{q-2}\alpha_q\,M_1^{q-1}\|\Phi_2\|^{q-2}_{L^p(\Omega,\omega_1)}\,\|u_{m_k}-u\|_X.$$

In the case $q=2$ we have $(Gu)(x)=\Delta u(x)$. Hence,

$$\|Gu\|_{L^2(\Omega,\omega_2)}=\|\Delta u\|_{L^2(\Omega,\omega_2)}\leq M_1\|\Delta u\|_{L^p(\Omega,\omega_1)}\leq M_1\|u\|_X,$$

$$\begin{aligned}
\|Gu_{m_k}-Gu\|_{L^2(\Omega,\omega_2)} &\leq M_1\|\Delta u_{m_k}-\Delta u\|_{L^p(\Omega,\omega_1)}\\
&\leq M_1\|u_{m_k}-u\|_X.
\end{aligned}$$

Therefore (for $2\leq q<\infty$), we obtain $\|Gu_{m_k}-Gu\|_{L^{q'}(\Omega,\omega_2)}\to 0$, that is, $Gu_{m_k}\to Gu$ in $L^{q'}(\Omega,\omega_2)$. By the Convergence Principle in Banach spaces, we have

$$Gu_m\to Gu \text{ in } L^{q'}(\Omega,\omega_2). \tag{6.108}$$

Step 5. We define the operator $H: X\to L^{z'}(\Omega,\omega_5)$ by

$$(Hu)(x)=\mathcal{H}(x,u(x),\nabla u(x)).$$

We also have that the operator H is continuous and bounded. In fact,
(i) Using $(H12)$, Theorem 2.1.5 (since $\omega_3 \in A_p$) and Remark 6.5.2(b) we obtain

$$
\begin{aligned}
&\|Hu\|^{z'}_{L^{z'}(\Omega,\omega_5)}\\
&= \int_\Omega |Hu|^{z'}\,\omega_5\,dx\\
&= \int_\Omega |\mathcal{H}(x,u,\nabla u)|^{z'}\,\omega_5\,dx\\
&\leq \int_\Omega \left(K_3 + h_3|u|^{z/z'} + h_4|\nabla u|^{z/z'}\right)^{z'}\,\omega_5\,dx\\
&\leq C_z \int_\Omega (K_3^{z'} + h_3^{z'}|u|^z + h_4^{z'}|\nabla u|^z)\omega_5\,dx\\
&\leq C_z \int_\Omega K_3^{z'}\,\omega_5\,dx + \|h_3\|^{z'}_{L^\infty(\Omega)}\int_\Omega |u|^z\,\omega_5\,dx\\
&+\|h_4\|^{z'}_{L^\infty(\Omega)}\int_\Omega |\nabla u|^z\omega_5\,dx\Big]\\
&\leq C_z\Bigg(\|K_3\|^{z'}_{L^{z'}(\Omega,\omega_5)} + \|h_3\|^{z'}_{L^\infty(\Omega)}M_3^z\|u\|^z_{L^p(\Omega,\omega_3)}\\
&+\|h_4\|^{z'}_{L^\infty(\Omega)}M_3^z\|\,|\nabla u|\,\|^z_{L^p(\Omega,\omega_3)}\Bigg)\\
&\leq C_z\Bigg(\|K_3\|^{z'}_{L^{z'}(\Omega,\omega_5)} + M_3^z(C_\Omega^z\|h_3\|^{z'}_{L^\infty(\Omega)} + \|h_4\|^{z'}_{L^\infty(\Omega)})\,\|u\|^z_X\Bigg),
\end{aligned}
\tag{6.109}
$$

where the constant C_z depends only on z. Hence, in (6.109), we obtain

$$
\begin{aligned}
&\|Hu\|_{L^{z'}(\Omega,\omega_5)}\\
&\leq C_z^{1/z'}\Big[\|K_3\|_{L^{z'}(\Omega,\omega_5)} + M_3^{z-1}(C_\Omega^{z-1}\|h_3\|_{L^\infty(\Omega)}
\end{aligned}
$$

$$+\|h_4\|_{L^\infty(\Omega)})\|u\|_X^{z-1}\Big].$$

(ii) Applying (H9), (H12), Remark 6.5.2(b), (6.96), (6.97) and (6.98) by the same argument used in Step 1(ii), we obtain analogously, if $u_m \to u$ in X then

$$Hu_m \to Hu, \quad \text{in} \quad L^{z'}(\Omega, \omega_5). \tag{6.110}$$

Step 6. Since $\dfrac{f_j}{\omega_3} \in L^{p'}(\Omega, \omega_3)$ $(j = 0, 1, ..., n)$ then we have

$$\begin{aligned}
&|\mathbf{T}(\varphi)| \\
&\leq \int_\Omega |f_0||\varphi|\,dx + \sum_{j=1}^{n} \int_\Omega |f_j||D_j\varphi|\,dx \\
&= \int_\Omega \frac{|f_0|}{\omega_3}|\varphi|\omega_3\,dx + \sum_{j=1}^{n} \int_\Omega \frac{|f_j|}{\omega_3}|D_j\varphi|\,\omega_3\,dx \\
&\leq \|f_0/\omega_3\|_{L^{p'}(\Omega,\omega_3)}\|\varphi\|_{L^p(\Omega,\omega_3)} \\
&+ \left(\sum_{j=1}^{n} \|f_j/\omega_3\|_{L^{p'}(\Omega,\omega_3)}\right)\|\,|\nabla\varphi|\,\|_{L^p(\Omega,\omega_3)} \\
&\leq C_\Omega\|f_0/\omega_3\|_{L^{p'}(\Omega,\omega_3)}\|\,|\nabla\varphi|\,\|_{L^p(\Omega,\omega_3)} \\
&+ \left(\sum_{j=1}^{n} \|f_j/\omega_3\|_{L^{p'}(\Omega,\omega_3)}\right)\|\,|\nabla\varphi|\,\|_{L^p(\Omega,\omega_3)} \\
&\leq \left(C_\Omega\,\|f_0/\omega_3\|_{L^{p'}(\Omega,\omega_3)} + \sum_{j=1}^{n} \|f_j/\omega_3\|_{L^{p'}(\Omega,\omega_3)}\right)\|\varphi\|_X.
\end{aligned}$$

Moreover, we also have

$$\begin{aligned}
&|\mathbf{B}(u,\varphi)| \\
&\leq |\mathbf{B_1}(u,\varphi)| + |\mathbf{B_2}(u,\varphi)| + |\mathbf{B_3}(u,\varphi)| + |\mathbf{B_4}(u,\varphi)| + |\mathbf{B_5}(u,\varphi)|
\end{aligned}$$

$$\le \int_\Omega |\mathcal{A}(x,u,\nabla u)||\nabla\varphi|\,\omega_3\,dx + \int_\Omega |\mathcal{B}(x,u,\nabla u)|\,|\nabla\varphi|\,\omega_4\,dx$$
$$+ \int_\Omega |\Delta u|^{p-1}|\Delta\varphi|\,\omega_1\,dx + \int_\Omega |\Delta u|^{q-1}|\Delta\varphi|\,\omega_2\,dx$$
$$+ \int_\Omega |\mathcal{H}(x,u,\nabla u)|\,|\varphi|\,\omega_5\,dx. \tag{6.111}$$

In (6.111) we have:
(i) By (H4) and Theorem 2.1.5,

$$\int_\Omega |\mathcal{A}(x,u,\nabla u)|\,|\nabla\varphi|\,\omega_3\,dx$$
$$\le \int_\Omega \left(K_1 + h_1|u|^{p/p'} + h_2|\nabla u|^{p/p'}\right)|\nabla\varphi|\,\omega_3\,dx$$
$$\le \|K_1\|_{L^{p'}(\Omega,\omega_3)}\|\,|\nabla\varphi|\,\|_{L^p(\Omega,\omega_3)}$$
$$+\|h_1\|_{L^\infty(\Omega)}\|u\|^{p/p'}_{L^p(\Omega,\omega_3)}\|\,|\nabla\varphi|\,\|_{L^p(\Omega,\omega_3)}$$
$$+\|h_2\|_{L^\infty(\Omega)}\|\,|\nabla u|\,\|^{p/p'}_{L^p(\Omega,\omega_3)}\|\,|\nabla\varphi|\,\|_{L^p(\Omega,\omega_3)}$$
$$\le \left(\|K_1\|_{L^{p'}(\Omega,\omega_3)} + (C_\Omega^{p-1}\|h_1\|_{L^\infty(\Omega)} + \|h_2\|_{L^\infty(\Omega)})\|u\|_X^{p-1}\right)\|\varphi\|_X.$$

(ii) By (H8), Theorem 2.1.5 (with $k = 1$, since $\omega_4 \in A_s$) and Remark 6.5.2(b)

$$\int_\Omega |\mathcal{B}(x,u,\nabla u)|\,|\nabla\varphi|\,\omega_4\,dx$$
$$\le \int_\Omega \left(K_2 + g_1|u|^{s/s'} + g_2|\nabla u|^{s/s'}\right)|\nabla\varphi|\,\omega_4\,dx$$
$$\le \|K_2\|_{L^{s'}(\Omega,\omega_4)}\|\,|\nabla\varphi|\,\|_{L^s(\Omega,\omega_4)}$$
$$+\|g_1\|_{L^\infty(\Omega)}\|u\|^{s/s'}_{L^s(\Omega,\omega_4)}\|\,|\nabla\varphi|\,\|_{L^s(\Omega,\omega_4)}$$
$$+\|g_2\|_{L^\infty(\Omega)}\|\,|\nabla u|\,\|^{s/s'}_{L^s(\Omega,\omega_4)}\|\,|\nabla\varphi|\,\|_{L^s(\Omega,\omega_4)}$$
$$\le M_2\|K_2\|_{L^{s'}(\Omega,\omega_4)}\|\,|\nabla\varphi|\,\|_{L^p(\Omega,\omega_3)}$$

$$+ M_2^{s-1} C_\Omega^{s-1} \|g_1\|_{L^\infty(\Omega)} \| \, |\nabla u| \, \|_{L^p(\Omega,\omega_3)}^{s-1} M_2 \| \, |\nabla \varphi| \, \|_{L^p(\Omega,\omega_3)}$$
$$+ M_2^{s-1} \|g_2\|_{L^\infty(\Omega)} \| \, |\nabla u| \, \|_{L^p(\Omega,\omega_3)}^{s-1} M_2 \| \, |\nabla \varphi| \, \|_{L^p(\Omega,\omega_3)}$$
$$\leq \Big[M_2 \|K_2\|_{L^{s'}(\Omega,\omega_4)} + \big(M_2^s C_\Omega^{s-1} \|g_1\|_{L^\infty(\Omega)}$$
$$+ M_2^s \|g_2\|_{L^\infty(\Omega)} \big) \|u\|_X^{s-1} \Big] \|\varphi\|_X.$$

(iii) By (H12), Theorem 2.1.5 (since $\omega_3 \in A_p$) and Remark 6.5.2(b)

$$\int_\Omega |\mathcal{H}(x,u,\nabla u)| \, |\varphi| \, \omega_5 \, dx$$
$$\leq \int_\Omega \Big(K_3 + h_3 |u|^{z/z'} + h_4 |\nabla u|^{z/z'} \Big) |\varphi| \, \omega_5 \, dx$$
$$\leq \int_\Omega K_3 \, |\varphi| \, \omega_5 \, dx + \|h_3\|_{L^\infty(\Omega)} \int_\Omega |u|^{z/z'} |\varphi| \, \omega_5 \, dx$$
$$+ \|h_4\|_{L^\infty(\Omega)} \int_\Omega |\nabla u|^{z/z'} |\varphi| \, \omega_5 \, dx$$
$$\leq \|K_3\|_{L^{z'}(\Omega,\omega_5)} \|\varphi\|_{L^z(\Omega,\omega_5)} + \|h_3\|_{L^\infty(\Omega)} \|u\|_{L^z(\Omega,\omega_5)}^{z/z'} \|\varphi\|_{L^z(\Omega,\omega_5)}$$
$$+ \|h_4\|_{L^\infty(\Omega)} \| \, |\nabla u| \, \|_{L^z(\Omega,\omega_5)}^{z/z'} \|\varphi\|_{L^z(\Omega,\omega_5)}$$
$$\leq M_3 \, \|K_3\|_{L^{z'}(\Omega)} \|\varphi\|_{L^p(\Omega,\omega_3)}$$
$$+ \|h_3\|_{L^\infty(\Omega)} M_3^{z-1} \|u\|_{L^p(\Omega,\omega_3)}^{z-1} M_3 \|\varphi\|_{L^p(\Omega,\omega_3)}$$
$$+ \|h_4\|_{L^\infty(\Omega)} M_3^{z-1} \| \, |\nabla u| \, \|_{L^p(\Omega,\omega_3)}^{z-1} M_3 \|\varphi\|_{L^p(\Omega,\omega_3)}$$
$$\leq \Big[M_3 C_\Omega \, \|K_3\|_{L^{z'}(\Omega,\omega_5)} + M_3^z \big(C_\Omega^{z-1} \|h_3\|_{L^\infty(\Omega)}$$
$$+ \|h_4\|_{L^\infty(\Omega)} \big) \|u\|_X^{z-1} \Big] \|\varphi\|_X.$$

(iv) We have

$$\int_\Omega |\Delta u|^{p-1} |\Delta \varphi| \, \omega_1 \, dx$$

$$\leq \left(\int_\Omega |\Delta u|^{(p-1)\,p'} \omega_1 \, dx \right)^{1/p'} \left(\int_\Omega |\Delta \varphi|^p \, \omega_1 \, dx \right)^{1/p}$$
$$= \|\Delta u\|^{p-1}_{L^p(\Omega,\omega_1)} \, \|\Delta \varphi\|_{L^p(\Omega,\omega_1)}$$
$$\leq \|u\|^{p-1}_X \, \|\varphi\|_X;$$

(v) By Remark 6.5.2(a),

$$\int_\Omega |\Delta u|^{q-1} |\Delta \varphi| \, \omega_2 \, dx$$
$$\leq \left(\int_\Omega |\Delta u|^{(q-1)\,q'} \omega_2 \, dx \right)^{1/q'} \left(\int_\Omega |\Delta \varphi|^q \, \omega_2 \, dx \right)^{1/q}$$
$$= \|\Delta u\|^{q-1}_{L^q(\Omega,\omega_2)} \, \|\Delta \varphi\|_{L^q(\Omega,\omega_2)}$$
$$\leq M_1^{q-1} \|\Delta u\|^{q-1}_{L^p(\Omega,\omega_1)} \, M_1 \, \|\Delta \varphi\|_{L^p(\Omega,\omega_1)}$$
$$\leq M_1^q \, \|u\|^{q-1}_X \, \|\varphi\|_X.$$

Hence, in (6.111) we obtain, for all $u, \varphi \in X$

$$|\mathbf{B}(u,\varphi)|$$
$$\leq \Big[\|K_1\|_{L^{p'}(\Omega,\omega_1)} + (C_\Omega^{p-1} \|h_1\|_{L^\infty(\Omega)} + \|h_2\|_{L^\infty(\Omega)}) \|u\|_X^{p-1}$$
$$+ M_2 \|K_2\|_{L^{s'}(\Omega,\omega_4)} + M_2^s (C_\Omega^{s-1} \|g_1\|_{L^\infty(\Omega)} + \|g_2\|_{L^\infty(\Omega)}) \|u\|_X^{s-1}$$
$$+ M_3 C_\Omega \|K_3\|_{L^{z'}(\Omega,\omega_5)} + M_3^z (C_\Omega^{z-1} \|h_3\|_{L^\infty(\Omega)}$$
$$+ \|h_4\|_{L^\infty(\Omega)}) \|u\|_X^{z-1} + \|u\|_X^{p-1} + M_1^q \|u\|_X^{q-1} \Big] \|\varphi\|_X.$$

Since $\mathbf{B}(u,.)$ is linear, for each $u \in X$, there exists a linear and continuous functional on X denoted by Au such that $(Au|\varphi) = \mathbf{B}(u,\varphi)$, for all u, $\varphi \in X$ (here $(f|x)$ denotes the value of the linear functional f at the point x). Moreover

$$\|Au\|_*$$

$$\begin{aligned}
&\leq \|K_1\|_{L^{p'}(\Omega,\omega_1)} + (C_\Omega^{p-1}\|h_1\|_{L^\infty(\Omega)} + \|h_2\|_{L^\infty(\Omega)})\|u\|_X^{p-1}\\
&+ M_2\|K_2\|_{L^{s'}(\Omega,\omega_4)} + M_2^s(C_\Omega^{s-1}\|g_1\|_{L^\infty(\Omega)} + \|g_2\|_{L^\infty(\Omega)})\|u\|_X^{s-1}\\
&+M_3C_\Omega\|K_3\|_{L^{z'}(\Omega,\omega_5)} + M_3^z(C_\Omega^{z-1}\|h_3\|_{L^\infty(\Omega)}\\
&+\|h_4\|_{L^\infty(\Omega)})\|u\|_X^{z-1} + \|u\|_X^{p-1} + M_1^q\|u\|_X^{q-1},
\end{aligned}$$

where $\|Au\|_* = \sup\{|(Au|\varphi)| = |\mathbf{B}(u,\varphi)| : \varphi \in X, \|\varphi\|_X = 1\}$ is the norm of operator Au. Hence, we obtain the operator

$$\begin{aligned}
A &: X \to X^*\\
u &\mapsto Au.
\end{aligned}$$

Consequently, problem (P) is equivalent to the operator equation

$$Au = \mathbf{T},\ u \in X.$$

Step 7. Using (H2), (H6), (H10) and Lemma 6.5.2(b), we obtain (for $u_1, u_2 \in X$, $u_1 \neq u_2$)

$$\begin{aligned}
&(Au_1 - Au_2|u_1 - u_2)\\
&= \mathbf{B}(u_1, u_1 - u_2) - \mathbf{B}(u_2, u_1 - u_2)\\
&= \int_\Omega \left(|\Delta u_1|^{p-2}\Delta u_1 - |\Delta u_2|^{p-2}\Delta u_2\right)\Delta(u_1 - u_2)\,\omega_1\,dx\\
&+ \int_\Omega \left(|\Delta u_1|^{q-2}\Delta u_1 - |\Delta u_2|^{q-2}\Delta u_2\right)\Delta(u_1 - u_2)\,\omega_2\,dx\\
&+ \int_\Omega \langle \mathcal{A}(x,u_1,\nabla u_1) - \mathcal{A}(x,u_2\nabla u_2), \nabla(u_1 - u_2)\rangle\,\omega_3\,dx\\
&+ \int_\Omega \langle \mathcal{B}(x,u_1,\nabla u_1) - \mathcal{B}(x,u_2,\nabla u_2), \nabla(u_1 - u_2)\rangle\,\omega_4\,dx\\
&+ \int_\Omega (\mathcal{H}(x,u_1,\nabla u_1) - \mathcal{H}(x,u_2,\nabla u_2))(u_1 - u_2)\,\omega_5\,dx\\
&\geq \beta_p \int_\Omega \left(|\Delta u_1| + |\Delta u_2|\right)^{p-2} |\Delta u_1 - \Delta u_2|^2\,\omega_1\,dx
\end{aligned}$$

$$+\beta_q \int_\Omega \Big(|\Delta u_1| + |\Delta u_2|\Big)^{q-2} |\Delta u_1 - \Delta u_2|^2\, \omega_2\, dx$$

$$+\theta_1 \int_\Omega |\nabla(u_1 - u_2)|^p\, \omega_3\, dx + \theta_2 \int_\Omega |\nabla(u_1 - u_2)|^s\, \omega_4\, dx$$

$$\geq \beta_p \int_\Omega \Big(|\Delta u_1| + |\Delta u_2|\Big)^{p-2} |\Delta u_1 - \Delta u_2|^2\, \omega_1\, dx$$

$$+\theta_1 \int_\Omega |\nabla(u_1 - u_2)|^p\, \omega_3\, dx$$

$$\geq \beta_p \int_\Omega |\Delta u_1 - \Delta u_2|^{p-2} |\Delta u_1 - \Delta u_2|^2\, \omega_1$$

$$+\theta_1 \int_\Omega |\nabla(u_1 - u_2)|^p\, \omega_3\, dx$$

$$= \beta_p \int_\Omega |\Delta u_1 - \Delta u_2|^p\, \omega_1\, dx + \theta_1 \int_\Omega |\nabla(u_1 - u_2)|^p\, \omega_3\, dx > 0.$$

Therefore, the operator A is strictly monotone. Moreover, we have by (H3), (H7) and (H11),

$$\begin{aligned}
(Au|u) &= \mathbf{B}(u,u) \\
&= \mathbf{B}_1(u,u) + \mathbf{B}_2(u,u) + \mathbf{B}_3(u,u) + \mathbf{B}_4(u,u) + \mathbf{B}_5(u,u) \\
&= \int_\Omega |\Delta u|^p \omega_1\, dx + \int_\Omega |\Delta u|^q \omega_2\, dx \\
&+ \int_\Omega \langle \mathcal{A}(x,u,\nabla u), \nabla u\rangle\, \omega_3\, dx \\
&+ \int_\Omega \langle \mathcal{B}(x,u,\nabla u), \nabla u\rangle\, \omega_4\, dx \\
&+ \int_\Omega \mathcal{H}(x,u,\nabla u)\, u\, \omega_5\, dx \\
&\geq \int_\Omega |\Delta u|^p \omega_1\, dx + \int_\Omega |\nabla u|^q\, \omega_2\, dx
\end{aligned}$$

$$
\begin{aligned}
&+ \quad \lambda_1 \int_\Omega |\nabla u|^p \, \omega_3 \, dx + \lambda_2 \int_\Omega |\nabla u|^s \, \omega_4 \, dx \\
&+ \quad \Lambda_2 \int_\Omega |u|^s \, \omega_4 \, dx \\
&+ \quad \lambda_3 \int_\Omega |\nabla u|^z \, \omega_5 \, dx + \Lambda_3 \int_\Omega |u|^z \omega_5 \, dx \\
&\geq \quad \int_\Omega |\Delta u|^p \omega_1 \, dx + \lambda_1 \int_\Omega |\nabla u|^p \, \omega_3 \, dx \\
&\geq \quad \gamma \left(\int_\Omega |\Delta u|^p \, \omega_1 \, dx + \int_\Omega |\nabla u|^p \, \omega_3 \, dx \right) \\
&= \quad \gamma \|u\|_X^p,
\end{aligned}
$$

where $\gamma = \min\{\lambda_1, 1\}$. Hence, since $2 < p < \infty$ we have

$$\frac{(Au|u)}{\|u\|_X} \rightarrow +\infty, \text{ as } \|u\|_X \rightarrow +\infty,$$

that is, A is coercive.

Step 8. We need to show that the operator A is continuous. Let $u_m \rightarrow u$ in X as $m \rightarrow \infty$. We have,

$$
\begin{aligned}
&|\mathbf{B_3}(u_m, \varphi) - \mathbf{B_3}(u, \varphi)| \\
&\leq \sum_{j=1}^n \int_\Omega |\mathcal{A}_j(x, u_m \nabla u_m) - \mathcal{A}_j(x, u, \nabla u)||D_j \varphi| \, \omega_3 \, dx \\
&= \sum_{j=1}^n \int_\Omega |F_j u_m - F_j u||D_j \varphi| \, \omega_3 \, dx \\
&\leq \left(\sum_{j=1}^n \|F_j u_m - F_j u\|_{L^{p'}(\Omega, \omega_3)} \right) \| \, |\nabla \varphi| \, \|_{L^p(\Omega, \omega_3)} \\
&\leq \left(\sum_{j=1}^n \|F_j u_m - F_j u\|_{L^{p'}(\Omega, \omega_3)} \right) \|\varphi\|_X,
\end{aligned}
$$

and, by Remark 6.5.2(b),

$$\begin{aligned}
&|\mathbf{B_4}(u_m,\varphi)-\mathbf{B_4}(u,\varphi)| \\
\leq &\sum_{j=1}^{n}\int_\Omega |\mathcal{B}_j(x,u_m,\nabla u_m)-\mathcal{B}_j(x,u,\nabla u)||D_j\varphi|\,\omega_4\,dx \\
= &\sum_{j=1}^{n}\int_\Omega |G_ju_m-G_ju||D_j\varphi|\,\omega_4\,dx \\
\leq &\left(\sum_{j=1}^{n}\|G_ju_m-G_ju\|_{L^{s'}(\Omega,\omega_4)}\right)\|\,|\nabla\varphi|\,\|_{L^s(\Omega,\omega_4)} \\
\leq &M_2\left(\sum_{j=1}^{n}\|G_ju_m-G_ju\|_{L^{s'}(\Omega,\nu_2)}\right)\|\,|\nabla\varphi|\,\|_{L^p(\Omega,\omega_3)} \\
\leq &M_2\left(\sum_{j=1}^{n}\|G_ju_m-G_ju\|_{L^{s'}(\Omega,\nu_2)}\right)\|\varphi\|_X,
\end{aligned}$$

and we also have

$$\begin{aligned}
&|\mathbf{B}_1(u_m,\varphi)-\mathbf{B}_1(u,\varphi)| \\
\leq &\int_\Omega \left||\Delta u_m|^{p-2}\Delta u_m-|\Delta u|^{p-2}\Delta u\right||\Delta\varphi|\,\omega_1\,dx \\
= &\int_\Omega |Fu_m-Fu||\Delta\varphi|\,\omega_1\,dx \\
\leq &\|Fu_m-Fu\|_{L^{p'}(\Omega,\omega_1)}\,\|\Delta\varphi\|_{L^p(\Omega,\omega_1)} \\
\leq &\|Fu_m-Fu\|_{L^{p'}(\Omega,\omega_1)}\,\|\varphi\|_X,
\end{aligned}$$

and by Remark 6.5.2(a)

$$\begin{aligned}
&|\mathbf{B}_2(u_m,\varphi)-\mathbf{B}_2(u,\varphi)| \\
\leq &\int_\Omega \left||\Delta u_m|^{q-2}\Delta u_m-|\Delta u|^{q-2}\Delta u\right||\Delta\varphi|\,\omega_2\,dx \\
= &\int_\Omega |Gu_m-Gu||\Delta\varphi|\,\omega_2\,dx
\end{aligned}$$

$$\begin{aligned}
&\leq \|Gu_m - Gu\|_{L^{q'}(\Omega,\omega_2)} \|\Delta\varphi\|_{L^q(\Omega,\omega_2)} \\
&\leq M_1 \|Gu_m - Gu\|_{L^{q'}(\Omega,\omega_2)} \|\Delta\varphi\|_{L^p(\Omega,\omega_1)} \\
&\leq M_1 \|Gu_m - Gu\|_{L^{q'}(\Omega,\omega_2)} \|\varphi\|_X,
\end{aligned}$$

and by Remark 6.5.2(b))

$$\begin{aligned}
&|\mathbf{B}_5(u_m\varphi) - \mathbf{B}_5(u,\varphi)| \\
&\leq \int_\Omega |\mathcal{H}(x,u_m,\nabla u_m) - \mathcal{H}(x,u,\nabla u)|\,|\varphi|\,\omega_5\,dx \\
&= \int_\Omega |Hu_m - Hu|\,|\varphi|\,\omega_5\,dx \\
&\leq \|Hu_m - Hu\|_{L^{z'}(\Omega,\omega_5)} \|\varphi\|_{L^z(\Omega,\omega_5)} \\
&\leq M_3 \|Hu_m - Hu\|_{L^{z'}(\Omega,\omega_5)} \|\varphi\|_{L^p(\Omega,\omega_3)} \\
&\leq M_3 \|Hu_m - Hu\|_{L^{z'}(\Omega,\omega_5)}\, C_\Omega \|\nabla\varphi\|_{L^p(\Omega,\omega_3)} \\
&\leq M_3\, C_\Omega \|Hu_m - Hu\|_{L^{z'}(\Omega,\omega_5)} \|\varphi\|_X,
\end{aligned}$$

for all $\varphi \in X$. Hence,

$$\begin{aligned}
&|\mathbf{B}(u_m,\varphi) - \mathbf{B}(u,\varphi)| \\
&\leq |\mathbf{B_1}(u_m,\varphi) - \mathbf{B_1}(u,\varphi)| + |\mathbf{B_2}(u_m,\varphi) - \mathbf{B_2}(u,\varphi)| \\
&+ |\mathbf{B_3}(u_m,\varphi) - \mathbf{B_3}(u,\varphi)| + |\mathbf{B_4}(u_m,\varphi) - \mathbf{B_4}(u,\varphi)| \\
&+ |\mathbf{B_5}(u_m,\varphi) - \mathbf{B_5}(u,\varphi)| \\
&\leq \Bigg[\sum_{j=1}^n \|F_j u_m - F_j u\|_{L^{p'}(\Omega,\omega_3)} \\
&+ M_2 \sum_{j=1}^n \|G_j u_m - G_j u\|_{L^{s'}(\Omega,\omega_4)} \\
&+ \|Fu_m - Fu\|_{L^{p'}(\Omega,\omega_1)} + M_1 \|Gu_m - Gu\|_{L^{q'}(\Omega,\omega_2)} \\
&+ M_3 C_\Omega \|Hu_m - Hu\|_{L^{z'}(\Omega,\omega_5)} \Bigg] \|\varphi\|_X.
\end{aligned}$$

Then we obtain

$$\begin{aligned}
&\|Au_m - Au\|_* \\
&\leq \sum_{j=1}^{n} \left(\|F_j u_m - F_j u\|_{L^{p'}(\Omega,\omega_3)} + M_2 \|G_j u_m - G_j u\|_{L^{s'}(\Omega,\omega_4)} \right) \\
&+\|F u_m - F u\|_{L^{p'}(\Omega,\omega_1)} + M_1 \|G u_m - G u\|_{L^{q'}(\Omega,\omega_2)} \\
&+M_3\, C_\Omega \|H u_m - H u\|_{L^{z'}(\Omega,\omega_5)}.
\end{aligned}$$

Therefore, using (6.99), (6.102), (6.107), (6.108) and (6.110) we have $\|Au_m - Au\|_* \to 0$ as $m \to +\infty$, that is, A is continuous and this implies that A is hemicontinuous.
Therefore, by Theorem 5.1.1, the operator equation $Au = T$ has a unique solution $u \in X$ and it is the unique solution for problem (P19).

Step 9. Estimates for $\|u\|_X$. In particular, by setting $\varphi = u$ in Definition 6.5.4, we have

$$\begin{aligned}
\mathbf{B}(u,u) &= \mathbf{B_1}(u,u) + \mathbf{B_2}(u,u) + \mathbf{B_3}(u,u) + \mathbf{B_4}(u,u) + \mathbf{B_5}(u,u) \\
&= T(u). \qquad (6.112)
\end{aligned}$$

Hence, using (H3), (H7) and (H11) we obtain

$$\begin{aligned}
\mathbf{B}(u,u) &= \mathbf{B_1}(u,u) + \mathbf{B_2}(u,u) + \mathbf{B_3}(u,u) + \mathbf{B_4}(u,u) + \mathbf{B_5}(u,u) \\
&= \int_\Omega \langle \mathcal{A}(x,u,\nabla u), \nabla u \rangle\, \omega_3\, dx + \int_\Omega \langle \mathbf{B}(x,u,\nabla u), \nabla u \rangle\, \omega_4\, dx \\
&+ \int_\Omega |\Delta u|^p\, \omega_1\, dx + \int_\Omega |\Delta u|^q\, \omega_2\, dx + \int_\Omega \mathcal{H}(x,u,\nabla u)\, u\, \omega_5\, dx \\
&\geq \lambda_1 \int_\Omega |\nabla u|^p\, \omega_3\, dx + \int_\Omega |\Delta u|^p\, \omega_1\, dx, \qquad (6.113)
\end{aligned}$$

and, since $\omega_3 \in A_p$,

$$T(u) = \int_\Omega f_0\, u\, dx + \sum_{j=1}^{n} \int_\Omega f_j\, D_j u\, dx$$

$$\begin{aligned}
&\leq \|f_0/\omega_3\|_{L^{p'}(\Omega,\omega_3)}\|u\|_{L^p(\Omega,\omega_3)} \\
&+ \left(\sum_{j=1}^{n}\|f_j/\omega_3\|_{L^{p'}(\Omega,\omega_3)}\right)\|\,|\nabla u|\,\|_{L^p(\Omega,\omega_3)} \\
&\leq \left(C_\Omega\,\|f_0/\omega_3\|_{L^{p'}(\Omega,\omega_3)} + \sum_{j=1}^{n}\|f_j/\omega_3\|_{L^{p'}(\Omega,\omega_3)}\right)\|u\|_X \\
&= M\,\|u\|_X, \qquad (6.114)
\end{aligned}$$

where $M = C_\Omega\,\|f_0/\omega_3\|_{L^{p'}(\Omega,\omega_3)} + \sum_{j=1}^{n}\|f_j/\omega_3\|_{L^{p'}(\Omega,\omega_3)}$. Hence in (6.112), using (6.113) and (6.114), we obtain

$$\begin{aligned}
\gamma\|u\|_X^p &= \gamma\left(\int_\Omega |\nabla u|^p\,\omega_3 dx + \int_\Omega |\Delta u|^p\,\omega_1\,dx\right) \\
&\leq \lambda_1\int_\Omega |\nabla u|^p\,\omega_3\,dx + \int_\Omega |\Delta u|^p\,\omega_1\,dx \\
&\leq M\,\|u\|_X.
\end{aligned}$$

Therefore,

$$\|u\| \leq \left(\frac{M}{\gamma}\right)^{1/(p-1)}.$$

□

Example 6.5.1. *Let* $\Omega = \{(x,y)\in\mathbb{R}^2 : x^2+y^2 < 1\}$, *the weight functions* $\omega_1(x,y) = (x^2+y^2)^{5/2}$, $\omega_2(x,y) = (x^2+y^2)^{3/2}$, $\omega_3(x,y) = (x^2+y^2)^{-1/2}$, $\omega_4(x,y) = (x^2+y^2)^{-1/3}$ *and* $\omega_5(x,y) = (x^2+y^2)^{-1/3}$ ($\omega_3\in A_5$, $\omega_4\in A_3$, $\omega_1\in A_5$, $\omega_2\in A_3$, $p=5$, $q=3$, $s=z=3$), *and the function*

$$\begin{aligned}
&\mathcal{A}:\Omega\times\mathbb{R}^2\to\mathbb{R}^2 \\
&\mathcal{A}((x,y),\xi) = h_1(x,y)\,|\xi|^2\,\xi,
\end{aligned}$$

where $h_1(x,y) = 2\,\mathrm{e}^{(x^2+y^2)}$, *and*

$$\begin{aligned}&\mathcal{B} : \Omega \times \mathbb{R} \times \mathbb{R}^2 \to \mathbb{R}^2\\ &\mathcal{B}((x,y),\eta,\xi) = g_2(x,y)\,|\xi|\,\xi,\end{aligned}$$

where $g_2(x,y) = 2 + \cos(x^2+y^2)$ *and*

$$\begin{aligned}&\mathcal{H} : \Omega \times \mathbb{R} \times \mathbb{R}^2 \to \mathbb{R}\\ &\mathcal{H}((x,y),\eta,\xi) = h_3(x,y)\,\eta,\end{aligned}$$

where $h_3(x,y) = 1 + \sin^2(xy)$. *Let us consider the partial differential operator*

$$\begin{aligned} Lu(x,y) &= \Delta\big[|\Delta u|^3 \Delta u\,\omega_1 + |\Delta u|\Delta u\,\omega_2\big]\\ &- \operatorname{div}\big(\mathcal{A}((x,y),u,\nabla u)\,\omega_3(x,y)\\ &+ \mathcal{B}((x,y),u,\nabla u)\omega_4(x,y)\big)\\ &+ \mathcal{H}((x,y),u,\nabla u)\,\omega_5(x,y). \end{aligned}$$

Therefore, by Theorem 6.5.3, the problem

$$(P)\begin{cases} Lu(x) &= \dfrac{\cos(xy)}{\sqrt{x^2+y^2}} - \dfrac{\partial}{\partial x}\left(\dfrac{\sin(xy)}{\sqrt{x^2+y^2}}\right)\\ &\quad -\dfrac{\partial}{\partial y}\left(\dfrac{\sin(xy)}{\sqrt{x^2+y^2}}\right) \text{ in } \Omega,\\ u(x) &= \Delta u(x) = 0 \text{ on } \partial\Omega \end{cases}$$

has a unique solution $u \in X = W_0^{2,5}(\Omega,\omega_1,\omega_3)$.

Bibliography

[1] Adams, Robert and Fourier, John. 2003. *Sobolev Spaces*, New York. Academic Press.

[2] Bélinan, Phillipe, Boccardo, Lucio, Gallouët, Thierry, Gariepy, Ron, Pierre, Michael and Vasquez, Juan Luis. "*An L^1-theory of existence and uniqueness of solutions of nonlinear elliptic equations*". Annali della Scuola Normale Superiore di Pisa, (4) 22, no 2, (1995) 241 - 273.

[3] Bensoussan, Alain, Boccardo, Lucio and Murat, François. "*On a nonlinear partial differential equation having natural growth terms and unbounded solution*". Annales de l'Institut Henri Poincaré - Analysis non lineáire, Tome 5, No.04, (1988) 347-364.

[4] Björn, Anders, Björn, Jana and Christensen, Andreas. "*Poincé inequalities and A_p weights on bow-ties*", Preprint, 2022, arXiv:2202.07491v1.

[5] Björn, Jana. "*Poincaré inequalities for powers and produts of admissible weights*". Annales Academiæ Scientiarum Fennicæ, vol.26 (2001) 175-188.

[6] Boccardo, Lucio, Murat, François and Puel, Jean-Pierre. "*Existence of bounded solutions for nonlinear elliptic unilateral problems*". Annali di Matematica Pura ed Applicata, 152, (1988) 183-196.

[7] Boccardo, Lucio, Murat, François and Puel, Jean-Pierre. "*L^∞-estimate for some nonlinear elliptic partial differential equations and applications to an existence result*". SIAM Journal of Mathematical Analysis, vol.23,No.02, (1992) 326-333.

[8] Bresch, Didien, Lemoine, Jérôme and Guíllen-Gonzalez,Francisco. "*A note on a degenerate elliptic equation with applications for lakes and seas*". Electronic Journal of Differential Equations, vol. 2004, no.42, (2004) 1-13.

[9] Cavalheiro, Albo Carlos. "*An Approximation Theorem for Solution of Degenerate Elliptic Equations*". Proceedings of the Edinburgh Mathematical Society, v.45, (2002) 363 - 389.

[10] Cavalheiro, Albo Carlos. "*The solvability of Dirichlet problem for a class of degenerate elliptic equations with L^1 data*". Applicable Analysis, v.85, (2006) 941 - 961.

[11] Cavalheiro, Albo Carlos. "*Existence and approximation of solutions for a class of degenerate elliptic equations with Neumann boundary condition*". Note di Matematica, 40, no. 2, (2020) 63-81.

[12] Cavalheiro, Albo Carlos. "*An approximation theorem for solutions of degenerate semilinear elliptic equations*". Communications in Mathematics, v.25, (2017) 21 - 34.

[13] Cavalheiro, Albo Carlos. " *Uniqueness of solutions for some degenerate nonlinear elliptic equations*". Applicationes Mathematicae, v.41, (2014) 93 - 106.

[14] Cavalheiro, Albo Carlos. "*Existence of solutions in weighted Sobolev spaces for some degenerate quasilinear elliptic equations*". Georgian Mathematical Journal, v.15, (2008) 627 - 634.

[15] Cavalheiro, Albo Carlos. " *Existence of solutions for some degenerate quasilinear elliptic equations*". Le Matematiche (Catania), v.LXII, (2008) 101 - 112.

[16] Cavalheiro, Albo Carlos. " *Existence of entropy solutions for degenerate quasilinear elliptic equations*". Complex Variables, Theory & Application, v.53, (2008) 945 - 956.

[17] Cavalheiro, Albo Carlos. "*Existence of Solutions for a class of Nonlinear Degenerate Elliptic Equations*". Journal of Inequalities and Applications, v.7, (2002) 367 - 377.

[18] Cavalheiro, Albo Carlos. "*Existence results for Dirichlet problems with degenerated p-Laplacian*". Opuscula Mathematica, v.33, (2013) 439.

[19] Cavalheiro, Albo Carlos. "*Existence and uniqueness of solutions for problems with weighted p-Laplacian and p-Biharmonic operators*". Electronic Journal of Mathematical Analysis and Applications, v.3, (2015) 215 - 226.

[20] Cavalheiro, Albo Carlos. " *L^∞-solutions for some degenerate quasilinear elliptic equations*". Electronic Journal of Mathematical Analysis and Applications, v.2(1), (2014) 149 - 163.

[21] Cavalheiro, Albo Carlos. "*Existence of a solution for a nonlinear degenerate elliptic equations having natural growth terms*". Journal of Abstract Differential Equations and Applications, v.7, (2016) 26 - 38.

[22] Chiarenza, Filippo. " *Regularity for solutions of quasilinear elliptic equations under minimal assumptions*". Potential Analisys, 4, (1995) 325-334.

[23] Cassano, Francesco Serra. "*A counterexample on the weighted Poincaré inequality*". Rendiconti del Seminario Matematico - Politecnico di Torino, vol.51, 1, (1993) 65-72.

[24] Cassano, Francesco Serra. "*On the local boundedness of certain solutions for a class of degenerate elliptic equations*". Bolletino dell' Unione Matemtica Italiana, (7) 10 B, (1996) 651-680.

[25] Chanillo, Sagun and Wheeden, Richard. " *Harnack's inequalities and mean-value inequalities for solutions of degenerate elliptic equations*". Communications in Partial Differential Equations, 11 (10), (1986) 1111-1134.

[26] Chipot, Michael. 2009. *Elliptic Equations: An Introduction Course*, Berlin, Advanced Texts, Birkhäuser.

[27] Colombo, Maria. "*Flows of Non-Smooth Vector Fields and Degenerate Elliptic Equations: With Applications to the Vlasov-Poisson and Semigeostrophic Systems*". Publications of the Scuolla Normale Superiore Pisa, 22 (2017), Pisa.

[28] Diening, Lars, Harjulehto, Petteri, Hästö, Peter and Ružička, Michael. 2011. "*Lebesgue and Sobolev spaces with variable exponents*", Lecture Notes in Mathematics, vol.2017, Springer-Verrlag, Heidelberg.

[29] Diening, Lars, Hästö, Petteri and Nekvinda, Alĕ. "*Open problems in variables exponent Lebesgue and Sobolev spaces*". (www.math. cas.cz/fsdona2004 /diening.pdf).

[30] Drábek, Pavel and Milota, Jaroseav. 2007. " *Methods of Nonlinear Analysis - Applications to Differential Equations*". Birkhäuser, Basel.

[31] Drábek, Pavel, Kufner, Alois and Mustonen, Ville. "*Pseudo-monotonicity and degenerated or singular elliptic operator*". Bulletin of the Australian Mathematical Society, 58, (1998) 213-221.

[32] Drábek, Pavel, Kufner, Alois and Nicolosi, Francesco. 1997. "*Quasilinear Elliptic Equations with Degenerations and Singularities*". De Gruyter Series in Nonlinear Analysis and Applications, 5, Walter de Gruyter, Berlin.

[33] Dunford, Nelson James and Schwartz, Jacob. 1958. "*Linear Operators, Part I: General Theory*". Interscience Publishers, New York.

[34] Fučik, Svatopluk, John, Oldrich and Kufner, Alois. 1997. "*Function Spaces*". Noordhoff International Publishing, Leyden.

[35] Fabes, Eugene, Kenig, Carlos, Serapioni, Raul. "*The local regularity of solutions of degenerate elliptic equations*". Communications in Partial Differential Equations, 7, (1982) 77-116.

[36] Franchi, Bruno and Serapioni, Raul. "*Pointwise estimates for a class of strongly degenerate elliptic equations: A geometric approach*". Annali della Scuola Normale Superiore di Pisa, 14, (1987) 527-568.

[37] Folland, Gerald. 1984. "*Real Analysis*". Wiley-interscience, New York.

[38] Garcia-Cuerva, José and Rubio de Francia, José. 1985. " *Weighted Norm Inequalities and Related Topics*". North-Holland Mathematics Studies 116, Amsterdam.

[39] Gilbarg, David and Trudinger, Neil. 1977. " *Elliptic Partial Differential Equations of Second Order.* " Springer-Verlag, New York.

[40] Gorban, Yuliya and Kovalevsky, Alexander. " *On the boundedness of solutions of degenerate anisotropic elliptic variational inequalities*". Results in Mathematics, 65, (2014), 121-142.

[41] Hajłasz, Piort and Koskela, Pekka. *Sobolev mets Poincaré*, Memmoirs of the American Mathematical Society, vol. 145 (2000).

[42] Halsey, Thomas. "*Electrorheological fluids*". Science, vol.258, Issue 5083, (1992), 761-766.

[43] Heinonen, Juha, Kilpeläinen, Tero and Martio, Olli. 1993."*Nonlinear Potential Theory of Degenerate Elliptic Equations*". Oxford Math. Monographs, Clarendon Press.

[44] Kinderlehrer, David and Stampacchia, Guido. 1980. " *An Introduction to Variational Inequalities and their Applications*". Academic Press, New York.

[45] Kovácik, Ondry and Rákosník, Jiří. "*On spaces $L^{p(x)}$ and $W^{1,p(x)}$*", Czechoslovak Mathematical Journal, 41 (116), (1991) 592-618.

[46] Kufner, Alois. 1985. "*Weighted Sobolev Spaces*". John Wiley & Sons, New Jersey.

[47] Kufner, Alois, John, Oldřich and Fučík, Svatopluk. 1977. " *Function Spaces*". Noordhoff International Publishing, Leyden.

[48] Kufner, Alois and Opic, Bohumir. 1990. " *Hardy-type Inequality*". Pitman Research Notes in Mathematics, Series 219, Longman Scientific Technical, Harlow.

[49] Leray, Jean and Lions, Jacques-Louis. "*Quelques reséultats de Višik sur les problèmes elliptiques non linéaires par les méthods de Minty-Browder*". Bulletin de la Société Mathématique de France, 93 (1965) 97-107.

[50] Muckenhoupt, Benjamin. "*Weighted norm inequalities for the Hardy maximal function*". Transactions of the American Mathematical Society, 165 (1972) 207-226.

[61] Zeidler, Eberhard. 1990. "*Nonlinear Functional Analysis and its Applications*". Vol.II/B, Springer-Verlag, New York.

[62] Zhikov, Vasilii Vasilievich. "*Weighted Sobolev spaces*", Sbornik: Mathematics 189:8, (1998) 1139-1170.

Index

[51] Rajagopal, Kumbakonam and Ružička, Marek. "*Mathematical modeling of electrorheological materials*". Continuum Mechanics and Thermodynamics, 13, (2001) 59-78.

[52] Samko, Stefan. "*On a progress in the theory of Lebesgue spaces with variable exponent: maximal and singular operators*". Integral Transforms and Special Functions, vol.16, No 5-6, July-September (2005) 461-482.

[53] Stampacchia, Guido. "*Équations elliptiques du second ordre à coefficients discontinus*". Séminaires de Mathématiques Supérieures, 16, Presses de l'Université de Montréal, Montréal, (1966).

[54] Talbi, Mohamed and Tsouli, Najib. "*On the spectrum of the weighted p-Biharmonic operator with weight*". Mediterranean Journal of Mathematics, 4, (2007) 73-86.

[55] Torchinsky, Alberto. 1986. "*Real-Variable Methods in Harmonic Analysis*". Academic Press, San Diego.

[56] Turesson, Benght Ove. 2000. "*Nonlinear Potential Theory and Weighted Sobolev Spaces*". Lecture Notes in Mathematics, vol. 1736, Springer-Verlag, New York.

[57] Xu, Xiangsheng. "*A local partial regularity theorem for weak solutions of degenerate elliptic equations and its applications to the thermistor problem*". Differential Integral Equations, 12, no.1, (1999) 83-100.

[58] Yosida, Kôsaku. 1980. "*Functional Analysis*". Springer-Verlag, Berlin.

[59] Zeidler, Eberhard. 1990. "*Nonlinear Functional Analysis and its Applications*". Vol.I, Springer-Verlag, New York.

[60] Zeidler, Eberhard. 1990. "*Nonlinear Functional Analysis and its Applications*". Vol.II/A, Springer-Verlag, New York.